# A Book of Reed

# Dedication

To all those pioneering research on *Phragmites.*

(A difficult list, but one that certainly includes:
E. Bittmann, S. Björk, D. Dykyjová, K. Fiala, H. Godwin, H. Hurlimann,
R. Kickuth, F. Klötzli, J. Květ, J.H. Mook, P. Ondok, M. Pallis, L. Rudescu,
H. Sukopp, J. van der Toorn, D.F. Westlake.)

(Frontispiece) The changing seasons

# A Book of Reed

(*Phragmites australis* (Cav.) Trin. ex Steudel,
*Phragmites communis* Trin.)

S.M. Haslam, M.A., Sc.D.
Department of Plant Sciences, University of Cambridge

Tina Bone UK
MMXXI

**Published by: Tina Bone UK (©)**
18 Harbour Avenue, Comberton, Cambridge CB23 7DD, UK.
Email: tina@tinasfineart.uk • Web: https://tinasfineart.uk

*AUTHOR: © SYLVIA MARY HASLAM 2021*
*EDITOR: TINA BONE*

**REPRINT 2021**
**ISBN 978-1-9168781-0-5**

[First Published by: Forrest Text 2010
Cardigan SA43 2JG, UK
ISBN 978-0-9564692-0-5: out of print]

PAPERBACK **264** pp.
*c.* **125** Illustrations *(Figures, Graphs and Photographs)* • **54** Tables



*Front Cover picture:*
Reedbed, Minsmere Nature Reserve, Suffolk, August 2020.
*Back Cover picture:* 'The Mighty Reed with Fragility', water colour by Tina Bone. In Sylvia M. Haslam & Tina Bone, *Reed—on the edge* (River Friend Series). 2020. Book Plate, Figure 1, page 3.

# Contents

Acknowledgements

Preface

List of Tables

Chapter 1 Introduction ....................1

Chapter 2 Seeds and young plants ....................21

Chapter 3 Plant pattern and growth ....................27

Chapter 4 The reedbed itself ....................55

Chapter 5 Variation: genetic and clonal ....................83

Chapter 6 Variation: environmental ....................99

Chapter 7 Mixed stands: competition, invasion, succession and decline ....................121

Chapter 8 Malta: decline and endurance ....................133

Chapter 9 Uses of reed ....................147

Chapter 10 On the roof: thatch ....................167

Chapter 11 Impact and distribution ....................191

Chapter 12 Can reeds diagnose habitat? ....................209

Chapter 13 Remarkable reeds in marvellous marshes ....................217

Bibliography ....................225

Index ....................241

# Acknowledgements

The span of those to whom I am indebted ranges from Dr A.S. Watt, F.R.S., my excellent Ph.D. supervisor, to Mrs R. Noy, who in 2006–7 drove me on *Phragmites* hunts in Malta and, as usual, Mrs Tina Bone, Artist and Publisher.

May I just say how grateful I am to all those who gave me permission to work on their land, who have assisted me in various ways (including Mrs S.M. Hornsey, chemical analyses), have provided advice, or have given expenses money.

# Preface

*Phragmites australis* is the dominant species in reedbeds and occurs on every continent except Antarctica and may be one of the most wide-ranging of all flowering plants. *Phragmites* grows under a variety of environmental conditions ranging from salt to fresh water, which probably has led to its ubiquitous occurrence around the globe.

Now as a result of a great deal of talk about the value of reedbeds (largely prompted by EU Directives) mostly in terms of their importance as a habitat for other species, some rehabilitation, creation or conservation is being undertaken. What has been lacking is sufficient, serious and long-term research on the plant itself.

There is a need for Universities and research organisations to appoint aquatic or wetland ecologists whenever possible. If, with a preponderance of botanists and low disturbance, we can get the vegetation right, the proper fauna will generally follow: while the reverse does not apply.

Can future research be done properly and how can further understanding be gained? If this is not, this book could be authoritative for a century when, properly, it should be overtaken by a new book in perhaps a quarter of that time.

I apologise to those whose work is omitted or barely referred to. Some cover the same ground (with publications mostly referred to by name); some add to knowledge in different parts of the globe, confirming general behaviour as the same, but not otherwise increasing knowledge; some describe what will be the understanding of how chemical processes determine behaviour, but the research is still ongoing; and conclusions are missing.

To provide evidence for *Phragmites* behaviour often requires complex tables which the general reader does not need or want. Tables are therefore gathered at the end of chapters, where they can be studied if desired.

S.M.H.
January 2010

# List of Tables


*CHAPTER 1*

**Table 1.1** Variation in Reed (Modified from Haslam (1995). a) Non-climatic. b) Climatic.

**Table 1.2** Formation and loss of peat in Britain: rates of formation vary with climate (Haslam 2003).

**Table 1.3** Climate and geomorphic changes of the post-glacial era in Britain (compiled by the author). a) General. b) Fenland.

**Table 1.4** History of (Eastern) Broadland, England (George 1992).

**Table 1.5** History of the Lužnice flood plain in the Czech Republic (tentative reconstruction of the floodplain vegetation development since the Late Glacial [Jankovska, 1998).

**Table 1.6** Recent history of some reedbeds of Broadlands (selected reedbeds from Parmenter [1995]. Chronologically oldest first). 1. Estuarine and coast. 2. Inland, Freshwater.

*CHAPTER 2*

**Table 2.1** Development of young plants in fair (typical English) conditions.

**Table 2.2** The effect of 15° C and 25° C night temperatures on the growth of young plants.

**Table 2.3** The effect of adding phosphate (sodium dihydrogenate phosphate) to peat-stunted seedlings (another test showed no response to nitrogen alone).

**Table 2.4** The effect of cutting on plants 6 months old (all cut at 6 months, re-recorded at 9 months).

*CHAPTER 3*

**Table 3.1** Stem types of *Phragmites* (*Phragmites* is extraordinarily variable. This table merely lists the most typical behaviour).

**Table 3.2** Length and position of the growing season.

*CHAPTER 4*

**Table 4.1** Early-emerged buds are more likely to grow taller (accidents excluded) (Haslam, 1970a).

**Table 4.2** Effect of exposure of *Phragmites communis* to frost in plots* in two Breck fens (Suffolk). (* The 'Cavenham' site was exposed, while 'Icklingham' was flooded and so more sheltered.)

**Table 4.3** Larger buds emerge earlier (Haslam, 1969b).

**Table 4.4** Damage to emerged shoots. a) Dense stand. Litter removed, frosts very light. All causes (England). b) Death or damage from late summer internal competition. Affected shoots may have dead tips, be dying, or be dead (Haslam 1970a).

**Table 4.5** To show that flowering is commoner on taller shoots of *Phragmites* (Haslam, 1970a).

**Table 4.6** Comparison of performance in the same stand in different years (Breck Fens, England 1m$^2$), aerial stems (Haslam, 1972).

**Table 4.7** Competition for soil nutrients, Cavenham, England (Haslam, 1970a).

**Table 4.8** Effect of cutting during and after emergence (Haslam, 1969b) (living shoots were cut, leaving the litter dead standing aerial stems. Plots 1 x 1 m. Woodwalton, England).

**Table 4.9** Influence of invertebrates on the production of *Phragmites australis* in Czechoslovakia (Dvorak & Imhoff, in Květ et al., 1998).

**Table 4.10** Effects of *Lipara* spp. galls (Dvorak & Imhoff, 1998). a) Influence of gall formation by *Lipara* species on the length of *Phragmites* stems at Bohemian localities. b) Changes in dimensions of internodes of *Phragmites australis* caused by *Lipara* spp. infestation (percentage of the dimensions of unaffected stems; internodes numbered from top.

**Table 4.11** Infestation of *Phragmites australis* by some stem miners and some characteristics of the affected stand at Neusidedlersee (Dvorak & Imhoff, 1998).

**Table 4.12** Mammals in Central Europe (species diversity and some semi-quantitative data on mammals in Central European littoral reed belts (Dvorak & Imhoff, 1998).

**Table 4.13** Mammals and birds of Broadland (characteristic species) (George, 1992).

**Table 4.14** The 'macrolepidoptera' recorded from Broadland over the 1960s to 1990s, and their status (George, 1992).

**Table 4.15** Transpiration and evaporation during summer from open water by reed in the Danube Delta (Królikowska *et al*., 1998).

## CHAPTER 5

**Table 5.1** Infra-red patterns in the Delaware marsh. The records are of the number of clones of one infra-red colour invading those of another colour in each of the main habitats (Haslam, 1979).

**Table 5.2** Characters independent of clone infra-red colour (from Haslam, 1979).

**Table 5.3** Chromosome number, shoot length, and habitat of *Phragmites australis* provenances in The Netherlands (van der Toorn, 1972).

**Table 5.4** Ecotypes of *Phragmites australis* (Květ & Westlake in Květ *et al*., 1998).

## CHAPTER 6

**Table 6.1** *Phragmites* performance in habitats of different nutrient status (1950s, 1960s data). a) Number of shoots, Modal height, Population type. b) The habitat types. Communities, nutrient status and *Phragmites* performance in a).

**Table 6.2** The effect of added nutrients (N, K and P) on *Phragmites* (Haslam 1995). a) Effect of fertiliser on a calcium-rich, East Anglian fen, deficient in phosphorus and other available nutrients. Records after fertilisers were added every three months for eighteen months. b) Effect of fertiliser on replacement crops growing after summer cutting in East Anglian nutrient-rich fens (one treatment).

**Table 6.3** Nutrient content of aerial shoots in June (results as per cent dry weight. Analyses kindly done by the Ministry of Agriculture, Fisheries and Food, Cambridge, 1960s). a) Breck fens. b) North-west Scotland.

**Table 6.4** Nutrient content of rhizomes in January (analyses as Table 6.1 [data comparable with other 1960s analyses, not necessarily with those of the 1980s] Haslam 1995).

**Table 6.5** Nutrient content of horizontal rhizomes of different ages (Haslam, 1995) (Norfolk samples from high-nutrient peat dyes; tank samples in mineral soil, Cambridge; Boarlan samples from nutrient-poor loch, north-west Scotland. Analyses as Table 6.1).

**Table 6.6** Variation in nutrient levels and reed strength (Haslam, 1995). (From marshes of apparently uniform habitat with conclusive evidence of clonal variation, Nutrient results expressed as per cent (oven) dry weight in internodes. Silica skeleton remaining after ashing: varies from powder to patterned tube recorded as 0–15 (high). Strength measured as fibrousness, lack of brittleness and hardness, from Poor to Good.)

**Table 6.7** Reed strength and nutrient content (of internodes at the butt, the part most exposed to weathering on the roof) (Haslam, 1995).

**Table 6.8** Comparison of nutrient contents of internodes, nodes and sheaths of dead reeds (Haslam, 1995). (Nutrient results expressed as per cent (oven) dry weight, standard deviations in brackets. Internode and node levels are all correlated to a significance better than 0.001. Internode and sheath levels of K, Ca, Mg and Ash are correlated to a significance better than 0.001; those of N and P to 0.05.

**Table 6.9** Nutrient content and strength of older reed (internodes) (Haslam, 1995). (Nutrient results expressed as per cent (oven) dry weight, standard deviation in brackets. Silica skeleton as number on scale (15 is high).

## *CHAPTER 9*

**Table 9.1** Some organic compounds taken up by the roots of land plants (from Shimp *et al*., 1993).

**Table 9.2** Examples of chemical removal in constructed wetlands (see, e.g., Athie & Cerri, 1987; Hammer, 1989; Reddy & Smith, 1987' Rubec & Overend, 1987).

**Table 9.3** Selecting best management practices by pollutant: rules of thumb (Haslam, 2003, after Novotny & Olem, 1994).

*CHAPTER 10*

**Table 10.1** Characters of reed unsatisfactory for (particularly) traditional thatchers, East Anglia (modified from Haslam, 2009). (These characters were unwanted by thatchers when reed was in surplus and choice was available.

**Table 10.2** Year-to-year variations in reed strength, East Anglian marshes.

**Table 10.3** Distribution of sample strengths within reedbeds.

**Table 10.4** Range of strengths in some East Anglian reedbeds.

*CHAPTER 11*

**Table 11.1** Causes for reed decline (Ostendorp, 1989).

**Table 11.2** Dominants and understory groups, Britain (Rodwell, 1995).

**Table 11.3** British communities with *Phragmites australis* according to the National Vegetation Classification (Rodwell, 1991a, b and 1995).

**Table 11.4** Extract from the description of the most complex *Phragmites*-mix Tall Herb fen, *Phragmites australis–Peucedanum palustre* Tall Herb fen of the (British) National Vegetation Classification. (Selected to show the influence of management and soil.)

*CHAPTER 12*

**Table 12.1** Reed features characteristic of different habitat.

**Table 12.2** Vegetation pattern in reedswamp, The Netherlands (Mook & van der Toorn, 1982).

**Table 12.3** Reedbed variation with habitat bands of different reedbed areas.

Artistic impression of the Bridge of Reeds, designed in Cambridge to go over the main (A14) road where this nearly touches the edge of The Fenlands. Intended to evoke a reedbed.

# Chapter 1
# Introduction

Reed, *Phragmites*, is one of the most widespread of plants, growing in all five continents, though most common and most variable in Europe, particularly eastern Europe, the eastern Mediterranean and Iraq. English 'Reed' is, rather surprisingly, not the same root everywhere: e.g., Schilf (German), Cana (Spanish), Canna (Italian), Riet (Dutch), Rør (Norwegian), Siv (Danish), Kanus (Turkish), Qasbet ir-Rih (Maltese).

Reedswamp is an ancient term, used to describe that shown on the frontispiece. This is reedswamp, swampy, wet or shallow-flooded, reedy, with tall tubular (grass-like) vegetation smoothly covering the swamp, with the reeds bending and soughing in the breeze.

The earliest English (Anglo-Saxon) poem on reed is:

*REED*

*Beside the shore and near the strand*
*Right where the sea beats on the land*
*I, rooted, dwelt in my old place*
*And few there were of human race*
*Who saw my solitary abode*
*But every morn the brown wave flowed*
*and with its watery clasp me caught.*

Here is one of the common habitats: shallow brackish water. Reedswamp is found at the back of saltmarshes, where salinity lessens, but that is not where sea beats on the land. It is also found behind low sand dunes or equivalent, where salt spray and perhaps some salt water can reach. But the habitat battered by the sea and with brown (silt-carrying) water is on the edges of estuaries. Now, in Britain, these are usually fairly small – but 1,200 years or so ago impact was far less, and the reedbeds could have been more. Brown sea water is that carrying silt down from the land, in rivers.

*And little had I any thought*
*That it should ever be decreed*
*That o'er the bench where men drank mead*
*I, mouthless once, should speak and sing*
*Or soon or late ...*

Reeds are not the best material for musical pipes as they are too brittle and difficult to handle. However, it is possible, and in climes too cold for the Great Reed (*Arundo donax*), it was used for centuries, or indeed millennia. This verse describes the Anglo-Saxon (wooden) hall, with benches for the thegns etc., and plenty of good alcoholic drink (mead), song (with instruments including reed pipes) and general merriment. (See, for a later wooden hall, Figure 10.4.)

*... a wondrous thing*
*To mind which cannot understand*
*How point of knife and strong right hand*
*The man's mind coupled with the blade*
*Pressed me purposely and made*
*Me give a message without fear*
*To thee, with no one else to hear*
*So that no other man e'er may*
*Tell far and wide the words we say.*

(Anglo-Saxon Poetry, transl. R.K. Gordon, Dent, London)

Reeds were used as pens! And made in the same way as, later, quill pens, then steel nibs. Cut obliquely, pressed down, then cut to points and separating into two. (Quill pens were used from at least the early seventh century. The development of the steel nib, in the early nineteenth century, slowly led to this replacing the goose quill.)

The Anglo-Saxon poem dates from when few were literate so a written message was much more secret than a text message is now.

*Phragmites* pens, however, far pre-date the Anglo-Saxons, being in use in Ancient Egypt (Täckholm & Täckholm, 1941). A larger variety of strong reed was used, from the Eighth Dynasty. However, these pens were not well suited to hieroglyphics, and were not common until the coming of Graeco-Roman script.

Reeds have always been useful to people (see Chapter 9); never a major crop, but continually of some value. There are now more efficient pens and wall-fillings, and much more efficient drinks, among others, so reeds are used less, but may still have iconic resonances. A 'Reed Bridge' was planned, which undoubtedly would have captured early twenty-first century design!

## Where does *Phragmites* grow?

The planet is largely covered by water, mostly saline, but including some fresh. Generally, wetlands occur anywhere at the interface between terrestrial and aquatic ecosystems where the soil is saturated most of the time, making them different from either, yet highly dependent on both. Reedswamps in fresh and brackish water are one of the commonest wetlands in temperate areas, as mangrove swamps are in salty water in tropical areas.

'Reedswamp' is so named provided the habitat is wetland, and the dominant plants are tall grasses or sedges, with tubular stems, flatter leaves, or leafy stems rising up *c.* 1–8 m. However, for the purposes of this book 'reedswamp' will be used to describe *Phragmites*, 'The Reed' in current parlance. Such reedswamps can cover vast areas of many hectares, both well inland, as in e.g., the Volga river marshes, and coastal, as now developing along the east (New Jersey to Georgia) and south coasts of the United States of America. The European ones have much declined in recent centuries, with drainage and disturbance, while in North America the coastal populations are spreading (but are in total much less, see below).

In contrast, reed populations may also barely reach a square metre in size. It is one of the strengths of *Phragmites* that it is so versatile. Freshwater and brackish, dominant in reedswamps, sparse in wet woodland, sedge bed, tall-herb marsh or poor fen, local in low-nutrient lake or bog, banded along lowland streams: *Phragmites* is common in all.

*Phragmites* can grow to nearly 70° N, and up to 17,600 m in Tibet, and 16,000 m in Kashmir. *P. australis* and *P. karka* occur together in some areas e.g., Pakistan and Australia. In the latter, *P. karka* is recorded in the tropical and subtropical north, *P. australis* in the east, from arid to temperate to montane regions, and also in the southwest, probably as an introduction (Roberts, 2000).

Bibby & Lunn (1982) and Lunn (1979 and 1980) estimate 2,300 ha of reedbed existed in Britain in 1980. An extra 1,200 ha was wanted by 2010 (Brookhouse, 1998). In 2021, Sussex Wildlife Trust reported 900 sites covering about 5,000 ha. British beds are discussed generally by Bibby & Lunn (1982) and Wood (1999). Reed grows up to *c.* 420 m (it was in pollen zone Vlc [see Fig. 1.4] at 670 m: when the climate was warmer).

Figure 1.1 shows the world-wide distribution of *Phragmites*. In fact, in terms of area covered and range of habitats, Europe exceeds Asia, America and Africa (the Sahara is somewhat short of reedswamps!). New Zealand is the only land mass with a temperate climate which is without *Phragmites*. In North America, *Phragmites* is less frequent than *Typha* spp. In Europe, it is the reverse. The habitat, in ways not yet elucidated, favours *Phragmites* in the one, *Typha* in the other. *Phragmites* has its centre of variation in eastern Europe, and even west Europe shows more variation than North America. Reedbeds are very common in China, covering large areas, with reed of similar size to that of Europe and North America (C. Zhou, pers. comm.). It could be that east Europe is the centre of origin, and the main genotypes radiated out from there, and that those further away, in Africa and the Americas, somehow diminished in plasticity and were unable to develop to compete with the other—often non-European —species. However, recent invasion of reed along the coasts of the USA suggests a common phenomenon: when impact becomes high, all too often traditional vegetation gives way to invaders better suited to the new conditions. (Increasingly salty conditions in China produce *Spartina*, replacing *Phragmites*. C. Zhou, pers. comm.)

Figure 1.2 shows the distribution within Britain. Lowland England has recorded occurrences in most squares. The highland and nutrient-poor areas have least. Figures 1.1 and 1.2, though, show the very widespread distribution of this species. (See also Armstrong, 2004; Ausden, 2005; Esselink, 2000; Gainey, 1997; Gilbert, 2003; Hoi, 2001; Humphreys, 2005; Nikolajevsky, 1971; Salmon, 2004; Ward, 1999 for regional and local distributions.)

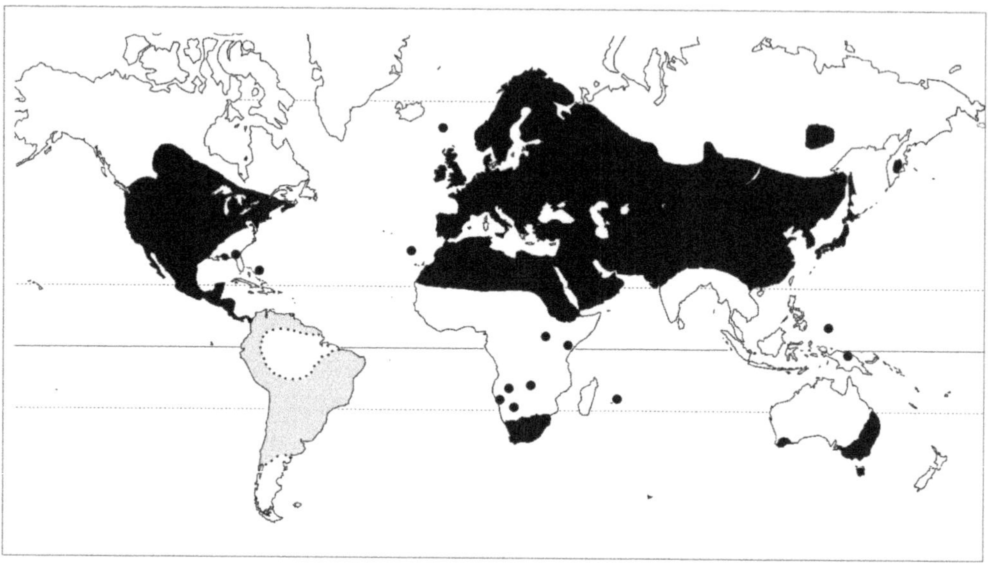

**Figure 1.1** (above) World-wide distribution of *Phragmites* (van der Toorn, 1972) Black: distribution fairly well known. Grey: probably occurring; distribution not well known. Broken line: indicates that the exact boundary of the distribution area is not known. Dots: isolated finds.

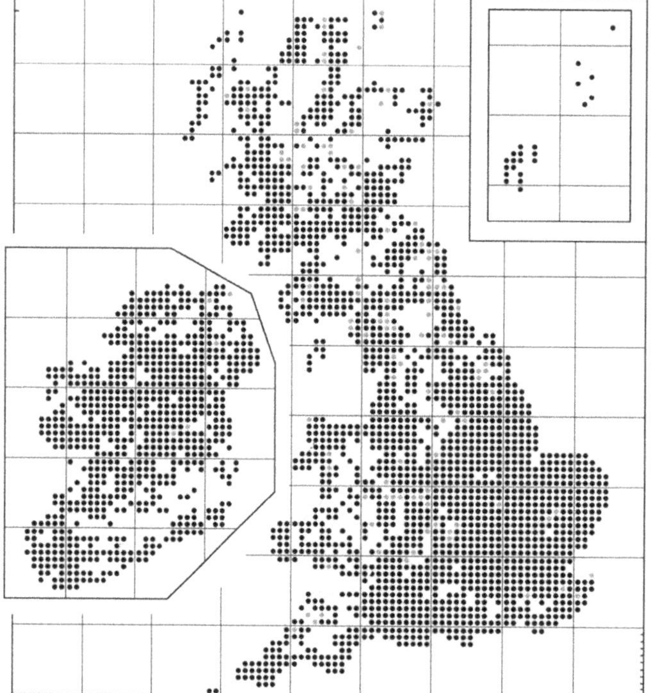

**Figure 1.2** (left) British distribution of *Phragmites australis* (data taken from the New Atlas of the British Flora, Preston *et al.*, 2002)

## Scientific names (and how many)

*Phragmites* is a very variable plant (e.g., Table 1.1), and DNA and reproductive evidence is, so far, minimal. That would be enough to complicate nomenclature, but *Phragmites* suffered a further, serious blow. The official technical description of *Phragmites* reads like this:

*Tribe oryzeae* [Rice]. A tall reed, with annual [not always] cane-like shoots often 1–8 m high, and an extensive perennial [living over two years] rhizome [underground stem] system. Leaf blades flat [may have central kink across, and are sometimes ribbed], up to *c.* 4 cm wide, tapering to long slender points, deciduous in winter. Sheaths [lower part of leaf, sheathing the stem], auricles [ear-like lobes at base of leaf blade] prominent. Ligule [a flap above the leaf sheath] replaced by a ring of hairs [this is a useful diagnostic character]. Panicle [a branched inflorescence] often (10–)30(–40) cm long, upright, yellow to dull purple, branches smooth, usually with scattered groups of a few long silky hairs. Spikelets [small units of the inflorescence, consisting of two glumes – bracts, flaps – and one or more florets – little flowers] *c.* 10–15 mm, (1–)3–(6) florets; rhachilla [stalk] hairy. Glumes shorter than first floret, lanceolate [lance-shaped], upper twice as long as lower. Lemma [lower glume] twice as long as upper glume, linear-lanceolate [very narrow lanceolate]. Palea [upper glume] shorter, ciliate [hairy] in upper parts. Florets, except for the lowest, with a tuft of long silky hairs, about as long as the lemma, at the base. Lodicules [scales outside stamen and ovary] 2, oblong. Stamens 1–3 in lowest floret; 3 in others, ovary glabrous [no hairs]; styles ['stem' between ovary and stigma, the part receiving pollen]. Helophyte [marsh plant: at least upper shoots above water].

And why is a plant centred and probably evolving in Europe, and sparse in Australia (Figure 1.1) known as *Phragmites australis*? When plant names became standardised in the eighteenth century (a movement led by Linnaeus – Carl von Linné, from Sweden), all were given Latin names, so that they could be recognised world-wide. (And see Arber, 1934.) The names 'Reed' and 'Schilf' are not immediately recognisable as the same! *Phragmites*, however, is, wherever it is found. Gradually these became standardised as a binomial, two names: *Phragmites*, a genus name, applies to all 'kinds' of similar reeds (*Salix* for willows, *Rosa* for roses, etc.), and *Phragmites communis* is the ordinary name, used up to *c.* 1970. This was derived from the original *Phragmites communis europaeus*, a trinomial, so later not considered valid, and the *europaeus* was dropped. There is a botanical rule which says the first Latin binomial name applied to a species, and properly authenticated and published, is the official name, however daft this becomes. It so happens that a botanist looked at Australian *Phragmites*, and, reasonably wanting to separate this from the European reed, named it (in 1799) *Phragmites australis*. Clayton (1967 and 1968) discovered that, unluckily:

1. *Phragmites australis* was given its binomial when the European reed still had the trinomial.
2. Studying the two, later, showed the differences between them were not sufficient to call them different species. (Here this author agrees. I – before DNA testing – could find no greater differences than there are within the European populations.)

3. Therefore, the European, American, etc., reed is officially *P. australis*. Which is confusing. But the rules do not allow exceptions, even when such would make sense. How many of those working on wetlands after 1970 think, or thought, reeds are an Australian import?

*Phragmites* has also suffered from the variation within the genus. There is a general tendency for taxonomists to want to separate many different species, subspecies, varieties, forms, etc.. That is their proper function: to decide what variation merits taxonomic status. There is an equally general tendency for ecologists to want to lump together as many types as possible. They are less interested in the, e.g., number of lodicules, and more interested in whether a type will vary in different habitats, will breed true from seed, or grow true from transplants, will be the same across a large reedbed, etc..

Hegi (1906) in his authoritative *Flora of Central Europe*, recognised, as well as the main species, two subspecies, three varieties, one sub-variety, and one form. Subspecies *pseudodonax* (Aschers et Graebner) is 8(–10) m high, with leaves up to 75 cm long and 6 cm wide. This occurs in north Germany. It is similar to the Giant Reed of the Danube Delta.

At the other extreme is subspecies *humilis*. This is the small form – Hegi puts shoots as not over 12 cm high, the one with small leaves hard and sharp enough to scratch skin. This, however, is due (where studied later) to a salty, rather dry and trampled habitat, and is definitely an ecological, not a taxonomic, entity.

Hegi's varieties are var. *flavescens*, which is pale, var. *effusa*, which is close to normal and var. *stolonifera*, which is mostly *legehalme* (long runners) and has violet on the stems. Its *legehalme* are here up to 10 m long, and grow in fairly dry places. The first two have not been further studied, but violet on stems is partly habitat-controlled, and *legehalme* development means both a type which potentially can develop *legehalme*, and a habitat suited to their development (see Chapter 4). Sub-variety *pumila* is a rather short reed (30–60 cm) in dry places: where 'normal' reed is liable to be short anyway (see Chapter 4). Sub-variety *pumila* is a rather short reed (30–60 cm) in dry places: where 'normal' reed is liable to be short anyway (see Chapters 3 and 4), and *forma picta* looks rather like *Phalaris arundinacea*, reed-grass.

Unfortunately, therefore, there are ecological as well as taxonomic considerations to sorting out types of *Phragmites*! Different lists have been prepared for, e.g., Romania (Rudescu *et al.*, 1965), The Netherlands (van der Toorn, 1972), Sweden (Björk, 1967). Cope (1982) recognised var. *stenophylla*, under 3 m, with pungent short convoluted sheaths, as environmental. Hubbard (1968) described var. *variegatum* with striped leaves, as a mutant used for ornament. In Egypt, Täckholm & Täckholm (1941) record many types: var. *isaica* very tall and broad-leaved (*cf.* Hegi's var. *pseudodonax*); var. *stenophylla* creeping and strongly smelling; mostly *legehalme* (*cf.* Hegi's var. *stolonifera*); var. *mauritianus* tall and thick, rare on the banks of the Nile and flowering very late in November (see below); and var. *striatipicta*, with white to variegated leaves, also flowering late (most October–December, some July–March) (*cf.* Hegi's var. *flavescens*, which, in central Europe, would be killed and unable to flower in winter). It was supposed there were other varieties, not located. In Lake Manzalor, reedswamps had large reeds of a narrow-leaved type, dormant in January, and also scattered clumps of a broad-leaved winter-green type. While dormancy variants like these are infrequent, the story of the versatility of *Phragmites* is repeated frequently. It certainly supports the view that there is a single entity, *P. australis*, *sensu lato* (in a broad sense).

There is still, however, the question of the more major taxonomic groups, the (usually three) species, *P. australis* (mostly Europe, America, Australia), *P. karka* (mostly Asia) and *P. mauritianus* (mostly Africa). Undoubtedly in Central Africa two forms occur, and can be separated by simple taxonomic (floral, etc.) characters. Equally undoubtedly both occur in South Africa: but the taxonomic characters such as smooth or hairy leaf blades, happen to be different (e.g., Gordon-Gray & Ward, 1971; Phillips, 1930)! One *Phragmites* entity? In the rest of this book, *P. australis* is used *sensu lato*, unless otherwise specified. However, nearly all the investigative research has been done in Europe, especially in the Czech Republic, The Netherlands, Sweden, Britain and Malta. That described in this book is primarily from the latter two.

Table 1.1 lists some of the investigated variables, most of which may be determined both by inheritance (genetic variation) and by habitat (phenotypic, environmental, variation). The range of these factors is remarkable! The types of variation listed are:

1. **Clonal** – this, as long as the evidence is good, means genetic inheritance.
2. **Annual** – factors varying from year to year, primarily the weather.
3. **Spatial** – the variation across a bed, which may include clonal, or variations in thickness of litter mat, local deep holes, etc.. This, therefore, is a mixture of genetic and habitat factors which can be very difficult to disentangle.
4. **Management** (for thatching reed) – effects can last for at least 90 years after management stops, but probably much longer. They may also disappear after 2–3 years (see Chapter 10).
5. **Transplant** – when the rhizome is cut, severing the connection with the main plant, and removed to a different, probably very different habitat, the new plant suffers double effects. First, re-growing from a short rhizome is different to growing on the parent. This effect has to be identified and eliminated before the change in the plant by moving it to other conditions. Such results may be valuable in interpretation.

The range of features is impressive. For instance, down south in Malta, where there is no frost, larger shoots, having lost their leaves, remain alive for 2–3 years, and bear side shoots from nodes. This can not happen in frosty winters (or in material adapted to these, except for short-lived branches after damage). When Maltese transplants are grown in Britain, they are frost-killed. British transplants in Malta die back as in Britain. This is genetic adaptation to the cold winter. However, if British material spends the winter in hot and humid conditions, shoots remain alive, and side-shoots grow. Similar unlikely phenomena lie behind most of the other listed variations. (See Chapters 5 and 6 for a fuller discussion of variation, and such DNA results as are available.)

How can different names be given to such a plant? It is much better to stick to '*P. australis, sensu lato*'.

## Peat

*Phragmites* is an important part of past as well as present vegetation and where it has been able to form peat. It is therefore the interpretation of the past. Reedswamp peat is formed under water, unlike bog peat, which grows on land, water being held within the (mainly *Sphagnum*) moss layer. Under water, though, leaves and stems drop down to the bottom of the lake or marsh. Peat is dead plants that have not decayed and vanished, but have accumulated in non-degrading conditions, such as under water, and over time have broken up and softened and formed humus. With more time, and the heavy weight of water and newer material above it, it turns to peat. Peat is soft to the touch, though firm enough to be cut into turves for (when dried) burning. Given even longer, and it becomes lignite, then with the weight of millions rather than tens of years it becomes coal.

Peat grows at a rate varying with the habitat. If the reedswamp is sparse, so the dead material is little, growth is slow. If water dries for part of the year, so the top forming peat is oxidised and lost, peat formation overall is slow: or none at all, if the oxidation is greater than the formation. At most, the peat seldom grows more than a few centimetres a year, and as it is compressed over time, this gets less (Table 1.2).

Not all reedswamps form peat. There may be currents and waves and other exposure that take away all the leaves and stems that fall into the water. There may be silt or sand or indeed gravel brought in by those same currents, and dumped on the lake floor. Peat is plant material, with maybe a little sediment in too. Accumulating sand with a little plant material is not peat!

*Phragmites* peat is one of the commonest underwater peats in Europe. It is made up of brown decayed humus, held together by the rootfelt of recent roots (wiry, binding threads pushing all through) and penetrated by the *Phragmites* rhizomes, both those growing along (horizontal) and those growing up (vertical). Other species present are incorporated into the peat too. Usually these are sparse, but occasionally the peat can be mixed, e.g., *Phragmites–Cladium mariscus*. Peat layers may be consecutive, as in Figure 1.3, or, in wet conditions, may be separated by layers of mud (Figure 1.4) or other water-deposited sediment. The peat core itself may be divided by layers of water, peat growing over holes (pits) in the layer below. In drained Britain, though these occur, there is much less water than in, say, The Netherlands. (Both in the Norfolk Broads area, and in Wicken Fen (Friday, 1997), heavy jumping causes the quiver which shows a floating layer below.)

Plant material humifies, and its pore space decreases. It becomes a highly active chemical factory (acrotelm), until covered and pushed down under the new peat, where it can remain little-changed for millennia (catotelm). Seeds and, particularly, pollen, blown in from the surroundings, are incorporated, and so are any human artefacts around at the time, such as axes or causeways. This means peat stores history, and is very valuable in that way. The study of pollen analysis reveals the history of the vegetation during the period (millennia?) that peat was made.

The peat is surprisingly pale, actually mid-brown. Once taken into the air, oxidation soon turns it darker, then nearly black. So it is easy to tell if a *Phragmites* peat has been exposed to the degrading influence of the air. While entirely suitable for burning, since turves are not in the air long enough to decay, when exposed and ploughed on fields the peat all-too-rapidly is lost. First, the highly-structured material becomes formless black powdery material, then it is further oxidised and blown away.

In order to form peat, a reedswamp must exist for a long time, and it must ***be*** a reedswamp. *Phragmites* peat does not come from sparse populations, as in wet woods or poor fens: the *Phragmites* is insufficient. It does not come from dry populations. Peat cannot form in these. Nor can it form in wave-beaten or nutrient-poor lakes where reedswamp is little and localised (Figure 1.5).

Reedswamp (all types) is a primary invader. It invades and spreads over water, as bog does over land. It occurs in any flooded habitat except bog, or on soils either very coarse or very unstable. *Phragmites* swamp ranges from those bogs with added nutrients (run-off, flushes) to the most nutrient-rich (eutrophic) waters found.

In Britain – and other regions glaciated in the last Ice Age – some fen peat (like bog peat) dates back to the late Glacial, *c.* 10,000 years BC. It is local in the last Inter-Glacial period in Europe (Tables 1.2 and 1.3).

Wet climates lead to more peat (both fen and bog). Dry climates and human impact lead to less being formed, and more being destroyed by drying, cultivation and, recently, large scale removal for power stations and horticulture.

Climates, landscapes and now human impact change. No reedswamp lasts for ever. For millennia, yes, frequently, but not over millions of years. Tables 1.3–1.6 and Figures 1.3 and 1.4 exhibit this pattern, in lakes, large and small. The now-Somerset Levels and Moors were reedbeds over 6,000 years ago, and *Phragmites* peat duly built up. Then sea level rose (relatively), the sea came in, and estuarine mud was deposited. This is incompatible with *Phragmites* peat (see above) and indeed the water was probably too deep, and *Phragmites* disappears from the record. (Any small populations were not incorporated, at the test sites.) Next, conditions became drier,

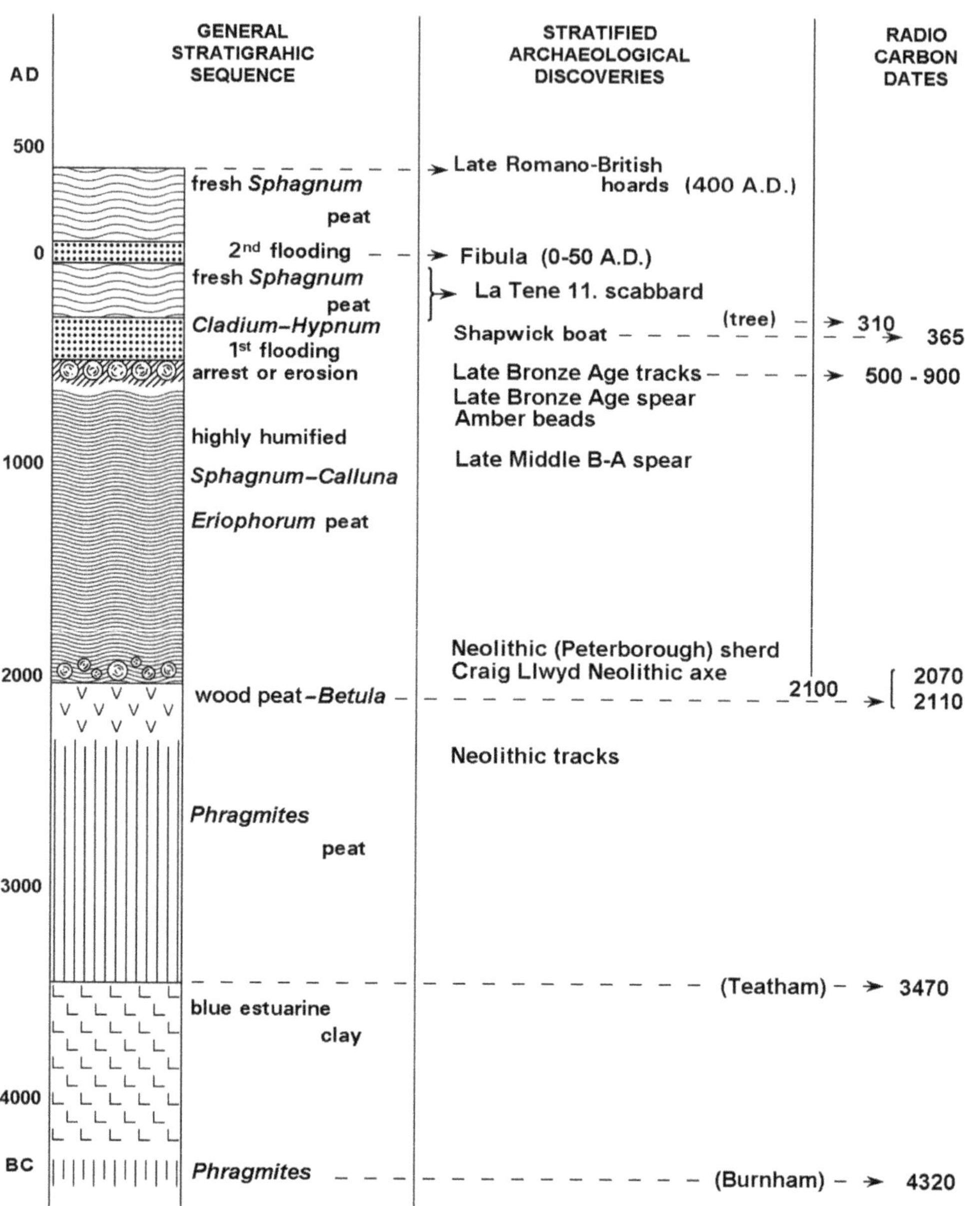


**Figure 1.3** Peat section. Somerset Levels, part (after Godwin, 1986).

From top to bottom: the most recently formed peat, to the first-formed, oldest.

From left to right: dates AD and BC, peat, names of principal peat-forming plants, archaeology, and radiocarbon dates. *Phragmites* is the earliest peat recorded. A sea incursion led to clay deposits before again reedbeds could invade, and this time they stayed for a thousand years. After drying led to birch wood, bog was able to invade (rainwater collecting on surface, see text). Finally, there was a freshwater incursion leading to fen (too few nutrients for reedbed?), followed by bog–fen–bog. Any peat formed after 500 AD has been removed.

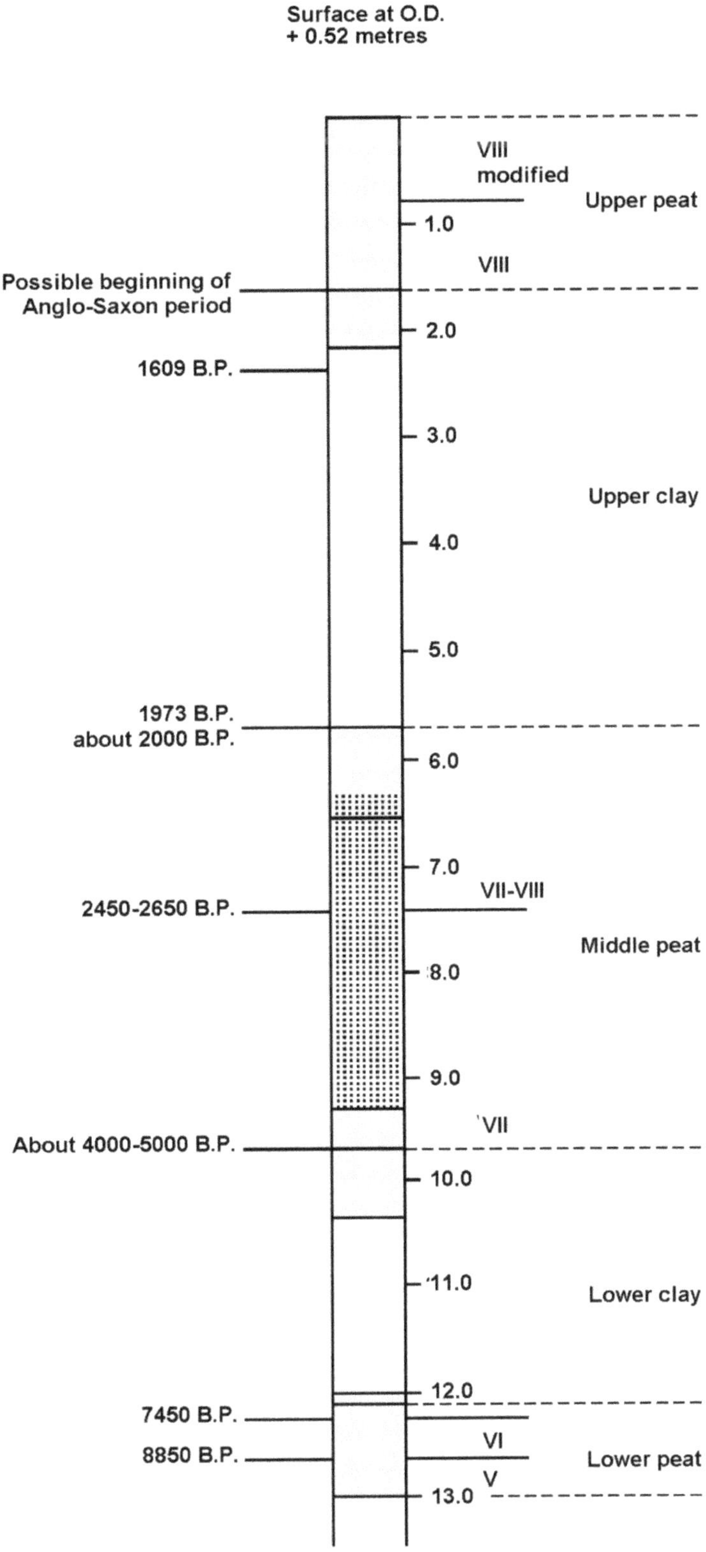

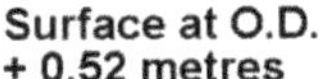


**Figure 1.4** Peat section, Broadland, part (after George, 1992).

The peat on the surface, the newest peat, is at the top, and the peat core goes down as far as possible. Once more, *Phragmites* peat is the lowest soil found (light grey), and it appears four times in the peat core, in total over about 1500 years. Any recent peat has been removed. These were two sea incursions (clay), and a drier period between (see Table 1.4). From left to right: dates, given as BP (Before Present, defined as before 1950); depth in m; pollen zones (V to VIII); and substrate zone.

a)

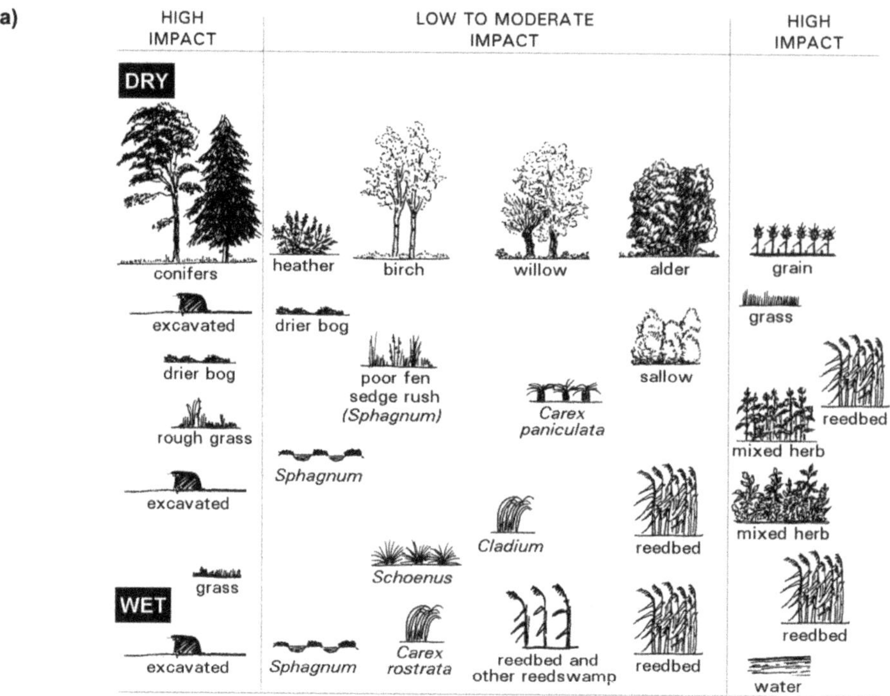


b)

| | BOG-INFLUENCED | LOW NUTRIENTS | HIGH NUTRIENTS |
|---|---|---|---|
| **Wetland type** | **Wet** | **Medium**[1] | **Dry**[2] **(wetland!)** |
| BOG | *Sphagnum* (cotton-grass, *Eriophorum angustifolium*, etc.) | Small shrub (e.g. *Calluna vulgaris*, *Erica tetralix*) cotton-grass, *Eriophorum vaginatum* | As medium, or with birch and pine woods. With increased nitrogen metabolism, *Molinia*-heath |
| CALCIUM-DOMINATED | Reedswamp, e.g. *Schoenus nigricans*, *Cladium mariscus* | *Cladium mariscus*, *Schoenus nigricans*, *Juncus subnodulosus*, short sedge/rush fen | Carr (alder, birch, buckthorn, etc., birch nearer springs, alder further off. Trees colonise tussocks. Tall herb communities; *Cladium*, tall-sedge vegetation. Short communities nutrient-poor or -rich |
| RICH FEN | Reedswamp, e.g. ***Phragmites australis***, *Glyceria maxima*, *Typha* spp. | Reedbed, large-sedge, short grass, etc., fen, tall herb | Carr (sallow, willow, alder, ash, etc.), tall-herb, short grass, etc. |
| MARSH | Reedswamp, e.g. ***Phragmites australis***, *Glyceria maxima*, *Typha* spp. | Reedbed, large-sedge, short grass, etc., fen, tall herb | Carr (as last, with more willow and poplar, oak), tall-herb, short grass, etc. |

[1] Tree invasion not possible in wetter places; above that, controlled by management (grazing, mowing, burning) as much as by water level.

[2] With further draining, mixed deciduous wood develop. On richer soils: oak, poplar, ash, elm increase.

**Figure 1.5** General distribution of British wetland types, including reeds and reedbeds (Haslam, 2003)

a) Pictorial (and incomplete) distribution.

b) Tabular (and incomplete) distribution.

Sparse *Phragmites* in all but the most dry and the most boggy habitats.

the sea retreated, and peat-forming reedswamp again flourished, for over a thousand years. The peat built up, the lake dried (with or without climatic influences). It was dry enough for birch seedlings to colonise gaps in the *Phragmites*, which gradually out-competed the (much-shorter) reed, and the record shows wood peat. Birch is a fairly low-nutrient species (sallow and alder colonise more eutrophic reedbeds). Unlike the eastern Fenland, the Somerset basin drains calcium-poor as well as some calcium-rich areas, and the nutrient status is lower (Figure 1.3).

After 2000 BC there is no *Phragmites* in the record, just variations of boggy (and sedge) peats (which include archaeological artefacts). This pattern is typical, first the open water, then the reedswamp, then either wood or bog (see Chapter 11; and Hardy, 1939). And, most often in Europe, the reedswamp is *Phragmites*.

There are various other patterns. Open water organic mud is a frequent precursor of *Phragmites* peat: a lake too deep for reedswamp has the slow drop of diatoms, other phyto- and zooplankton, organic drop from such vegetation and larger animals as may be present, plus some sediment from run-off. When the lake is filled enough to allow reedswamp (or a sea level change dries the lake), this invades, and *Phragmites* or other peat develops. The duration of the reedswamp is usually between hundreds and thousands of years (see Godwin & Mitchell, 1938, and Godwin & Newton, 1938).

While *Phragmites* is one of the commoner peat-formers, it is not the only one, either in the large British fen wetlands of the Fenland and the Somerset basin, or in small areas (Tables 1.4–1.6). They, at relevant times, did not meet the habitat criteria of water depth, nutrient status and disturbance.

*Phragmites*, like other peats, may be shallow or deep. It may reach several metres in ancient reedbeds: provided they have not been drained, cultivated or excavated down to just a little remaining peat!

## Living in the reedbed, Iraq

People living by reedbeds, or on islands within the beds, have used reeds since time immemorial. Most, on the smaller beds, have been variants of land cultures: fish and fowl from the marshes, dead reeds for farm and domestic use, with willow, alder and land farming (even in a small way) also available.

One of the largest beds is that in Iraq, estimated as 120,000 ha in the 1960s (compared with 110,000 ha in the Volga, Rudescu *et al.*, 1965). Here a full culture developed within the beds, as is to be expected from the size. This was well described by Thesiger (1964). Unfortunately, the Iraq government decided to drain these, the inhabitants having proved dissident to the then government. After the 2001 Iraq war, re-wetting the marshes began. In that *Phragmites* can re-grow easily, and putting water on to dried marshes is a technical matter, this is simple (e.g., Jaweir & Al-Kenzawi, 2009, record good re-invasion). However, when a people is taken out of a special habitat, where life is hard, and customs strict and traditional, and is taken to a city, culture changes. The older, in general, wanted to return home, while the younger were attracted by twenty-first century gadgets, and the wider world. The outcome is, at the time of writing, uncertain (see also Ward, 1999).

The canoes used in the reedbeds have remained the same for at least 5000 years. It is possible, therefore, that other aspects of culture, and artefacts, have done so, also. (The people are Muslim, so that, anyway, was a change not over 1300 years ago.)

The reeds are giant (Figure 1.6, and above), and covered the marsh except for lagoons and creeks, and the drier islands and islets where they have been removed. Some of the reed is floating (see Chapter 4). While the village houses are fairly simple, the sheikh's halls of government (*mudhif*), which vary by tribe, are magnificent (Figure 1.7) – not at all what is expected of primitive life! A strong culture! Houses are on ground, or soggy piles of reed like swans' nests. Farm fences are made of reed. Reed is cut as fodder for water buffalo, and moved by boat. It is sucked, for sugar. Once, rice was grown in parts nearer the delta, but not by the 1950s. Wild animals include jackals, wolves, snakes, eagles, a great quantity of wildfowl (over-exploited and decreasing) and fish. Fully enough for survival!

**Figure 1.6** Giant reeds.

**Figure 1.7** A typical Sheikh's hall built by the Marsh Arabs.

The reed dies back in winter, in climates with cold winters. In Africa it is likely to remain green.

## AND ...

*Phragmites* the plant, and *Phragmites* populations or reedbeds are dynamic systems. The plants grow and change, responding to habitat and the passing of time. The reedbed does likewise. The reedbed is not stable. It arrived, and, after years or millennia it will be lost. Within that there are variations by the year and by the century. Because of the long time-span, observers may miss the dynamism, and the changes in *Phragmites*. The acreage of dominant *Phragmites* is vast, but there are also co-dominant communities (see Chapter 7), and ones where *Phragmites* is subdominant, sparse, or local. Habitats change, slowly or rapidly, and the vegetation changes with it, between *Phragmites* and other plants, and those with habitat changes.

Because of its cosmopolitan distribution, its fascinating biology, its uses to people and its responses to so many habitat factors, *Phragmites* is indeed a plant worth studying: and reading about.

**Table 1.1** Variation in reed

Modified from Haslam (1995).

**a) NON-CLIMATIC**

| Character | Type of variation | | | | |
|---|---|---|---|---|---|
| | Genetic/ Clonal | Spatial | Annual | Management of reedbed | Transplant |
| **Stem, summer** | | | | | |
| height | + | + | + | + | + |
| width | + | + | + | + | + |
| density | + | + | + | + | + |
| colour | + | ? | + | + | + |
| shoot anatomy | + | + | + | ? | ? |
| **Leaf** | | | | | |
| size | + | + | + | + | + |
| texture | + | + | + | + | + |
| shape | + | + | + | + | + |
| number | + | + | + | + | + |
| autumn colour | + | + | + | ? | ? |
| position | + | + | + | + | ? |
| **Inflorescense** | | | | | |
| density | + | + | + | + | + |
| shape | + | + | + | + | + |
| size | + | + | + | + | + |
| seed germination | + | +(?) | + | ? | ? |
| **Timing** | | | | | |
| emergence | + | + | + | + | ? |
| growth rate | + | + | + | + | ? |
| flowering | + | -(?) | little | ? | ? |
| fruit dispersal | + | -(?) | + | ? | ? |
| leaf fall pattern | + | + | + | ? | ? |
| ***Legehalme*** | | | | | |
| production | + | + | + | n/a | + |
| **Rhizome** | | | | | |
| density | + | + | + | ? | n/a |
| cross-section shape | + | + | ? | ? | + |
| depth in soil | + | + | ? | ? | n/a |
| branching of verticals | + | + | +(?) | + | n/a |
| stimulation of buds on upper verticals | + | + | + | + | + |
| **Infra-Red colour** | | | | | |
| (in late summer) | + | - | -(?) | -(?) | ? |
| **Rhizosphere/root, etc.** | | | | | |
| factors in competition with other species | + | + | ? | ? | n/a |
| breakdown of pollutants, etc. | + | ? | ? | ? | ? |
| tolerance of salt | + | -(?) | - | - | -(?) |
| **Reed (dead)** | | | | | |
| straightness | + | + | + | + | + |
| strength | + | + | + | +(?) | + |
| wall thickness | + | -(?) | + | ? | ? |
| wax, sheen | + | ? | + | ? | + |
| colour | + | ? | + | ? | + |
| appearance | + | ? | + | + | + |
| lignification | + | + | + | ? | ? |
| nutrient content | + | + | + | ? | ? |
| **Pests and diseases** | | | | | |
| e.g. *Archanara* spp. (reed bug), *Ustilago* spp. (rust) | + | + | + | + | - |

n/a not applicable

Annual variation from year to year from any cause (e.g. summer temperature, grazing intensity, flood)

Spatial variation across the population, from any cause (e.g. competitive species, water depth, unseen clonal variation)

Management management for reed production only

**Table 1.1** Continued

| Character | Type of Variation | | | | |
|---|---|---|---|---|---|
| | Genetic/ Clonal | Spatial | Annual | Management of reedbed | Transplant |
| **b) CLIMATIC** | | | | | |
| Variation between Britain (approximate latitude 50°–58°, longitude 6° (6° W–0°) and Malta (latitude 36°, longitude 14° E) | Character varies? | | Can mimic, by altering habitat conditions, using material from only one country? | | |
| **Timing** | | | | | |
| emergence | + | | + | | |
| duration of leafy season | + | | + | | |
| flowering, fruiting seasons | + | | -(?) | | |
| autumn colour, leaf fall | + | | + | | |
| longevity and branching of shoots | + | | + | | |
| nutrient content (winter) | + | | + | | |
| (see also Table 6.1) | | | | | |

**Table 1.2** Formation and loss of peat in Britain: rates of formation vary with climate (Haslam, 2003)

| | Peat formed | | Human destruction | |
|---|---|---|---|---|
| | from | to | before 1900[1] | after 1900 |
| RAISED BOG | end Ice Age | now (local) | much | much |
| BLANKET BOG | at least from 8000 BC (from end of Ice Age?) | now | little | considerable |
| FEN | end Ice Age | now (local) | much | much |

[1] The rate of peat loss is greater after 1900.

**Table 1.3** Climate and geomorphic changes of the post-glacial era in Britain
Compiled by the author.

**a) General**

| | | | |
|---|---|---|---|
| AD | 1900 | Warmer | |
| | 1600 | Maximum cold, climatic upturn | |
| | 1400 | Broads peat pits flooding | |
| | 1200 | Cooler, moister | |
| | 1000 | Warm | |
| | 500 | Wetter | Anglo-Saxon |
| BC | 55 | Drier, warmer, hospitable climate | Roman |
| | 500 | SUB-ATLANTIC, wetter, warmer. Much peat development | Iron Age |
| | 2500 | Climate much as now. *Sphagnum* covering sub-boreal bogs | Bronze Age |
| | 3000 | SUB-BOREAL, cooler, drier, *c.* 2° C. Warmer than now, more continental. Settled agriculture. Peat decline, pine spreading over bog | Neolithic |
| | 5500 | ATLANTIC, warm (2–4° C above present), wet blanket bog spreading over hills, replacing pine and birch<br>Raised bog growing in lowlands<br>Sea level rise, Britain an island | |
| | 7000 | BOREAL, cool, dry. Pine forest spreading over peat bog | Mesolithic |
| | 9000 | PRE-BOREAL, bogs spread (e.g. Teesdale) | |
| | 10 000 | Post-glacial. Cold | Upper Palaeolithic |

**Table 1.3** Continued

**b) Fenland**

| | |
|---|---|
| Anglo-Saxon | Huge fens |
| Roman | Man-made watercourses, drainage |
| Iron Age | Build-up of silt, giving the present division of Silt and Peat Fens |
| 1000 BC–AD 0 | Sea incursion, extensive waterlogging, open sedge fen |
| To *c.* 2000 BC | Sea level drop, freshwater fen spread, and fen passing via carr to wood. Peat increase, trees invaded from the margin. Raised bog developing where alkaline flooding least (Middle Fens, and fen edge) |
| *c.* 2500–2000 BC | A vast brackish lagoon, 1–2 m deep. Coastal silt deposited, followed by *Phragmites* then inland sedge and woods far inland |
| *c.* 2500 BC | Rapid peat development, sedge fen, recent black peat. Sedge fen leading to carr then woods |
| *c.* 3000 BC | Sea incursion, waterlogged, freshwater ponded inland. Black peat developed in this |
| 4000–3000 BC | Dry acid peat, south fens remained alkaline, middle, to raise bog (only marginally affected by fen clay) |

**Table 1.4** History of (Eastern) Broadland, England (George, 1992)

| **Summary of events** | |
|---|---|
| *c.* 30 BP (i.e. 1920 AD)–*c.* 1000 BP | The town of Yarmouth develops from a small fishing settlement established on the shingle spit, and the peat laid down in the middle and upper sections of the valleys is exploited as a source of fuel. Between *c.* 1000 and *c.* 600 BP the excavations often extended through the Upper Clay into the Middle Peat and these relatively deep pits subsequently flooded to form the broads. Peat continued to be cut from extensive, shallow workings in the fens until about 1920 AD. |
| *c.* 1000 BC–*c.* 1610 BP | Tidal penetration up-valley progressively decreases, as a result of a combination of circumstances, including the development of a new shingle spit across the mouth of the embayment. Fen communities develop once again in the lower valleys, but the resultant 'Upper Peat' has subsequently wasted away following the embankment of the rivers, and the drainage of the adjoining marshland. Further up-valley, the Upper Peat directly overlies the Middle Peat formed previously. |
| | The shingle spit disintegrates (possibly as a result of a major storm surge), thus re-creating an open embayment. The increased tidal action in the valleys, allied to a continuing rise in relative sea level, leads to the deposition of a thick layer of 'Upper Clay' on top of the Middle Peat. Although the lower valleys were occupied by mudflats and saltings during this period, open fen communities predominated near the valley margins (owing to run-off from the higher ground bordering the latter) and in mid-valley. The Upper Clay sediments become thinner further upstream, and are absent from the upper sections of the valleys, where alder, carr and other fen communities persisted throughout the Transgression. The latter reaches its maximum extent in about 1609 BP. |
| *c.* 2250 BC–*c.* 5000 BP | A shingle spit forms across the mouth of the open embayment, deflecting the River Yare southwards, and reducing tidal penetration, thus allowing fen vegetation to replace the estuarine communities which had existed previously. The 'Middle Peat' thus formed, consisted initially of the remains of open (i.e. non-woody) communities dominated by *Phragmites*, but in the middle and upper valleys these soon gave way to alder-dominated woodland, which formed brushwood peat. Towards the end of the period, rainfall, and therefore fluvial flows, increase, and these find, or create, a breach in the shingle spit, leading to increased tidal penetration, and the re-development of estuarine conditions in the lower valleys. |
| *c.* 5000 BP–*c. 7500 BP* | The relative sea level continues to rise, and a thickening layer of sandy, silty or clayey sediments, collectively known as the 'Lower Clay', is laid down under open estuarine conditions in the lower and middle sections of the valleys. Brackish and freshwater conditions prevail further upstream and lead to the formation of 'Lower Peat' in these areas. |
| *c.* 7500 BP–*c.* 8500 BP | The once forested bed of the North Sea submerges as a consequence of a progressive rise in sea level relative to the land (i.e. eustasis and isostasis). Tidal influence begins to be apparent in the embayment into which the rivers discharged, with the consequent formation of the 'Lower Peat' in their lowermost reaches. |

**Table 1.5** History of the Lužnice flood plain in the Czech Republic

Tentative reconstruction of the floodplain vegetation development since the Late Glacial (Jankovska, 1998).

Because the analyzed profile covered only the last two and half thousand years, extrapolated results from other localities in the Třebon Basin were used for the reconstruction of floodplain vegetation development during the Holocene. The main periods of vegetation development are outlined below.

*Late Sub-Atlantic (13th century until the present)*
The massive colonization of the whole landscape which commenced at the end of the 12th century affected the floodplain as well. It was largely deforested and converted to meadows and pastures, partly even to arable land. The amount of open grassland and marsh expanded. Human activities changed not only the vegetation cover but also water regimes and sedimentation processes. Deforestation of the catchment enhanced erosion and the amount of transported material. This resulted in substantial sedimentation and deposition of silt deposits. The water regime was also affected by the construction of many fishponds and canals. The intensity of human alteration of the landscape increased rapidly after the 16th century. Since the end of the 18th century, *Picea abies* and *Pinus sylvestris* have been planted, displacing almost completely the original forests.

*Early Sub-Atlantic (800 or 500 BC–13th century AD)*
During this period, there were no substantial changes in vegetation of the floodplain itself. Species typical of open grassland survived sporadically. At this time, humans penetrated to the landscape of the Třebon Basin and especially to the flood.

*Sub-Boreal (2500–800 or 500 BC)*
The slight decrease of temperature experienced during this period did not probably affect vegetation cover to a large extent, and it remained comparable to the previous period in the floodplain. Human activity in the landscape is seen for the first time in the pollen diagrams, principally due to deforestation.

*Late Atlantic (4000–2500 BC)*
In this still warm and wet period, the floodplain was covered by dense, tall woodland mixed oak. The open sites of ancient grasslands had almost disappeared. The surrounding landscape was covered by dense oak woodlands (which attained their optimum at this time), by waterlogged spruce forests and alder carrs. Peatlands were more extensive than pine forests and very limited in their distribution. *Fagus sylvatica* and *Abies alba* began to penetrate the region. The first ancient settlements were located in or near the floodplain and the river corridor was probably an important communication routeway at this time.

*Early Atlantic (5500 or 5000–4000 BC)*
In this warm and wet period, the Lužnice River created a landscape characterised by widespread meanders, oxbows and pools, the river reworking the deeper silty sediments by now accumulated in large amounts on the floodplain. Terrestrialization of the water bodies was more rapid. Willow scrub, open marshes, and alluvial woodlands, with mixed oak over the largest area, were common in the floodplain. In the surrounding landscape, pine forests were restricted to dry sandy soils. Vast areas were covered by peat bogs. While the Lužnice River corridor probably served as the least difficult communication route for migrating tribes at this time.

*Boreal (6800 or 6500 or 5500–5000 BC)*
Substantial increases in temperature were accompanied by a rapid expansion of vegetation cover. Biomass accumulated and terrestrialization decreased the extent of lakes and marshes. The Lužnice River formed probably a deeper bed cut down into silty sediments accumulating widely in the floodplain. Willows, reed belts, and tall-sedge communities are expected to have prevailed along the river at this time. *Alnus* species would have been present, and the formation of alluvial woodland dominated by *Quercus* started. *Pinus sylvestris* was still common; *Picea abies*, *Alnus*, and species typical for mixed oak woodland started to expand.

*Pre-Boreal (8300–6800 BC)*
Increases of temperature and precipitation in the beginning of the Holocene Period would have caused higher river discharges. While the floodplain was still formed mostly of sandy and gravel sediments, these deposits became more and more covered by silty material in areas where overbank sedimentation frequently occurred. These silty sediments supported an expansion of ancient grass lands over the floodplain. Beside the plant species mentioned as occurring also in the Late Glacial Period, *Polygonum bistorta*, *Valeriana officinalis*, *Geranium*, *Geum*, *Sanguisorba officinalis*, *Succisa pratensis*, etc. would have been expected to have occurred at this stage. *Filipendula ulmaria* and *Phragmites australis* especially expanded. Willows dominated along the river bed. Outside the floodplain, *Pinus sylvestris* and species of *Betula* formed a sparse woodland cover. As in the preceding period, wetlands would have been a common feature of the Pre-Boreal landscape.

*Late Glacial (12,000–8,300 BC)*
In this still cold period, the Lužnice River probably had the character of a subarctic river, flowing through an open landscape with vegetation cover composed mainly of scattered trees and shrubs of *Pinus sylvestris*, *Betula pubescens*, *B. verrucosa*, *B. nana*, *Juniperus communis*, *Populus tremula*, and diverse *Salix* species. Ericaceae and *Vaccinium* sp. (?) in the herb layer. There were a good number of smaller lakes. Between these water bodies, the Lužnice River would have occupied a narrow bed in the sandy and gravel sediments. Silt deposits would have been much less common than now.

**Table 1.6** Recent history of some reedbeds of Broadlands

Selected reedbeds from Parmenter (1995). Chronologically oldest first.

**1. Estuarine and coast**

a) *Whitecast Marshes*

1838 Commercial reedbeds, some grazing perhaps

1884 Fen or marsh

19th – early 20th century
Commercial reedbed, without management scrub invasion and reedbed deterioration

1902 Swamp, reeds and water. Quagmire. Crop used for litter. Bullocks

1906 Fen or marsh

1990 Reedbeds, species-poor, with some dominant *Phalaris arundinacea* and, with most tidal influence, *Glyceria maxima*. Scrub. Some quagmire

b) *Belton Ronds*

Saltmarsh vegetation and reedbed, some still harvested. The ronds originated in the 13th and 14th centuries when the Broadland rivers were embanked, providing washland for flood alleviation. Most reedbed or grazed. *Phragmites* dominates in the least saline areas

1886 Marsh

1907 Marsh

**2. Inland, Freshwater**

a) *Sprats Water*

1995 Drained by pumps near the river wall for several centuries
In the 20th century, abandonment led to coarse vegetation: *Phragmites* and *Carex riparia*, carr and mown fen meadow
Areas of spring-fed mixed *Carex riparia* and *Phragmites* and others. Reed cutting has led to a very rich flora
*Phragmites* is present in two communities
*Phragmites australis–Peucedanum palustre* Tall Herb Fen
*Salix cinerea–Betula pubescens–Phragmites australis* woodland
In the former, *Phragmites* is co-dominant in the highest layer. In the latter, subdominant within a taller, carr community

b) *Fritton Decoy*

A 13th and 14th century peat cutting, cut down to the gravel substrate
Swamp fringed the broad, most of which has been lost through recreation and pollution. It was rich in medium-nutrient species. A little reedswamp remains

1797 'Fritton Decoy'

1838 Decoy

1935 Fringing damp oak wood

1985 Damp scrub, goat willow carr

Note:
In this (River Waveney) area major reedbeds are where there is more inorganic (richer) and brackish water habitat. Both suit *Phragmites* well. In the lower-nutrient peat habitat, *Phragmites* is less often dominant. Changed conditions, e.g. from reed harvesting, will, however, lead to dominant reed (as, more recently, at Old Buckenham Fen, with soil disturbance and fertiliser).

# Chapter 2
# Seeds and young plants

## SEEDS AND DISPERSAL

The seed (technically, fruit) of *Phragmites* is small, with hairs that mean it is easily blown by the wind (Figure 2.1). In view of the usually low germination rate and difficulty of seedling establishment (see below), it might be thought propagation by seed is unimportant to the plant. After all, reedswamps can last thousands of years (peat records), and the plants in a reedswamp have been estimated as at least a thousand (Rudescu *et al.*, 1965).

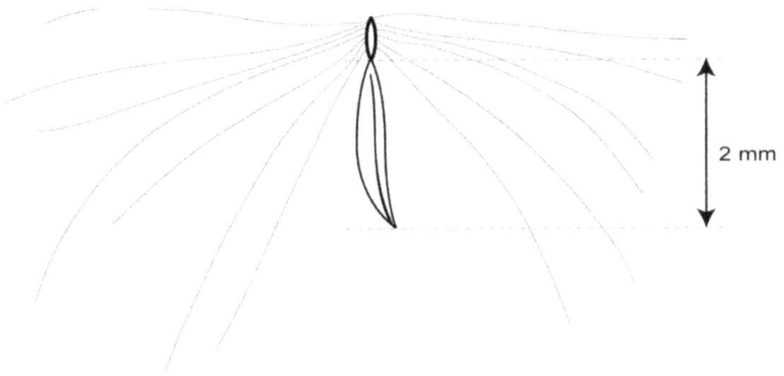

**Figure 2.1** Fruit of *Phragmites australis*

The ring (pappus) of hairs can carry the seed for scores of miles. (This one is probably infertile as it is thin.)

However, firstly, even long-lived reedswamps die out (from their success in building peat, from sea level changes, human impact, or other catastrophes), and then, plenty of reed stands in less favourable habitats live only years, decades or centuries. Propagation by bits of rhizomes, etc., washed away from stands to new places, is effective but limited to those places with water dispersal. Therefore seed propagation is, in the long term, necessary for reed survival.

Seed dispersal and germination, where it takes place, must in fact be very efficient, since within 2–10 years of a new, suitable site becoming available in Britain, it has tens to hundreds of young reed plants. These sites are abandoned gravel pits and suchlike: empty places, with a variety of water levels, and mineral soil. Disturbed (brackish) saltmarshes are another suitable habitat.

Twenty-five miles of sea, crossed in 14 years, is noted by Ridley (1923). Further distances are probable. While it is conceivable that young plants could be found in a reedbed, it is unlikely. If bad weather brings down fruiting reeds before the seeds have been blown away, seeds may germinate, and the seedlings start to grow on matted dead reeds. Soon, though, the new reeds of the bed emerge and the shade will kill them – if floods or frost have not done so first. Large openings in reedbeds, after some catastrophe, provide better opportunities for their growth (also see Ridley, 1930).

## Germination

Pollination is by wind, and dispersal by wind, birds' nests, people, and water (Bittmann, 1953).

Enormous numbers of fruit are recorded: up to 1,000 fertile fruits per panicle (Bittmann, 1953). Both the viability of the seeds, and their germination, is variable. Curran (1969), Djerbrouni (1992), Gustafsson & Simak (1963), and others find abnormalities preventing germination, and indeed in some populations germination is very poor indeed – virtually none. On the other hand, 100% was recorded in Switzerland from 1944 seed, though 1–55% was more usual, varying with year (Hürlimann, 1951). Similar results came from The Netherlands and Sweden (van der Toorn, 1972), and US coastlands (Pellegrin & Hauber, 1999). Mauchamp *et al.* (2001) records that too much drowning kills seedlings. When millions of seeds are deployed, even a 1% germination rate gives numerous seedlings. Genetic diversity is discussed by Krzakowa & Drapikowska (2000).

The first large-scale commercial use of *Phragmites* seeds was in The Netherlands, in the post-Second World War damming and draining of much of the Zuider Zee, converting land under sea to new polders for farming and (lesser) settlement. *Phragmites*, with its very high transpiration rate (Chapter 4), was used for the main soil drainage (after lake water had been removed). Sowing in May was found best, as germination then takes 8–10 days. Earlier sowing, and delayed germination, mean more competition from other species (Bittmann, 1953). In such drained polders, the soil is inorganic, and the water level is controlled (see below). Numerous seeds germinated and grew well on wet mud, and in good summer established on inorganic, nutrient-rich mud. Seed setting of 3–17% was recorded, giving plenty of seedlings (von der Toorn, 1972)!

The fastest germination reported for Britain and Malta is at 25° C, in *c.* 5 days, and the highest proportion of seeds germinating (from the same material) at 25° C day, and 15° C night temperatures: April to May in Malta, a warm May or June in Britain.

The first two leaves (Figure 2.2) are supported by the food in the seed: one narrow strap leaf, followed by a second, nearly twice the height, reaching perhaps 1.5 cm high. At this point the root is *c.* 1.5 cm long, and there is little further growth (though perhaps another leaf or two) if no nutrients are provided. This stage is supported by the food in the fruit. The seedling's next stage can be called 'Development' (below).

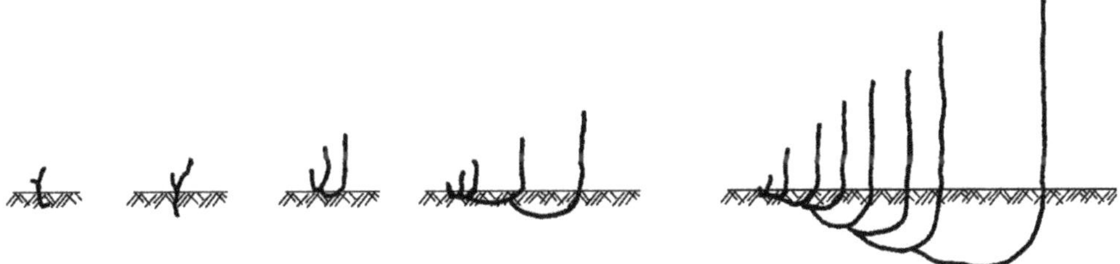

**Figure 2.2** Seedling growth (diagrammatic) (Haslam, 2003).
Food from the seed supports the plant to the first (or second) stage, which sometimes has a third leaf. In unsuitable conditions rapid growth can stop here. The leaf probably turns purplish or brownish, and remains at this stage, years, months, weeks or indeed no significant time. Once the seedling has the food from the habitat to reach the third stage, it can go on to form a proper plant, each shoot being further from, and taller than, the last. The final stage shown here has genuine horizontal rhizomes.

Germination is hindered by external (habitat) as well as by internal factors. Flood lowers and ultimately (if persistent) stops germination. So do dry conditions, although here there is also the possibility the seedling died young instead. Germination also decreases in salt. Different concentrations are quoted, germination certainly occurs up to 3% salinity, possibly 4%. However, differences could have come because the tests were done before genetic differences in salt tolerance were demonstrated; also, tests from different labs often show different results. The same may apply to nutrients and indeed to other characteristics (see, e.g., Bittmann, 1953; Björk, 1967; Chapman, 1960; Haslam, 1972; Hürlimann, 1951; Rudescu *et al.*, 1965; van der Toorn, 1972).

## Development

To have a new reed plant:

1. seeds must develop;
2. and be viable;
3. and be transported to where they can germinate;
4. and lastly, the tiny seedling must take the chances and survive the changes that all the habitat can do until it becomes the tough adult plant.

Figure. 2.2 shows the stages until the young plant is fairly resistant.

New shoots develop, each larger than the one before, with more leaves, and more distance from its predecessor. As the Figure shows, this is how the horizontal rhizomes develop, each being deeper as well as longer than the one before. Eventually they reach the depth of the mature rhizomes (Chapters 3–6). In the meantime, the shoots are becoming taller. The first three are close together, and mainly blade. Fast-growing plants usually have the typical leaf sheath on the fourth or fifth shoots. The longest blade reaches *c.* 7 cm in about the seventh shoot. The leaves now resemble those on the smallest adult shoots (*c.* 1 mm diameter).

The first side shoot develops when the first stem is *c.* 3.5–6.5 cm (height to top of upper sheath). Each side shoot is slightly wider, so able to grow longer, than the one before. European reed, i.e., that with a 'winter', has leaves and, except in the far south, stems dying in winter. The young plants in their first to third years have no dormancy. They remain leafy, unless killed by frost.

By about 10 shoots, the plant can resist frosts (because the horizontal rhizomes have now the adult quality of frost-resistance and are well below ground, so insulated; Chapter 3). It can resist the normal flooding of a shallow marsh, because some shoots are over 30 cm tall, and such deep flooding is usually temporary. (Armstrong *et al.*, 1992, describe effects of submergence on seedlings, and oxygen loss from roots.)

It can resist moderate shade, since if the plant can reach this size, in such shade, it can continue to tolerate it. The same applies to substrate. If this is unstable, the very young plant will die. If nutrients are severely deficient, the seedling will either not develop or will develop very slowly and feebly.

The 10-shoot, viable young plant can grow in three months from germination. However, this is by no means usual (Table 2.1)!

Within a year, in optimum conditions (warm, water level near surface, good soil, well-lit, etc.) the tallest shoots may be at least 1 m tall, with a maximum basal width of at least 5 mm, a maximum leaf width of 10 mm, and with maybe *c.* 70 shoots or more. Long horizontal rhizomes may be present after two months (Bittmann, 1953). This is like a small adult plant (Table 2.1). After two to three years, the plant is undoubtedly adult, above and below ground.

However, life is not always like that. At the other extreme are seedlings at the 1.5 leaf stage (Figure 2.2) which become discoloured, but otherwise do not change in at least two years, and probably more. Obviously these are vulnerable to anything from trampling to shading, and the longer a seedling remains in this state the more likely it is to be killed. Many young plants are in between, following the pattern described, but reaching each stage more slowly. Slow development means death from all causes is more likely. As an open habitat is invaded, bare, sunlit patches become smaller and fewer: competition with other species is more severe. Death by flood and drowning and drought are more likely the longer the seedling is vulnerable. The faster the growth, the better the survival, so the better the chance of dominant reedbed developing.

Growth is much faster in warm temperatures, so good survival is less likely in the north and, more important, in cooler summers. Growth of the same material to the 4-leaf stage was reached after *c.* 6 weeks in Britain, and in *c.* 3 weeks in Malta (Table 2.2).

Nutrient status, the nutrients available to the young plant, is of primary importance. There are two particularly low-nutrient habitats: those with very low nutrient concentrations, as on sparse silt in coarse substrates, where the silt is derived from low-nutrient rock (e.g., gneiss, acid sands); and those wet and with very high calcium, normally fen peats, where the calcium precipitates out the phosphorus, etc. (Haslam, 2003). Such peat is calcium-rich, other-nutrient poor: unless and until it is dried or disturbed or both. Then the held nutrients become available, and suddenly those species which have been excluded or, like *Phragmites* seedlings, kept small, start to grow and, if discoloured by deficiency, turn green. Such seedlings may have been just sitting on peat, browning, for years. Given a little phosphate, normal growth resumes (Table 2.3).

Balanced nutrients are necessary for good growth, of course, and in different habitats different nutrients can be deficient. No tests have been done on calcium deficiency, but (see Chapter 6 for the adult plant) that also is likely to delay growth.

Table 2.4 illustrates response to cutting (which would be the same as a single grazing), which at six months is good.

It is indeed surprising that, given a suitable and open habitat, young plants grow so well! No researcher has yet, however, recorded young plants within dominant reedbeds.

Dried lake edges after drought may, in spring, make suitable habitats for germination and establishment. The fate of the young plants varies with the subsequent conditions. If the young plants can form a shading canopy before the native, or other new plants do so, and be tall enough before the shore is again drowned, permanent colonisation may take place. In Sweden (Weissner, 1987 and 1991) some lakes dry like this occasionally (e.g., 1932–33), as also in Uganda (P. Denny, pers. comm.). Such episodes may spread *Phragmites* very rapidly and alter the dominance pattern.

**Table 2.1** Development of young plants in fair (typical English) conditions

| Age (weeks) | Number of shoots (av.) | Sheath height (cm) (av.) |
|---|---|---|
| 1 | 1 | – |
| 3 | 1 | – |
| 5 | 2.25 | 4.5 |
| 7 | 1.5 | 6 |
| 9 | 2.75 | 11 |
| 11 | 2 | 16 |
| 13 | 5 | 20 |
| 15 | 8 | 25 |
| 17 | 10 | 34 |
| 25 | 46 | 67 |

At 25 weeks probably, at 17 possibly, the young plant is viable in poor conditions. Four months is from about mid-May, when seeds typically germinate, to mid-August. Up to *c.* 1980 killing frosts in the Fens in September were frequent. In their present absence, more seedlings can become viable.

**Table 2.2** The effect of 15° C and 25° C night temperatures on the growth of young plants

| | Temperature ° C | Number of shoots (av.) | Sheath height (cm) (av.) |
|---|---|---|---|
| **Malta** | | | |
| | 15 | 2.5 | 20 |
| | 25 | 3 | 30 |
| **English fens, Icklingham** | | | |
| | 15 | 2.5 | 5 |
| | 25 | 3 | 10 |
| **Wangford** | | | |
| | 15 | 1 | 1 |
| | 25 | 2 | 2 |

Stems grow faster with hotter night temperatures, unless, as in the Wangford plants, they are stunted.

**Table 2.3** The effect of adding phosphate (sodium dihydrogenate phosphate) to peat-stunted seedlings

Another test showed no response to nitrogen alone.

| | Weeks | Number of shoots | Sheath height (cm) | Colour |
|---|---|---|---|---|
| **a)** | 6 (+$PO_4$) | 1 | 0.5 | Brown |
| | 10 (+$PO_4$) | 1 | 1 | Green |
| | 16 | 2 | 10 | Green |
| | 24 | 4 | 40 | Green |
| **b)** | 6 | 1 | 0.5 | Brown |
| | 10 | 1 | 0.5 | Brown |
| | 16 | 1 | 0.5 | Brown |
| | 24 | 1 | 7 | Green-Brown |

**Table 2.4** The effect of cutting on plants 6 months old

All cut at 6 months, re-recorded at 9 months.

| | Number of shoots | Maximum shoot height (cm) |
|---|---|---|
| Before | 42 | 70 |
| After | 41 | 85 |
| **Control** | | |
| 6 months | 41 | 80 |
| 9 months | 45 | 85 |

At 6 months, young plants recover well from being cut.

# Chapter 3
# Plant pattern and growth

## PLANT PARTS

*Phragmites* plants are highly complex! So this description will start with a small plant where there are sparse reeds in a dry (water *c.* 75 cm down), Tall Herb community. This is not in the reedbed (below), but is, for *Phragmites*, simple.

Figure 3.1 shows this plant, vertically and horizontally. For very obvious reasons of water and unpleasantness, such near-whole plant excavations are rare! This plant is about 4 m long from front to back, plus a *c.* 1 m branch, which failed to develop properly, and another branch which broke. Most is growing in a fairly straight line. This is the shortest way to try to get away from a poor habitat and try to reach a good one: and is usual in such places. Branches occur, and enable the plant to advance in different directions. *Phragmites* can live for long periods as a sparse species within another community, but is obviously less flourishing than where it dominates.

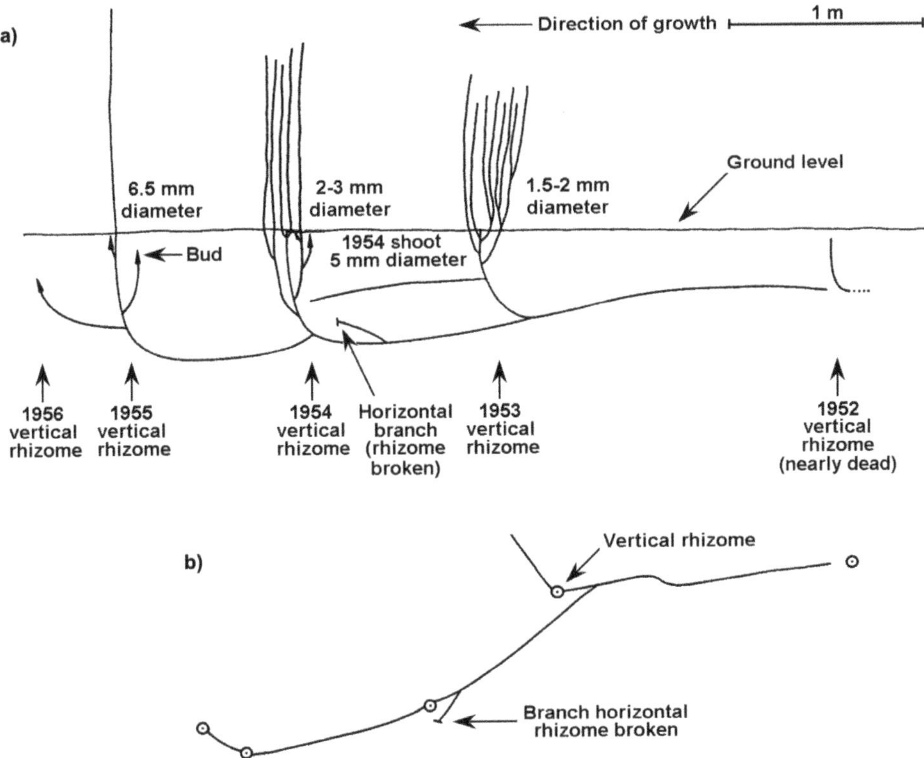


**Figure 3.1** Small *Phragmites* plant from dry fen, with strong apical dominance and restricted growth, Cavenham, England (Haslam, 1969a).

a) Vertical plan. (All aerial shoots of the same year; older ones mostly dropped.)
b) Horizontal plan.

The first point to notice in Figure 3.1a, is that lines go horizontal, across the page, and vertical, the latter are partly underground, partly in the air. This gives three types of stem:

1. horizontal rhizomes
2. vertical rhizomes
3. aerial shoots, reeds.

There is, in fact a fourth, the long runners, or *legehalme*, not here present (Figures 3.2 and 8.10).

Any grown plant has the first three, and may also have (Figure 3.3) curved and oblique rhizomes rising from the horizontal ones and becoming vertical by the soil surface. These are, except in position, like vertical rhizomes and occur in well-developed stands where divisions are less rigid than in Figure 3.1. Like any stem, leaves are borne on the nodes. The inflorescence panicle is on the tip of the aerial stem.

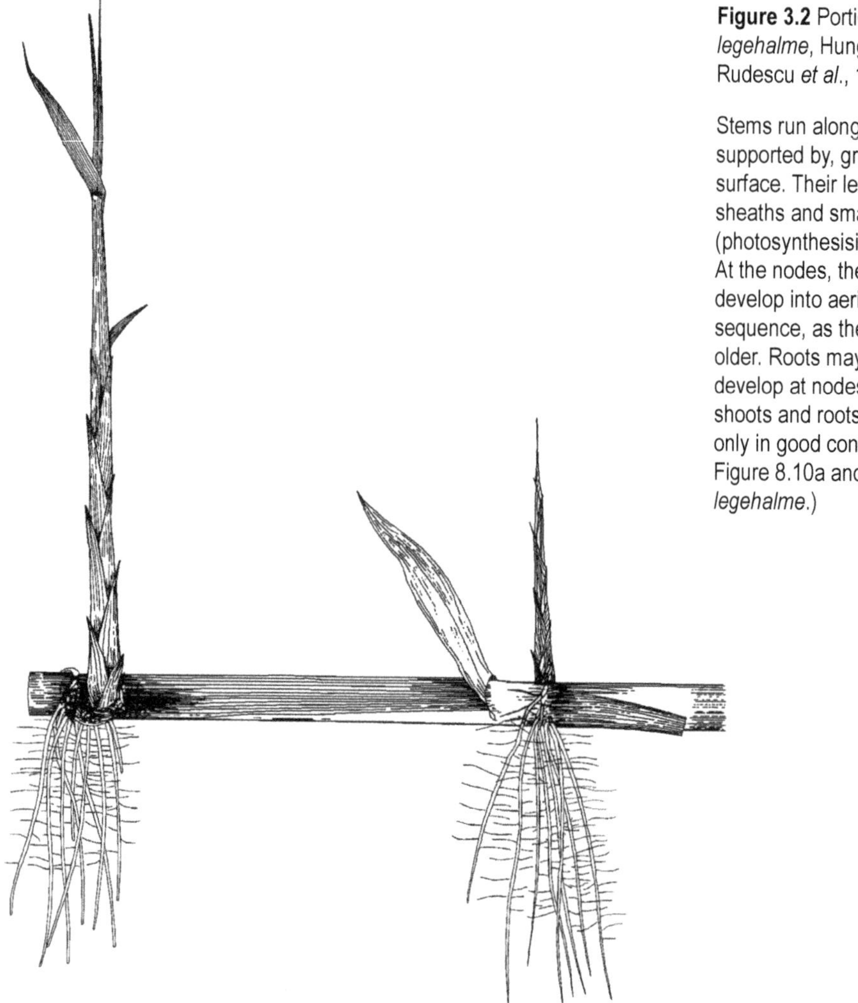

**Figure 3.2** Portion of *legehalme*, Hungary (after Rudescu *et al.*, 1965)

Stems run along, and are supported by, ground surface. Their leaves have sheaths and small (photosynthesising) blades. At the nodes, the buds may develop into aerial shoots: in sequence, as the stems grow older. Roots may also develop at nodes. Both shoots and roots develop only in good conditions. (See Figure 8.10a and b for whole *legehalme*.)

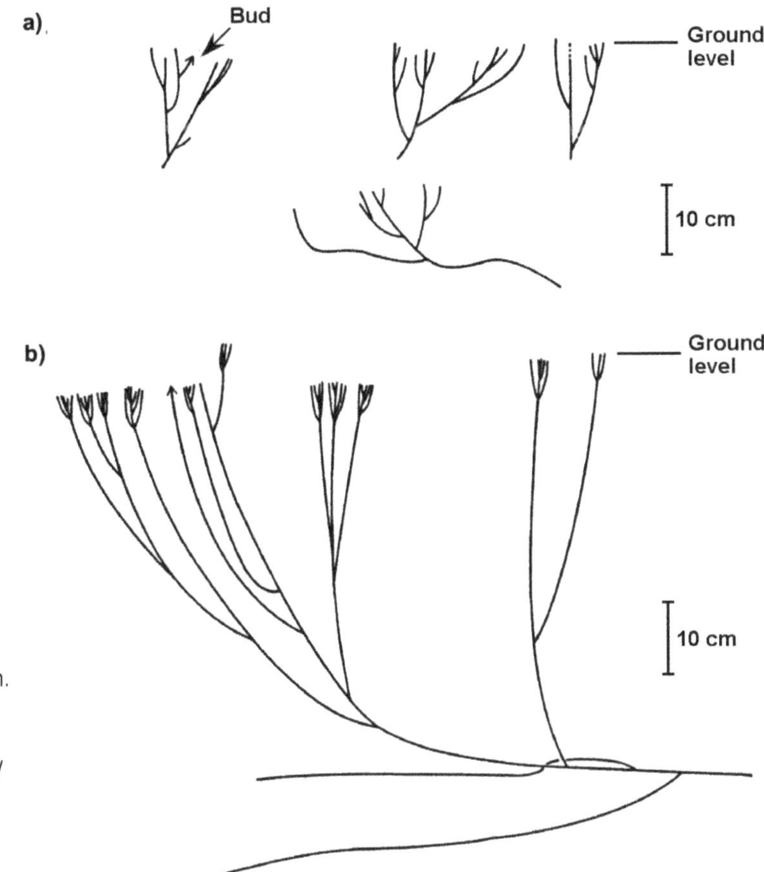


**Figure 3.3** Vertical rhizome systems from vigorous stands, Icklingham, England (Haslam, 1972)

a) Winter-flooded valley fen
b) Rather drier valley fen. Unlike Figure 3.1, these vertical patterns are complex, and, especially in wetter places, are slanting and curving as well as upwards.

## Stem types

The differences between stem types are summarised in Table 3.1. Basically the horizontal rhizomes stay horizontal, at the level at which they branched off from the parent (in developing or changing habitat they may first slant up or down). They also maintain their width. All will be much the same width, even when arising from a bud on a much narrower vertical rhizome. Figure 3.4, from the careful research of Rudescu *et al.* (1965), illustrates the plant structure better than words. These come from the superb reed habitat of the Danube Delta. In less clement habitats the rhizome internodes are less sausage-shaped. They are relatively longer, less wide, and keep this width through the nodes.

Horizontal rhizomes in poor conditions may die before four years old, as that in Figure 3.1. They may, though, live for at least seven years. Longer life is usually in more flourishing stands.

**Figure 3.4 a–k.** Detailed drawings of *Phragmites*, Romania (after Rudescu *et al.*, 1965)

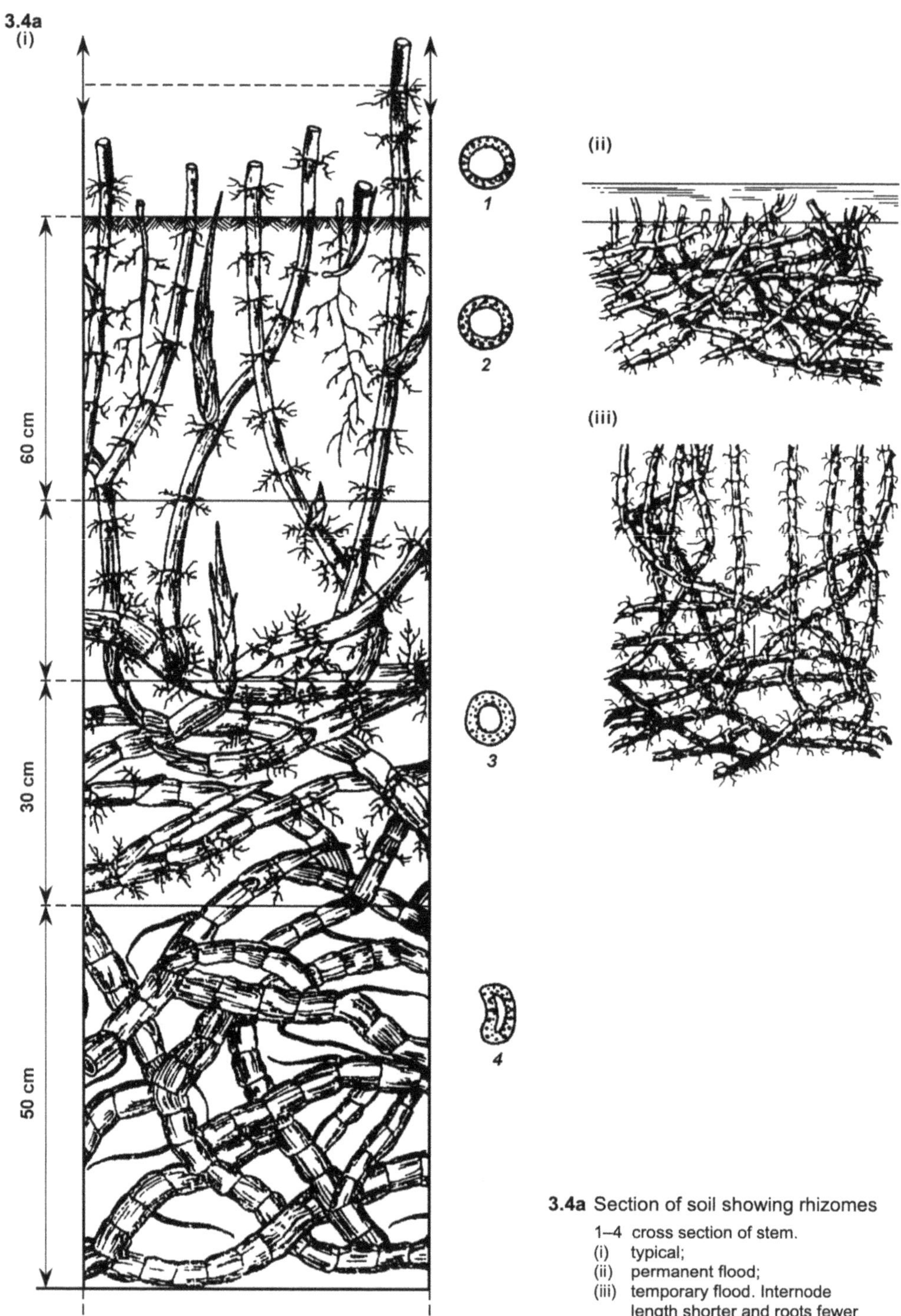


**3.4a** Section of soil showing rhizomes

1–4 cross section of stem.
(i) typical;
(ii) permanent flood;
(iii) temporary flood. Internode length shorter and roots fewer and shorter than in England.

3.4b

**3.4b** Emerging bud.

Note, especially, the stem tip is near the surface while the leaf tip is well above it. The roots are 'water roots' and as yet none have grown from the bud.

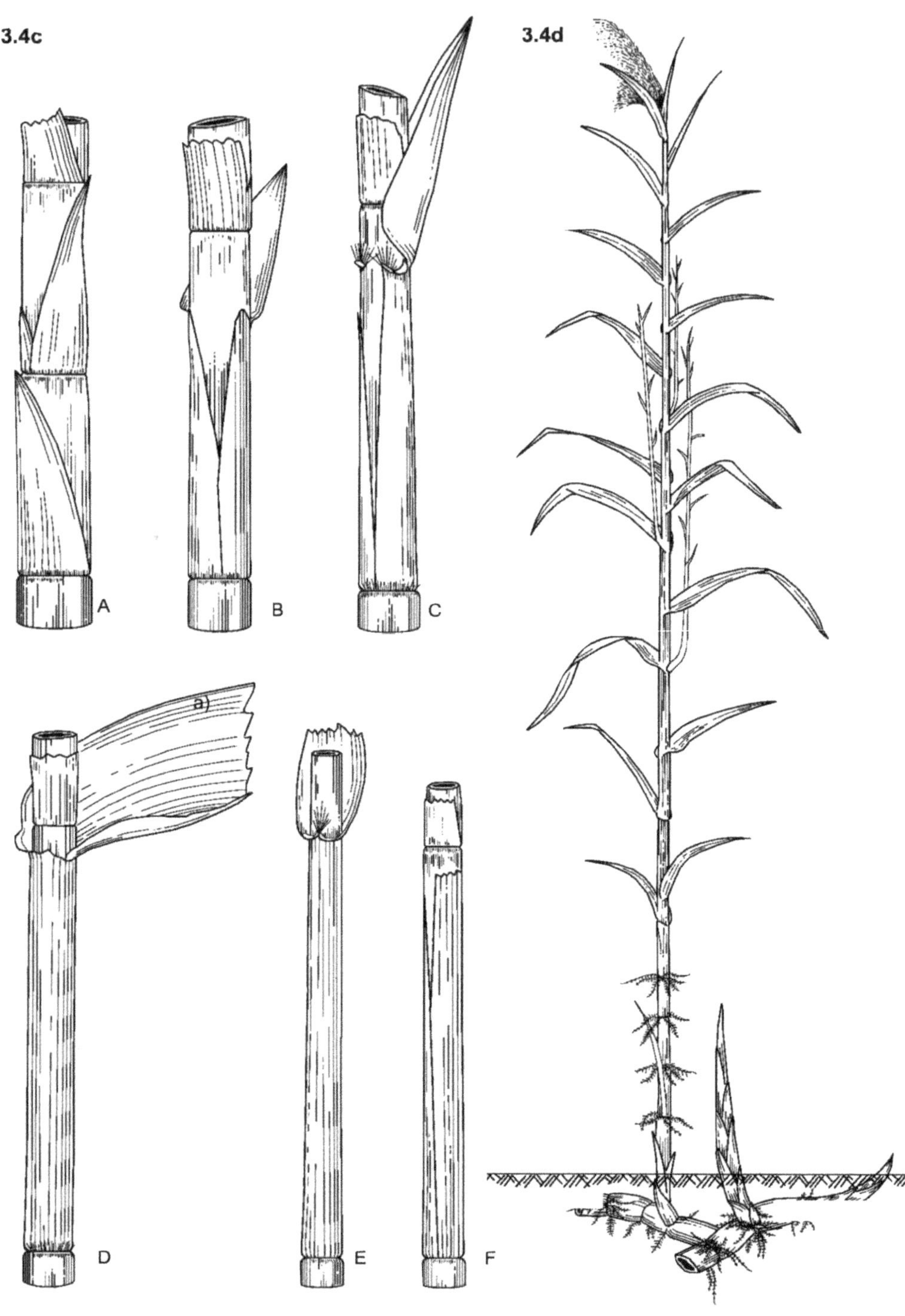


**3.4c** Consecutive internodes going up the aerial shoot.

A–F showing transition from rhizome leaves to sheaths and long blades.

**3.4d** The adult shoot.

'Water roots' present under water. Side shoots on 3 nodes: unlikely, except in damaged shoots, Britain, Malta.

**3.4e** Details of water roots, buds and rhizomes.

Rhizome internodes still much shorter than in England.

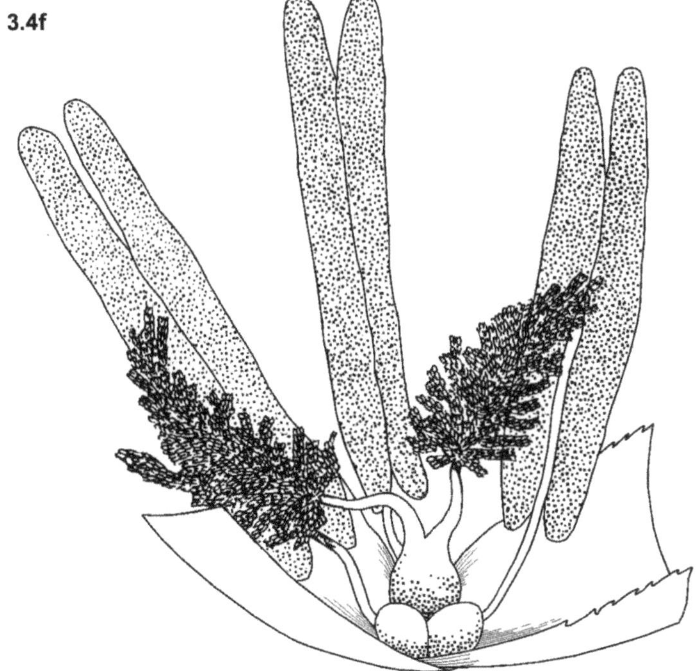


**3.4f** Flowers Detail.

**3.4g** Panicle development (pictures).

II Emerging.
IX Flowering, pollen dispersing.
X Fruit dispersing.

**3.4h**

(i)

(ii)

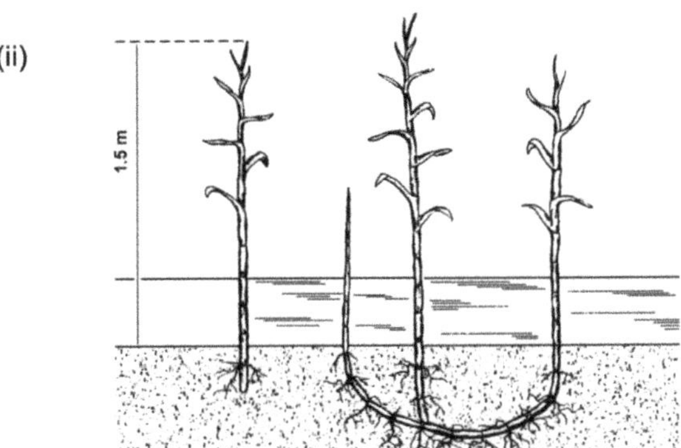


(iii)

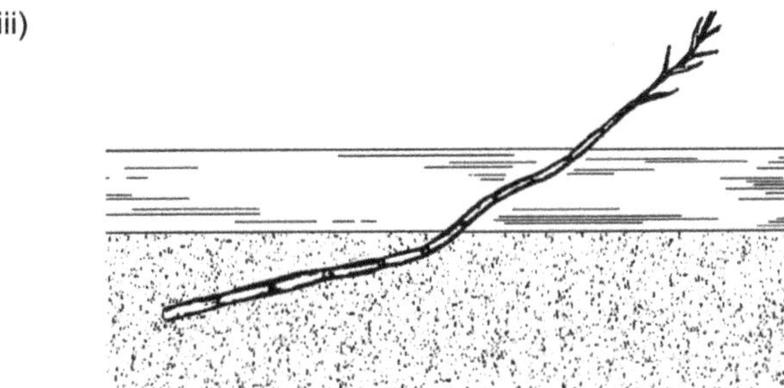

(iv)

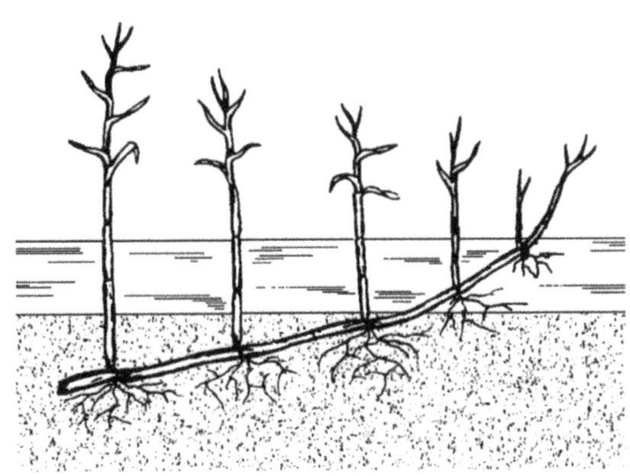

**3.4h** Propagation.

(i) from rhizome plus shoot;
(ii) from shoot, upright;
(iii) and (iv) development from shoot plus rhizome, oblique.

While these drawings illustrate development and growth, propagation in Britain is now mainly by rhizome 'plugs', that is, larger amounts of rhizome, where successful establishment is easier.

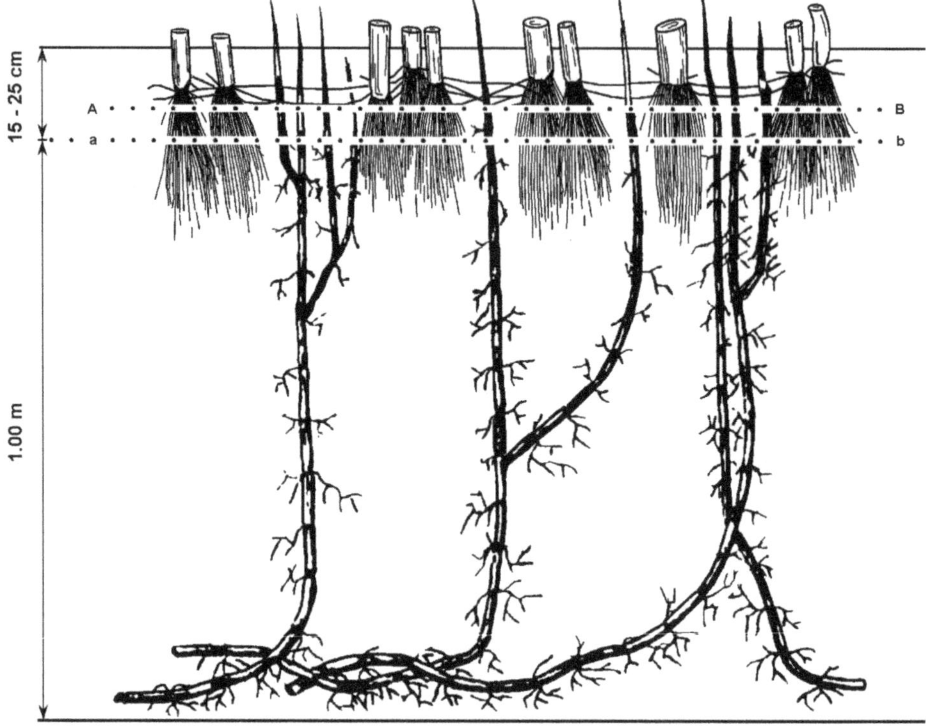


**3.4i** Competition underground avoided by species occupying different parts of the soil.

A/B The *Phragmites* here being unaffected – not narrowed – by another's root mat.
a/b Internode length much as in Britain and Malta.

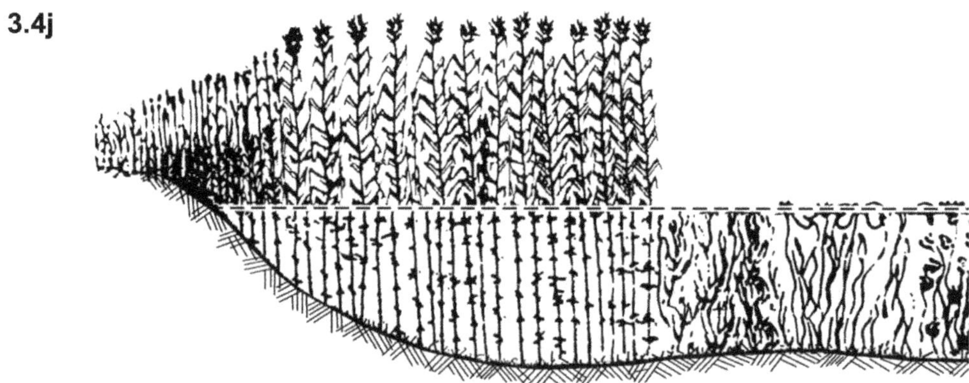


**3.4j** Typical position of *Phragmites* at the water's edge. Note that, without the East Anglian management of the marshland, *Phragmites* is absent there.

**3.4k**

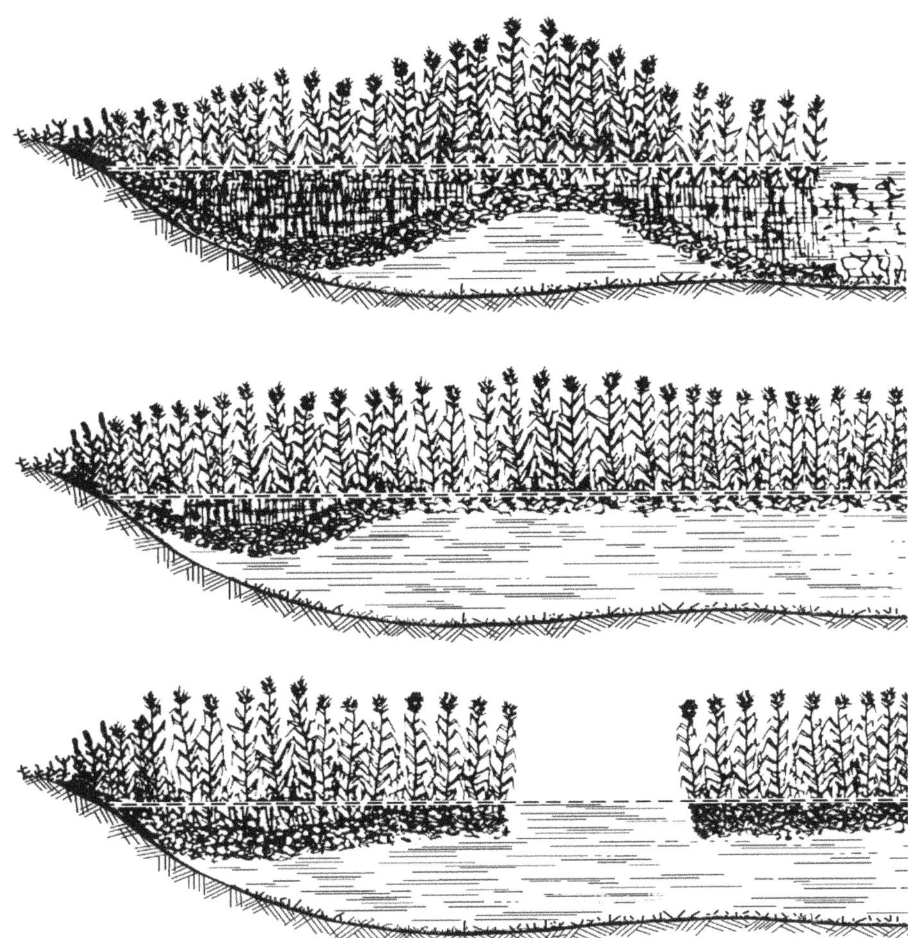

**3.4k** The development of plav (plaur, hover) from attached reedswamp in the Danube Delta.

Horizontal rhizomes are spread through maybe 20 cm–2 m of substrate. Such a narrow band as 20 cm occurs when this is the depth of soil over rock! Deeper ranges are normal, but vary with genotype and habitat (Chapter 5). To be large, pale, oval ('squashed') and juicy is a sign of good development.

Down in the soil, away from the seemingly important (and indeed necessary) aerial shoots, it is the horizontal rhizomes which control the plant. It is here that it is decided which of its buds shall grow, and whether to become horizontal or vertical rhizomes; and which of the buds on vertical rhizomes should grow into reeds, both this year, and in years to come. The potential size of the reeds, the flowering and fruiting, which small shoots should live, and which die, the root pattern, the movement of food up in spring, down in summer: all these are set into the plant. Unless, that is, the connecting rhizomes are severed, in which case all control is lost, or extreme conditions supervene above ground level (grazing, strong competition, etc.) when the pattern is modified. Although horizontal rhizomes turn green if exposed to light (artificially), photosynthesis is not their function, and they cannot therefore exist without the leafy shoots above. Curiously, putting reed under extra stress by late summer cutting seems to lead to more growth, so more death, of rhizomes.

Horizontal rhizomes are potentially endless. Perpetually in the state of being able to grow, and grow at the same size and in the same way, there is no reason they should stop. Certainly the implication of a reedbed staying in the same place for thousands of years (Chapter 1), in conditions near-impossible for new plants to become established, is that the original invading plant or plants stay in place. So, unless stopped by earthquake, fire or flood (more mundanely by drying, grazing, bog and suchlike), once there, a *Phragmites* horizontal rhizome remains there until conditions become unsuitable, years or millennia later.

Vertical rhizomes end. They narrow, and head for ground surface. However, they do have the ability to bear new horizontal rhizomes from their branch buds, so have the potential to be endless (Figure 3.4h shows development of rhizomes from transplants).

The relative proportions of each rhizome type vary. The more favourable the habitat, the larger and more complex are the vertical rhizome systems, vertical growth predominating over horizontal, the two types finally almost intergrading. In some stands (as in Switzerland, Hürlimann, 1951) vertical growth in favourable conditions results in an increased number of small vertical rhizome systems arising laterally from horizontal rhizomes, rather than increasing the complexity of the terminal vertical rhizome system. The pattern can also be intermediate. The rhizome pattern can vary with water depth, seasonal water supply and movement, and with genotype (see Chapters 5 and 6).

While most *Phragmites* is firmly anchored in the land, it is possible for mats to become detached. These are termed 'plav', 'plaur' or 'hover'. Their usual formation is shown in Figure 3.4, though in wet peaty areas it is possible to have lenses of water within the peat, and if they are near the surface, and covered in reed, the weight of the reed plus the movement of water or air may be sufficient to detach the reed mat with this upper soil. Plav have been recorded from large riverine reedbeds like the Danube Delta, small ones in the English Lake District (Pallis, 1916) and of course plenty in between (e.g., Grundling *et al.*, 2008, records ones growing in huge peat marshes in South Africa).

(Biomass is described in Krasovskii, 1962.)

The longevity of *Phragmites* plav depends on habitat. If with little soil and being driven around lakes with eroding scour, it is not long. But if merely gently moving from alongside one reedbed to another, and with ample nutrients from soil or water, it can be for years.

The vertical rhizome's main function is to connect horizontals with the surface, and to bear the prospective reeds. Unlike horizontals, but like aerial stems, vertical rhizomes narrow along their length. In both, the narrowing may be little or much (necessarily, a 5 m high, 8 mm wide reed narrows very slowly towards its tip!). These, therefore, do not have the ability to continue the plant – narrowing must end in tips – and are therefore not independent entities. However, there can be, once the shoot is above the soil area, fairly even narrowing along the length. This is the typical pattern. There can also be excess narrowing in the first *c.* 50 cm, which is usually seen where food is short (e.g., grazed stands) or growth is inhibited from outside (root toxins in the upper soil). It has been considered that water depth affects rhizome length (Weissner, 1991). Longer rhizomes may have less ability to tolerate sudden increases in depth. Oxygen transport did not change when tested, but could be worse in deeper water.

The bud may be halted around soil level, either by intrinsic dormancy (see below) or because it is too cold. If so, it continues growing very slowly, so there is a bunch of close nodes (Figure 3.5). If it started growth during the spring emergence, there will be little delay, and internodes are much longer.

Stems like this grow from two meristems or growing areas. There is the tip, the apical meristem (the more important one), which forms the organs, there is only one per shoot or rhizome. There are also the rings just above the nodes, the intercalary meristems, which elongate the main stem length, which is the internodes, the lengths between each leaf and root.

The aerial stems are given a great boost of food from the rhizomes, and push up, and grow faster than the horizontal or vertical rhizomes, though the longer *legehalme* grow even faster. (Also see Gorham & Pearsall, 1956, for shoot production in relation to habitat; and Isambaev, 1964, for rhizome in relation to habitat.)

The aerial shoots end. They end that part of the plant. In larger shoots and in better populations, the tip turns into a panicle, flowers emerge (Figure 3.4), seeds set, and the fruit is dispersed. They are essential but disposable.

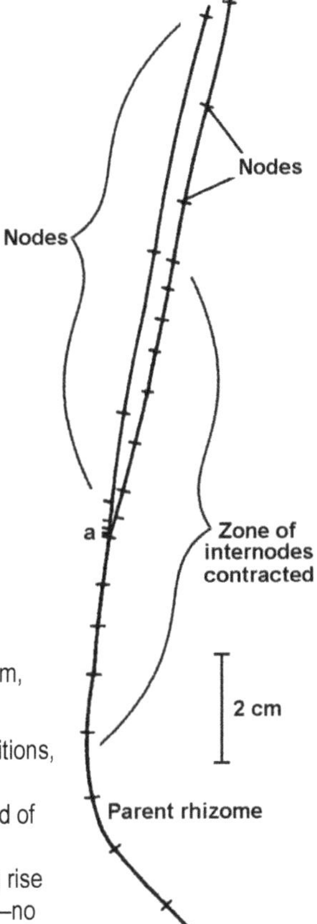


**Figure 3.5.** Speed of shoot growth and internodes around ground level (Haslam, 1969b).

Shoot growth continues, whether conditions are bad or good, but in poor conditions, while nodes develop (apical meristems), internodes remain short (intercalary meristems). Therefore the presence of contracted internodes indicates a period of near-dormancy. The parent bud was formed and grew near the surface in the autumn, where dormancy occurred, and apical growth continued slowly, giving rise to many short internodes. In late spring an auxiliary bud at (a) developed and—no dormancy—grew quickly into an early shoot.

Most reeds are only *c.* 2 mm wide at their tips, though larger reeds may be much wider at the base, so taper. Narrowing typically starts about half way up, while early tapering is encouraged by some competition, shortage of water or nutrients, cutting: and is more in some genotypes anyway.

In frosty climates reeds are killed in their first winter. Any branching comes because of damage: if the tip is damaged (grazed, gall, etc.), side shoots may develop. They are, of course, narrower, so shorter than the original one.

The modal height of reed stands can be anything from *c.* 10 cm to *c.* 10 m (one of the variable things about this plant): a substantial variation.

The height of reed in a stand is directly proportional to its width at emergence (Figures 3.6 and 3.7), though the actual angle of slope varies – a modal bud width of 7 mm can lead to a stand between under 1 m modal height to around 2 m, depending on habitat and genetics. Roughly speaking, the number of internodes is proportional to both height and basal width (Figure 3.8).

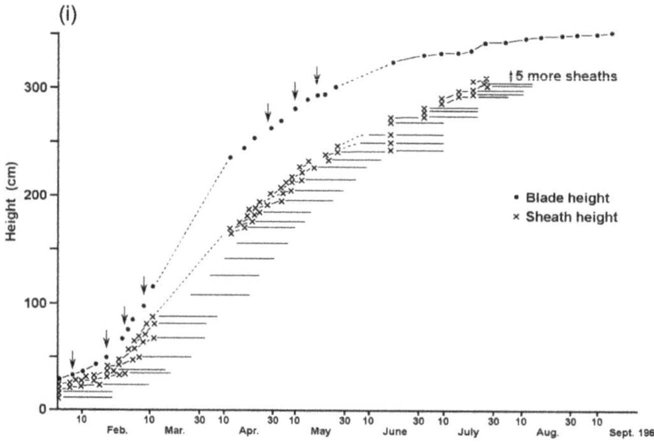


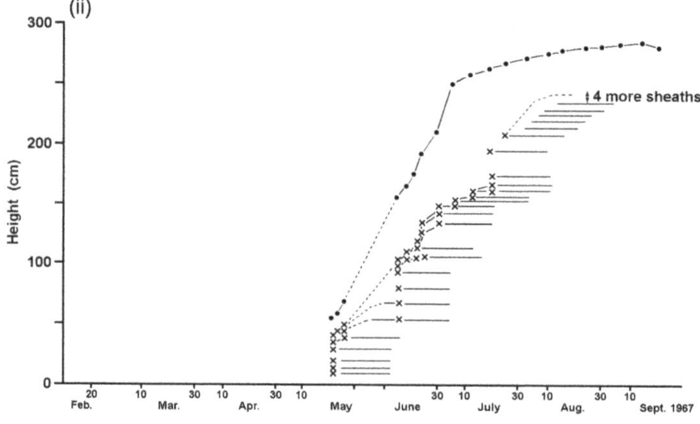


**Figure 3.6** Growth and development of shoots (Haslam, 1969c)

(i) Large (15 mm basal diameter) shoot in Malta. Emerged in February as part of the early emergence of mainly large buds. Growth is fast, the shoot is over 3 m high.

(ii) Fairly large (12 mm basal diameter) late-emerging shoot in Malta. Emerging late, growing even faster than (i): the left hand part of the sigmoid growth curve is missing. Being narrower, the shoot ends shorter than (i).

**Figure 3.6** Continued

(iii) Small (4 mm – a; 5 mm – b) shoots in Malta. Narrow, short shoots, which succumb to internal competition as too much of their food is removed. At 4–5 mm wide, in a typical English population, they would be large and flourishing shoots!

(iv) Large (for Braughing, England!) 7 mm shoot. This English shoot is one of the earliest and largest of its habitat, and emerges early: April. The growth curve is much like that of (ii), though nodes are fewer.

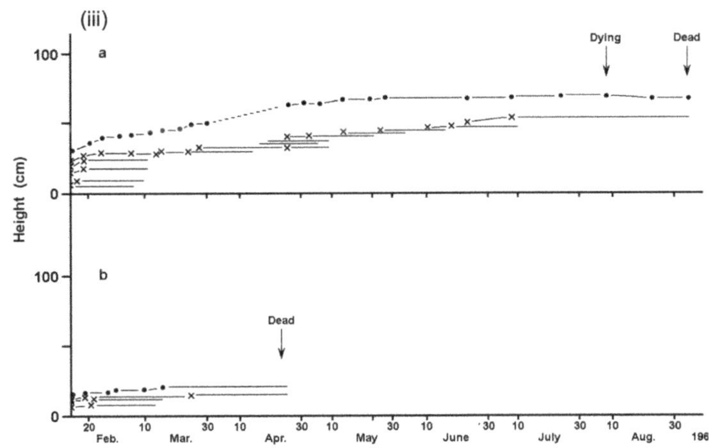


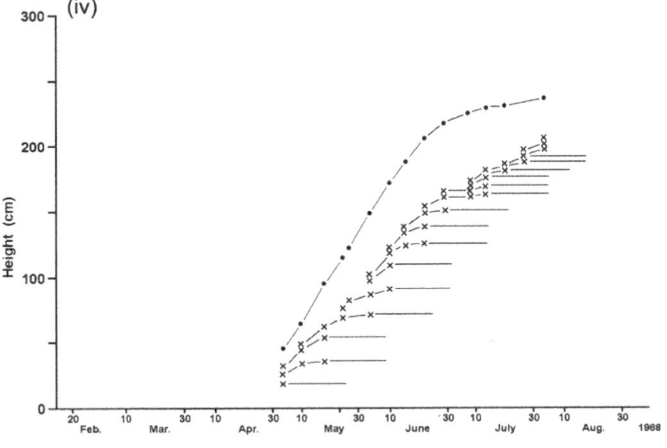


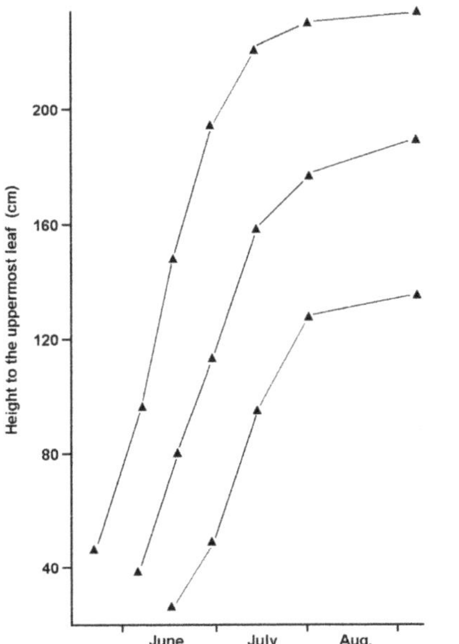


**Figure 3.7** The effect of (stem basal) width on growth (Haslam, 1972).

The tall, medium and short shoots are respectively wide, medium and narrow.

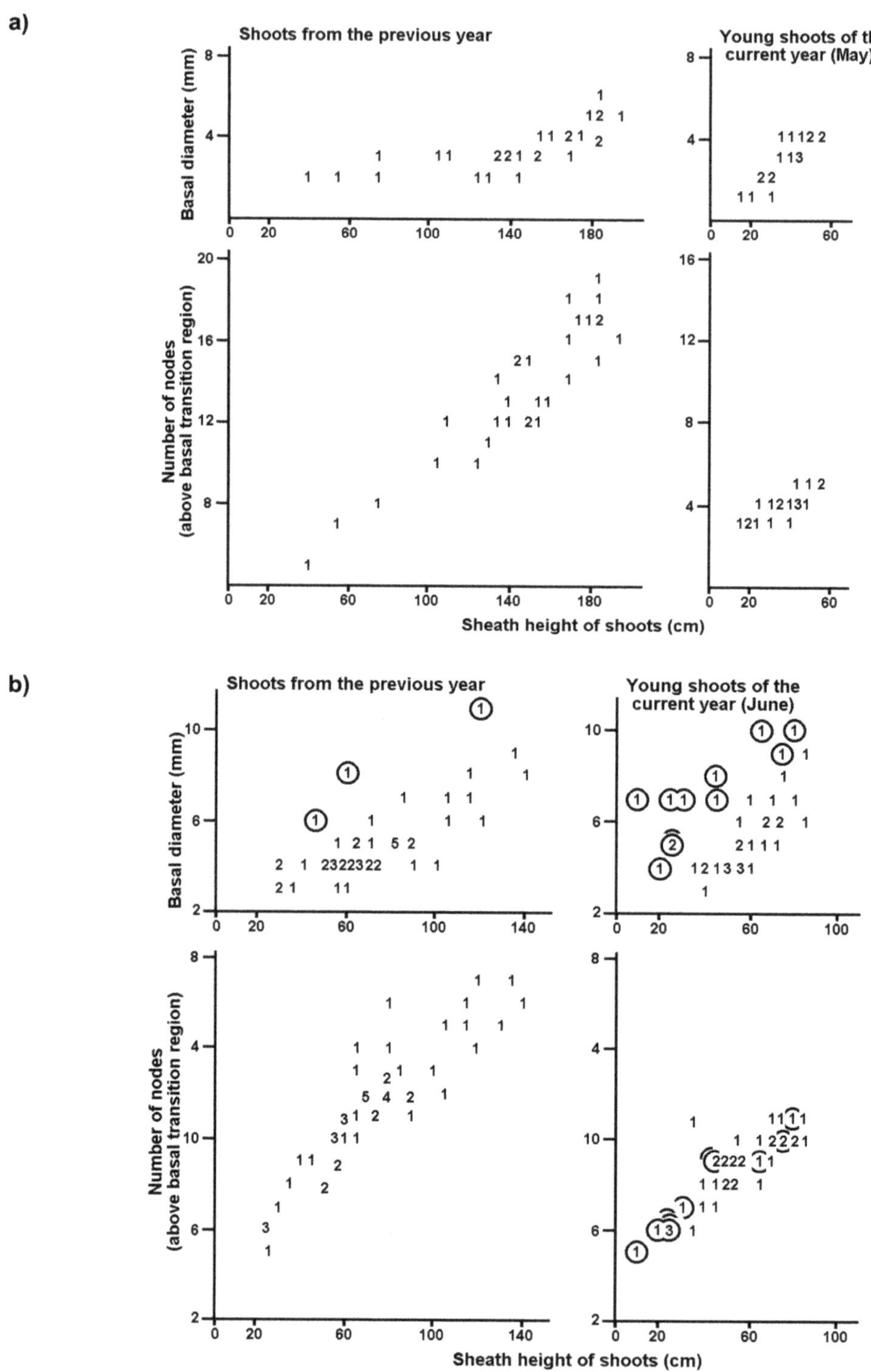


**Figure 3.8** Reed height in relation to (stem basal) width and number of internodes (Haslam, 1969c)

**a)** Dunfermline, Scotland. **b)** Lochluichart, Scotland. Circled shoots are summer shoots, comprising nearly a third. These are late, very wide for their height, but with the expected node number.

It is the taller stems in a stand which are most likely to flower (Figure 3.9). Flowering, though, as in the Malta genotypes and climate, may be on shoots as short as 20 cm. All reeds may flower, or panicles may be totally absent due to unfavourable circumstances. In English reedbeds, reeds too short for flowering often have short panicles which have not emerged. With a longer growing season, as further south, would they do so? (See Chapter 2 for fruiting and dispersal.)

Vertical rhizomes reach the surface. Each node bears a bud, and each one of the rhizome buds can develop into a shoot. Bearing these buds is also a vital process. If only the tip develops, the stand will be sparse. If a dozen or more, it will be dense (if there are enough verticals). The number of shoot-buds which develop depends on:

- the signals below about the state of the habitat reaching the rhizome;
- the conditions around soil surface, which may encourage or discourage shoot development (Chapter 4), especially during and just before emergence;
- the later history, whether reeds grow or are, e.g., grazed in May.

In Figure 3.1 the shoot buds are close to the original vertical rhizome. In Figure 3.3 some are more spread. This shows a more flourishing population.

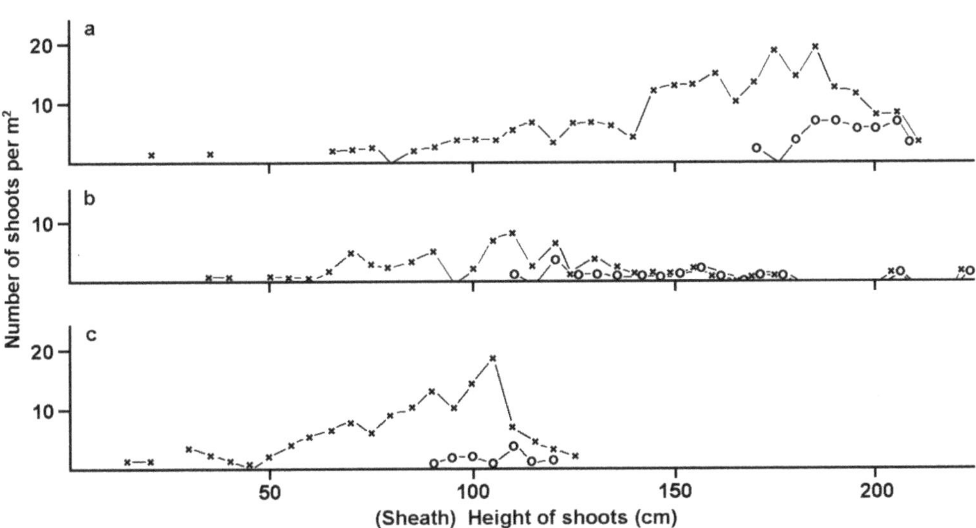


**Figure 3.9** Taller shoots are more likely to flower (Haslam, 1972).

Circled shoots are flowering.

a) Good population, Icklingham, England.
b) Restricted population, Eriswell, England.
c) Depauperate population, New Forest, England.

Lake stands – though equally monodominant – are usually sparser than marsh ones, even half the density. There, temperature fluctuations are hardly felt under the insulating layer of the water, and so bud stimulation from this cause is absent (see Matyuk, 1960).

*Legehalme* (Figures 3.2, 3.4–9 and Figure 8.10) are intermediate between rhizome and aerial shoot, and commonly have buds leading to both (also see Table 3.1). They are most often *c.* 1–2 m long, but can be over 10 m, and even reach 50 m in optimum conditions (enough water, soft soil, enough nutrients, no shading or damage, etc.) (see van der Toorn, 1972). Aerial shoots, well-grown, falling into water (around July in Britain or May in Malta) may resume rapid growth: more frequently in brackish than in fresh water. *Legehalme* may also grow out of the soil, from buds on the vertical rhizomes. This also is more common in a brackish habitat, and in advancing or young populations (it is also under genotypic control, Table 1.1). These grow together with the aerial shoots, and may be very dense, surrounding a young clone with a *legehalme*, and form new horizontal rhizomes (which gradually grow deeper). If so, the population advances rapidly. If not, as in Figure 8.10, they repeat yearly the unfortunately-futile attempt to colonise near the tip of the runner. A rare population composed entirely of *legehalme* is described by Bromfield (1841–2).

It is the experience of being out of the soil, in air or in water, which turns a rhizome into an aerial stem (except that the directive to become a *legehalme* overrides this). A vertical rhizome growing up into a tussock sedge (e.g. *Carex paniculata*) instead of, like its fellow, into shallow water and air, will remain a rhizome until it reaches the tussock top and comes into the air. It can even have some buds developing into (narrow) horizontal rhizomes within the tussock. (See Fiala, 1973, for further details on food movement, and van der Toorn & Mook, 1982, and Mook & van der Toorn, 1982, for performance and habitat factors.)

## At the stem nodes

At every node there is a leaf, with a bud in its axil, some of which grow (potentially all in soil and even some, at 10–30 cm above water). (See Rejmankova, 1973, for chlorophyll content.)

Rhizome leaves are sheathing, smooth, overlapping to form a tool which can bore through soil when young, becoming membranous behind, and are often shed young. This continues up the vertical rhizome, which, after all, still has to bore its way upwards, perhaps through very soft soil, perhaps through firmer material. Once above the soil, the leaf origins receive light, the sheaths become larger, fully covering the stem, and the blades develop. There is an absciss layer between blade and sheath and the blades are shed from the sheath by winter. The sheath remains, and forms part of the mature reed.

The first leaf blades are small and triangular, the later ones are larger and tapering, widest a quarter to a third the way along, and largest in all rather over half way up the stem (Figures 3.4 and 3.10).

**Figure 3.10** Growth of reeds (Sizes, stem patterns, leaf blade positions should be noticed. Compare Figure 3.4).

a) Early emergence. Some buds submerged, some other buds at ground level.

b) Middle emergence. Note small leaf blades: still the wider, earlier buds.

c) Rapid growth. First shoots have large blades bending out.

d) Nearly grown.

e

f

**Figure 3.10** Continued

e) Flowering.

f) Late fruit dispersal (December), most, not all leaf blades fallen. Reeds near-mature.

The difference in development between the bud which is wide and will potentially grow high, and that which is narrow, which cannot do so, is shown in Figure 3.11. At a similar height, the narrow one is adult, nearly as tall as it will be, with adult leaves, etc.. The wide one is juvenile, and has much potential.

The leaves on the rhizomes may disappear when their function as borers is over, but they may also remain while the rhizome lives, short and membranous. On the aerial shoot, the blades – even in lower latitudes – live under a year, dying in the autumn or early winter after they were formed (Table 3.2). The sheaths – and in the rhizome there is no division between sheath and blade – also die, but are not usually shed. They retain their function of surrounding and protecting the stem from damage, wear and tear, etc.. They may cover the whole internode, or more often they end just below, so the upper internode is exposed. (In favourable conditions, in some genotypes, this exposed part becomes pink.) When dead, they are as resistant to decay as the stem, though more easily torn away. While initial decay often seems more on the sheath in recently formed peat, leaves are more commonly near-intact than are aerial stems.

Aided by stems and sheaths, dead leaf blades decompose slowly, in air and on the ground, so may form litter mats. In water, decomposition is much faster, and in large water bodies there are currents which move debris, so in lake stands while there may be some debris on the soil, it is hardly a litter mat. In shallow-flooded and damp reedbeds, the litter may build up: particularly if the reed is not harvested, leaving the bed bare! At about 20–30 cm thick, the mat insulates the surface from temperature fluctuations (for effects, see Chapters 4 and 6). It also protects the bed from invasion by other species: few can grow by seed in heavy shade, and no short plants can come in vegetatively, without breaks in the litter mat. When it becomes really thick, however, 50 cm and more, the new reeds become very sparse, and shorter than those in a thinner litter mat beside. This, when tested in England, is due solely to the physical presence of the mat, as if it is removed, new buds emerge in a fortnight, too quick to leach toxins from the soil. Čížková *et al.* (1996), however, in the Czech Republic, think it may also be auto-toxicity, that leachate from the litter mat is poisoning the rhizomes below, inhibiting them from directing rhizome buds to develop into new shoots.

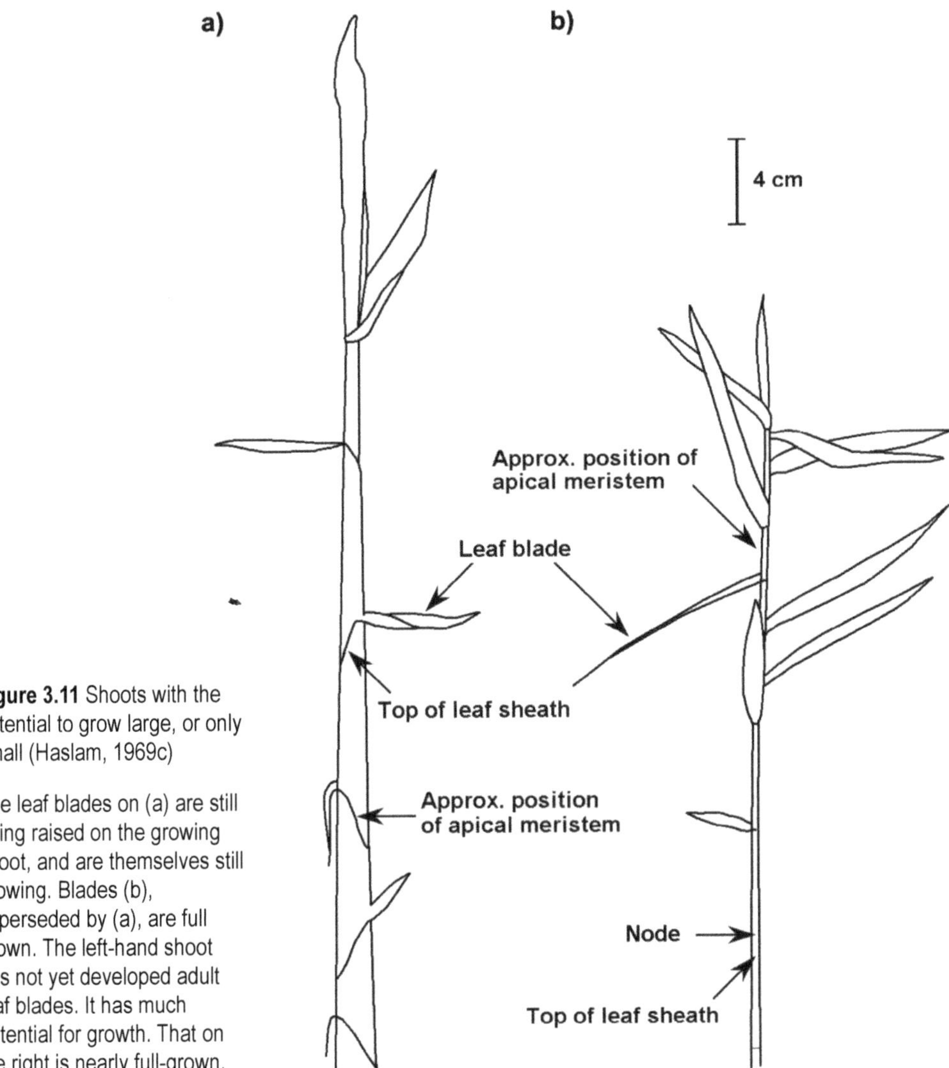


**Figure 3.11** Shoots with the potential to grow large, or only small (Haslam, 1969c)

The leaf blades on (a) are still being raised on the growing shoot, and are themselves still growing. Blades (b), superseded by (a), are full grown. The left-hand shoot has not yet developed adult leaf blades. It has much potential for growth. That on the right is nearly full-grown.

In general, the better the conditions for good *Phragmites* growth, the denser the roots (e.g., Weaver & Himmel, 1930). Roots are sparse on horizontal rhizomes, where many nodes bear none or two, 2–4 mm wide, arising from the underside of the rhizome and growing down *c.* 10 cm to well over 1 m, perhaps to >2 m. A second pair of roots may rise from the upper side, which are thinner, shorter and grow slanting down, not near-vertically.

This means such roots soon reach water down to 2 m below the horizontal rhizome, which can itself be at least 2 m below ground level. This gives a reach of 4 m to water, measured. *Phragmites*, that world-wide and hugely variable plant, is usually more variable than its few researchers have been able to quantify. It is not surprising (see Chapter 6) that *Phragmites*, in fact, grows further above water than that. Its main habitat is, of course, where water is available throughout the soil profile.

Roots also are essential to the plant. The tip is the main absorption site, and minerals, other solutes (good, bad and neutral) are all absorbed, as well as water. Water is essential for all life on earth, and is the largest component of all. Minerals are essential to make the new structure of the plant – stem, leaf, etc. – and the food to power these. *Phragmites* rhizomes have a good outside covering, and it is unlikely much of either can be absorbed through the undamaged skin. (Pieces of rhizome can absorb.) Each part of the *Phragmites* plant has its own, and irreplaceable part to play in the survival and well-being of the whole. Given this, *Phragmites* can manage very well on small pieces of rhizome or lower reed (or viable fruit) and reconstitute the plant!

Roots from horizontal rhizomes spread through the lower soil, including reaching where no other part goes. Going up towards the surface, the rhizomes, of course, become vertical, without an upper and underside. Roots are therefore equal around the rhizome.

In wet, high-nutrient places, roots are plentiful on the upper vertical and lower aerial stems, and there may be more than four at a node. These roots are quite different (Figure 3.4b) and are usually called water roots, as if initiated by the water here. In fact they are a developmental stage, as they do not occur in wet, calcium-dominated places where growth is much restricted (Chapters 5–7 and 11), and do occur, though sparsely, within the soil in dry, high-nutrient fens.

These water roots are short, narrow, often horizontal, and branch when several centimetres long, forming very dense mats.

*Phragmites* peat, when young, is fibrous, and this comes from these fibrous mats, in which dead leaves and to a lesser extent, stems, are incorporated. *Phragmites* is the chief peat-forming species in European (non-bog) wetlands. It is also the chief root-mat former. (Species such as *Cladium mariscus*, fen sedge, may form thick litter mats with their thick, tough leaves, but these are mostly in air, and only seldom pass into peat.)

## GROWTH

Figure 3.1 illustrates growth in this small, poorly-grown plant. In about July, when food from the reeds above has been pouring down into the rhizomes, and has already replenished the stores depleted by the effort of pushing so many reed buds up into the bed, horizontal rhizomes start to grow. They grow horizontally normally, of course. The typical length (in accessible stands) is about 1 m, and this, in frosty climates, is reached by near November. The boring tip then turns upwards and becomes a vertical rhizome, proceeding to the surface. Depending on various factors, including the temperature when it does so, it ceases growth below, in or above the ground surface. In lakes the latter is probably safe, but where it is frosted it dies and has to be replaced by a side bud developing. This aerial shoot, terminating a horizontal rhizome, is the one supplied with most food from the rhizome. Therefore the bud is the widest and (barring accidents) the reed will be the tallest and the most likely to flower. In thatcher's terms, it is a 'King reed'.

Meanwhile the rest of the rhizome has not been idle. Other horizontal rhizomes may have been developing, at the same level as the parent horizontal, or higher from a vertical rhizome bud, or, wherever it started, growing lower. All end in wide buds, as having long, subtending rhizomes. In addition, the vertical rhizome from last year's King reed is producing buds. The shorter the subtending rhizomes, the narrower the emerging bud and the shorter the final shoot. The better the conditions, generally, the more buds develop low on the vertical, giving more subtending rhizome and larger shoots.

The older the vertical rhizome, the less food is supplied to it from below, and the shorter and narrower the buds, and so also, the final reed. On the other hand, the older the vertical rhizome (before it starts to decay) the greater the number of (side) shoots it is likely to produce.

In Figure 3.1, as is typical for such a plant, the difference between the shoots on different-aged verticals is pronounced. In better conditions, though, there are both many intermediate (see Figure 3.3), and enough food to supply the back of the plant almost as well as the front, and bring those buds up almost as big. The King reed, of course, still remains the largest.

The most uniform beds are the commercially-harvested ones of, for example, the English East Anglia, cut for centuries. (Though management can create the same in wild beds within 2–3 years, see Chapter 10.) Here the management of cutting plus exposure to frost leads to dense, semi-uniform buds. (A bud frost-killed will have (1–)2–3(–4) replacements, sharing the same food, so each has less, and is smaller. Since the large buds tend to emerge first (Figures 3.6 and 3.7; Table 4.1) this reduces very tall shoots, and aids uniformity.) During the winter all to-be-emerging buds are dormant (except as described in Chapter 1). They still grow, but internodes remain very short (Figure 3.5).

By autumn, roughly half the final tally of emerging buds are present and near ground surface and, by the start of emergence, three quarters (the proportions vary). The internal or intrinsic damage imposed by the rhizomes is relaxed and the buds can grow up – if it is warm enough. In England intrinsic dormancy is relaxed by late February, but in cold springs and frost pockets emergence may not start until April. (Table 3.2 gives seasonal patterns in different countries.)

Emergence is normally spread over about a month. With frosty winters, short emergences are necessary, since as soon as the weather allows, reeds must start photosynthesising, making food. The other roughly half of the buds develop during this time, and grow straight up, without halting at the ground surface.

The young buds are reminiscent of rhizome tips, with shut leaves protecting them, and doing so for longer the wider the bud (Figure 3.4b). The green blades become larger up the stem. The green sheaths (and any stem receiving light) also photosynthesise. Only after about a month do the young shoots, or, rather, their upper rhizomes, send down roots. This means that they can obtain their own water and minerals, and are no longer totally dependant on rhizomes and their reserves. Somewhere, probably after this time, when leaves are reaching over half their number and have full-size blades, they are no longer drawing food from the rhizomes. No one can look at the thrust and power of emergence and think this energy can come from these tiny green things! Emergence is powered by, as well as organised (mostly) by, the rhizomes below.

This collective emergence also means a shading canopy, decreasing or preventing competitors, which is reached in hardly over a month. Figures 3.5 and 3.10 show the growing shoots.

This thrust to push food and buds upwards lasts until it is no longer needed, or until the food runs out. It is no longer needed if there is a good shading canopy growing with its own-made food, and merely receiving water and minerals from below. It runs out under two circumstances. Firstly, if rhizomes are themselves short of food (over-low nutrients, shade, etc.; Chapters 6, 11 and 12), they cannot support a dense, large stand. Secondly, if the emerged shoots are destroyed (grazed, frosted, cut, too heavily shaded, etc.) while the thrust is still upwards, meaning also that there is food to thrust, destroyed shoots are replaced. Supposing the rhizome system is satisfactory, June is the turnover month in England and the crop is replaced. In July it is not: unless minerals, fertilisers, are added. Is such replacement just because minerals were in short supply, and in a better-fed community July crops would emerge naturally? Maybe, but also, shoots take several months (Table 3.2) to reach their full height, flower and fruit. In countries with non-frosty winters the horizontal rhizomes' plans assert themselves after crop loss. A new crop comes, at any time of year, but growth is halted before the general emergence period, when up comes the 'proper' crop, as usual. In countries with frosty winters, the still-growing immature reeds will be killed, perhaps even before they have been able to refund the food needed for their growth, so being a net loss. So the absence of a replacement crop in July and August may be a protection to the plant.

During early growth, the apparent height of the reed is in fact that of the leaf sheaths. The stem tip is well down, protected by the leaves. It only emerges in late summer (panicle variation is described in Krzakova *et al.*, 2003) if flowering. The apical meristem at the tip forms nodes. The intercalary meristems in the ring above the node give the length. Nodes are short; internodes are long.

Seasonal growth varies genetically more than environmentally (Tables 1.1 and 3.2), and along a climatic gradient more than within genotypes in one area (see, however, Täckholm & Täckholm, 1941, for Egypt, in Chapter 1).

Reeds in non-frosty winters (including English reed in tropical glasshouses) do not die, or at least only the smaller ones die. The blades fall, but in spring the reeds produce a second crop. Some buds on the aerial stems develop into new green shoots, if plentiful giving the stand a bushy appearance. Naturally enough, these are narrower than the parent reed, so are short. Third-year reeds may produce yet more side shoots.

Similar branching occurs in the first year if the tip is damaged. If frost kills tips of shoots over *c.* 60 cm high, it may be only the upper part which dies, and side shoots arise from the lower part. Stems with galls at the tip (see below), if prevented from growing while still short, may also grow side shoots.

Flowering is in late summer, fruiting in the autumn. By fruiting much of the food has gone down to rhizomes, though of course the structural components (stem, leaves, panicle) remain. Fibres complete their mature length and density in August (when the final strength is determined), become hard in November, though their composition changes until December (Rudescu *et al.*, 1965). This is why thatching reed is not cut until after this (Chapter 10).

Leaf blades fall between late summer and late winter, depending on climate, genotype and temperature.

A blade falls after a breaking (absciss) layer forms. By September often only the upper half of the reed bears blades. Fall is hastened but not initiated by frost. It is earlier in a cold autumn, but if immature reed is killed (by frost or harvest) the dead blades remain attached to the stem.

Stems fall between their first and (usually) third spring as dead reeds, though they may remain longer (see Chapter 10). Stumps remain longer, so are available for aeration in shallow water or drier beds.

Reed fibres, whose length and number was complete in August, harden in November and contain their final chemicals in December. Food is moved from the shoots to the rhizome, leaving little more than the structural materials. The average mineral content decreases by 80–90% (Rudescu *et al.*, 1965). The mature reed rattles when handled on the bed: it is now hard.

Water is most crucial between +5 cm and -15 cm, where temperature fluctuations are received. A dry spring in a normally flooded bed may delay emergence and flowering by a month, and lower height by 30–50 cm. A dry autumn hastens shoot death. A wet autumn in a normally dry marsh delays maturation by up to a month (and, because of pending winter, may make the crop unusable for thatch). Unexpected deep floods in late spring may prevent any emergence, or allow just a few short shoots. The crop has normally recovered after four years.

The growth cycle is complete after about 16 months in frosty climates, the end of the reed's life overlapping with the new growth underground.

(This and the next chapter give only a very small fraction of the information available. See also the other publications of the author; Hürlimann, 1951; Rudescu *et al.*, 1965; van der Toorn, 1972.)

**Table 3.1** Stem types of *Phragmites*

*Phragmites* is extraordinarily variable. This table merely lists the most typical behaviour.

| | **Horizontal rhizome** | **Vertical rhizome** | **Aerial shoots** | ***Legehalme*** |
|---|---|---|---|---|
| *Direction of growth* | horizontal | slanting or straight upwards | upwards | horizontal |
| *Length determined by* | plant, habitat | ground surface | plant; bud width at emergence; habitat | habitat |
| *Length (annual, approximate)* | 0.75–2 m | 5 cm–2 m | 5 cm–10+ m | 1–50 m |
| *Width* | most; maintains width along length | usually less; narrowing upwards | least; narrowing upwards | width as lower aerial shoot; not narrowing |
| *Longevity* | 3–7+ years | same or less | one season, with cold winters | one season, with cold winters; branch shoots may survive |
| *Function* | food store; receive and respond to signals; controls; spread | bud density; as ← but less | photosynthesis; fruit production | spread |
| *Rigidity* | no | no | yes | no |
| *Adult leaves* | membranous, no blades | membranous, no blades | sheathing below, wider blades above | sheaths and small (or no) blades |
| *Roots* | downwards, no or little branching | upper, branched, fibrous, horizontal | below only, and as vertical | downwards, for establishment, where possible |
| *Branches may be* | horizontal rhizomes; vertical rhizomes | most bear terminal buds, i.e. verticals; also horizontals and sometimes *legehalme* | *legehalme* (in water) | aerial shoots; horizontal rhizomes |
| *Frost tolerance* | yes | yes, below; emerged buds are sensitive, least so in winter | no | not much |
| *Occur* | all viable stands | ← as this | ← as this | in some clones in some habitats |
| *Hormone content* | most | less | less | ? |

**Table 3.2** Length and position of the growing season

| | Main emergence | Rapid growth to | Growing season (months) | Fruiting |
|---|---|---|---|---|
| Britain | April | July | 6–7 | Oct/Nov |
| Canada (Central) | May | June | 5 | Oct/Nov |
| Czech Republic | April | July | 6 | Oct/Nov |
| Majorca | March | July | 8+ | Nov |
| Malta | February | July | 8 | Nov |
| Poland | May | June | 5 | Oct/Nov |
| Romania | February/April | June/July | 7–8 | Nov |

Flowering takes place July/August throughout, so may be fixed by photoperiod. Fruiting follows a couple of months later, and it may take several months for fruits to be dispersed.

A significant, and genetic difference is in emergence. This starts early in the warm winters of the south (February in Malta) and is not until the soil gets warm, May in Canada and Poland. This means growth can be taken leisurely in the south but must make haste in the north.

Transplant tests in England confirm Maltese and African material cannot grow enough in summer to flower. The autumn cold comes too soon.

# Chapter 4
# The reedbed itself

## Introduction

The reedbed, that mysterious vision of green waviness of the reeds moving in the wind, deserves study, too. It looks so simple, like an overgrown field of wheat or barley, like the pictures of old cereals, 1–2 m high. They are just the same, both are grasses.

Except, though, that this is *Phragmites*, an extremely complex plant! With such a complicated plant pattern (see Chapter 3), a reedbed must itself be complicated. The reeds, unlike wheat stems, are connected below ground, and this complex pattern is vital to the behaviour of the reed; whereas wheat stems are single, and consequently, in comparison, simple: not that wheat researchers would consider wheat simple! It is merely more simple than *Phragmites*.

## Emergence

In spring, buds gather near the surface (Table 4.1) having been formed through most of the year. There is then a period of rapid emergence, during a few weeks in frosty climates, and a few months, in warmer ones in Europe. The buds become ready at much the same time, and up they come.

Below ground a bud tip is normally 7–8 cm above the apical meristem (except in small lateral buds), and this tip is, as when underground, a boring tool to get the bud up: it can emerge through *c.* 10 cm of tarmac. The leaf sheaths are first. Emerged buds up to *c.* 20 cm may still have their apical meristems in the soil. Dormancy lasts until the apical meristems come above ground (however, though more often under water, some dormant buds have their tips above ground). Buds remain in the soil for between about ten months (summer-formed) and about two weeks (formed in late spring). Few buds for next year are formed in June and July.

The gathered buds are pushed up (Tables 4.1–4.3; Figures 4.1 and 4.2) and the reedbed is 'awake' once more. During the autumn and (warmer periods in) winter, occasional buds do come up, but are killed: except in warmer countries like Malta, where they survive. Generally, older, vertical rhizomes bear more and smaller buds, until 'old age' means decrease. The cycle may be only three years, or up to about seven. More are, of course, formed in favourable conditions, and indeed in optimum stands even first year rhizomes (Figure 3.1) can bear up to *c.* 6 side buds.

The start of emergence is well-defined, occurring over a very few days, and many buds come up initially. Then the rate decreases. Tables 3.1, 4.2 and Figures 4.1, 4.6, 4.7 show climatic and weather patterns. Onset in non-frosty Malta is close to 1 February, independently of temperature. In The Netherlands, early February is possible. The earliest recorded in Britain is late February, in an unusually frost-free site, and in plants kept warm. Normally, cold weather prolongs dormancy for a month or two more. There can be, in East Anglia, a month's difference within 100–200 m between emergence in a warm (by a spring) and a cold (Tall Herb) one (also see Chapter 6).

Spring and summer (June) frosts have greatly decreased in the past three decades, so frost-killed buds and young shoots have also decreased. Temperature fluctuations in the upper soil increase bud formation (bare soil more than litter mat, in equal habitats). Frost-killing increases it more.

The largest buds tend to emerge first (Table 4.3). These are those on longest rhizomes (Chapter 3) and on those formed earliest (generally speaking). (No such statements cover 100% of behaviour.) Since frost-killing leads to 1–3 replacement shoots (and indeed burn-killing, more) density increases and, since food has to be spread between more buds, the buds and shoots are smaller. With little frost, English stands have a 2–4-week emergence, though if there are many frost deaths, particularly from frosts over a period, it may last two months.

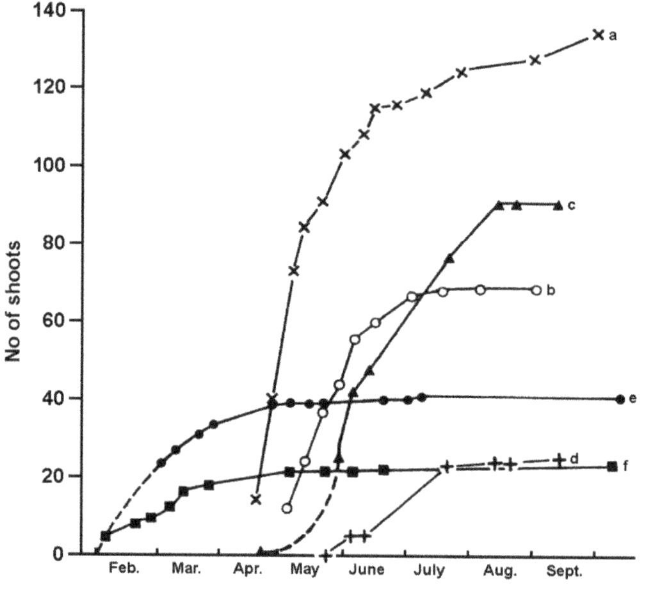


**Figure 4.1** Emergence in relation to temperature (Haslam, 1969b)

Malta sites, (e) and (f), emerge first, February, and the main emergence is spread over two and a half months. There is no temperature constraint. The English sites (Cavenham and Icklingham) have bud stimulation from frost, for both number and date, (a) has most frost, (d) least.

Each from 1 m$^2$.

At the end of general emergence, buds are small (Table 4.3) as are stragglers after the main event. 'Summer shoots', however, form an exception. They are wide, even if arising high on the rhizome. These emerge May–August in England and The Netherlands (van der Toorn, 1972), later in Malta (Figure 3.6). They may comprise up to a third of the total reeds, just some, or none at all. They develop while food is being produced by the reeds, the first-forming shoots growing quite tall. However, when food patterns change, and food is being withdrawn to the rhizome, then summer shoots often die. If they survive, they may well be too young, immature, to flower and mature properly in autumn, and end as little more than dead (soft) grass. The fact that, after fire, summer shoots may come up a month earlier suggests that those vertical rhizome buds which will produce summer shoots are designated early. Malta's pattern is similar, but less prescribed, emergence being more spread out, and there being more narrow early buds and wide late ones (of emergence: not summer shoots, which come later).

Burning breaks intrinsic dormancy. So does cutting a rhizome (e.g., by ploughing), and a rapid emergence follows over perhaps a month (in warm weather). Giving earlier (as well as denser) emergence by burning is another reason why it improves the crop: a shading canopy develops earlier. (Over-scorching burning can lead to delayed emergence, and six times the buds, so a dense, but short and late stand.)

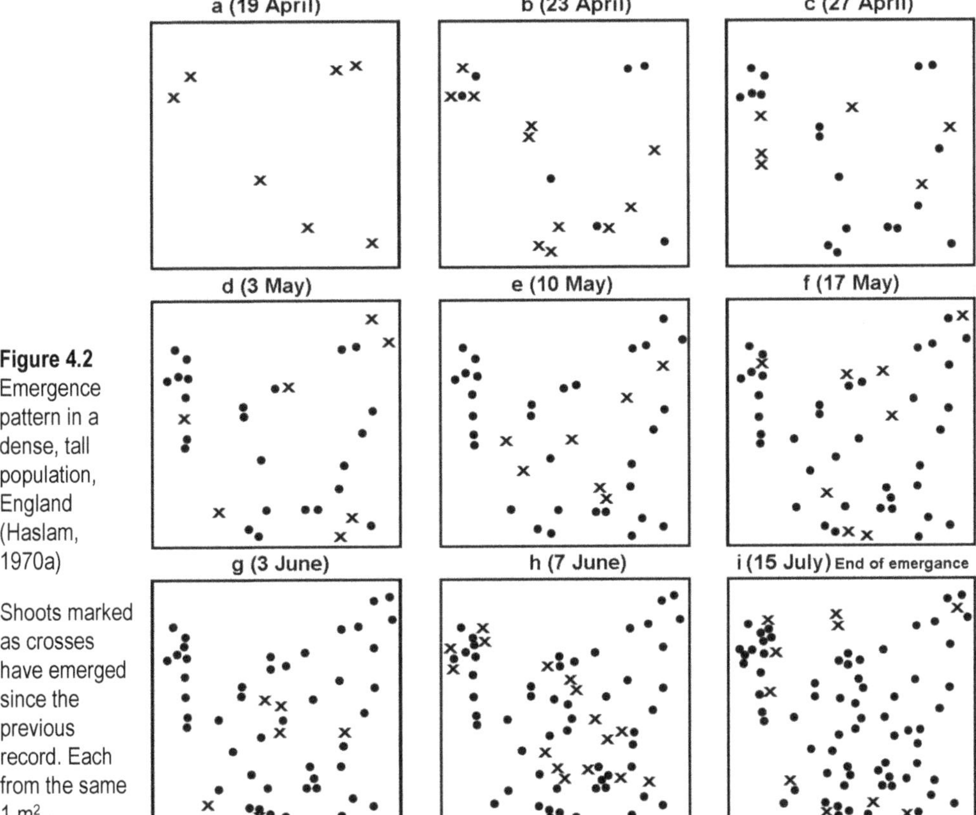


**Figure 4.2** Emergence pattern in a dense, tall population, England (Haslam, 1970a)

Shoots marked as crosses have emerged since the previous record. Each from the same 1 $m^2$.

Cutting and grazing during emergence do not lead to denser replacement shoots, and after emergence ends, there may be no further replacements (Table 4.5, and see Chapter 3). Some nutrient-rich stands can persist under regular July cutting, but are in a poor state.

Shallow flooding is good for bud development, provided rhizomes are aerated through standing stems at the time. By providing insulation, flooding allows earlier emergence—but not increased density.

Emergence can be delayed because of factors acting at least 2–3 years before. With an average rhizome age of four years, really bad frosts, severe burning, unexpected spring drought or deeper flooding, etc., all affect all living rhizomes. Emergence is delayed: until the majority of rhizomes are new ones, no longer influenced by the disaster. Really long-term effects on reedbeds are described in Chapters 10 and 11. That there are 'carry-over' effects on the bed from events likely to be unknown to most researchers, is important, and should not be forgotten in interpreting a bed.

## Development

In a sparse stand shoots may emerge in clusters (Figure 3.1), while in denser ones (e.g., Figure 3.3) they are more scattered. Figures 4.2 and 4.11 show a representative pattern of a reedbed. Since larger buds, on average, emerge first, these also grow fastest and end tallest (Table 4.3). Where the earliest emergence is of buds formed last summer, though, the final shoots are slightly shorter than the next group, those ending long rhizomes. Figures 4.3, 4.4, 3.7 and 3.8 illustrate various aspects of growing shoots in connection to relative time of emergence, basal width growth, node number and season, effect of internal competition and other causes of shoot death (also see Table 4.4).

Shoots emerging in spring may flower, and they mature and sclerify in autumn. Most English reeds are 4–5 months old at flowering, but in Malta, with a longer growing season, they may be up to 8 months. Because it is winter-warm there,

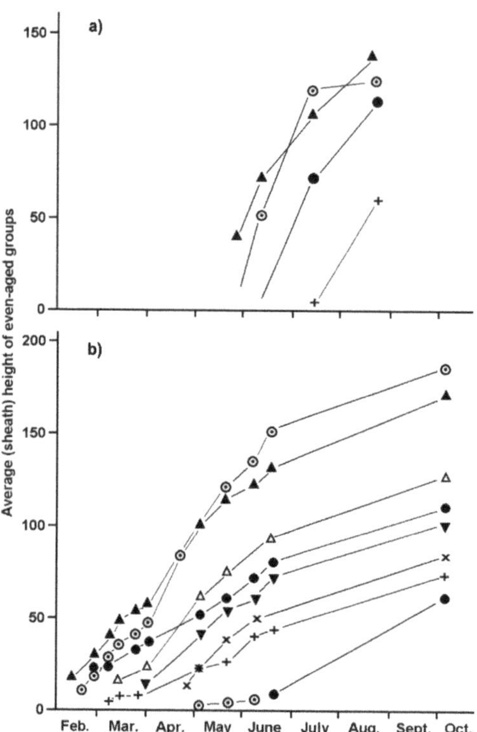


**Figure 4.3** Growth of even-aged shoots in relation to their time of emergence (Haslam, 1970a)

a) Icklingham, England.

b) Malta. Early shoots grow taller (unless damaged).

a)

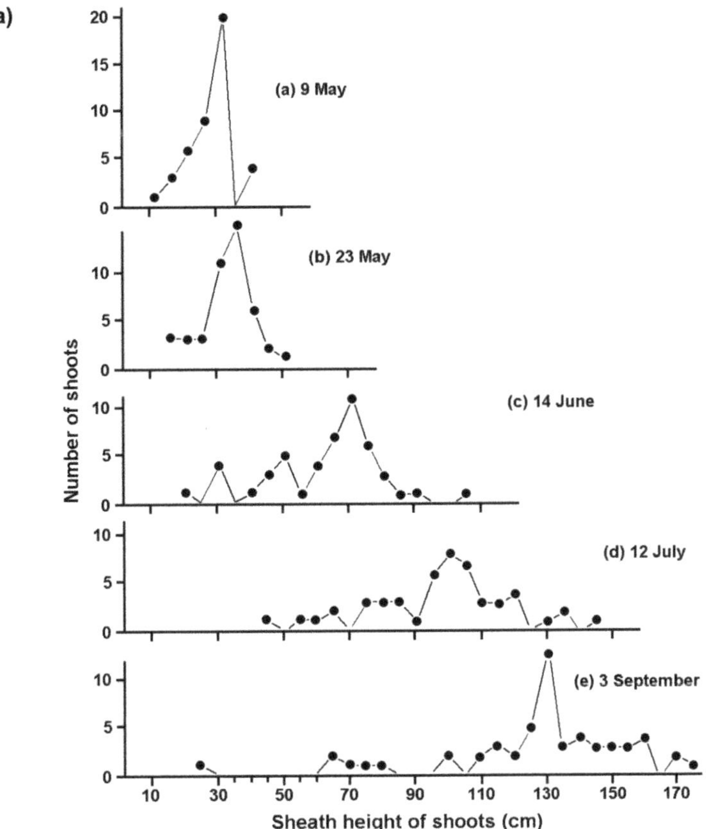


b)

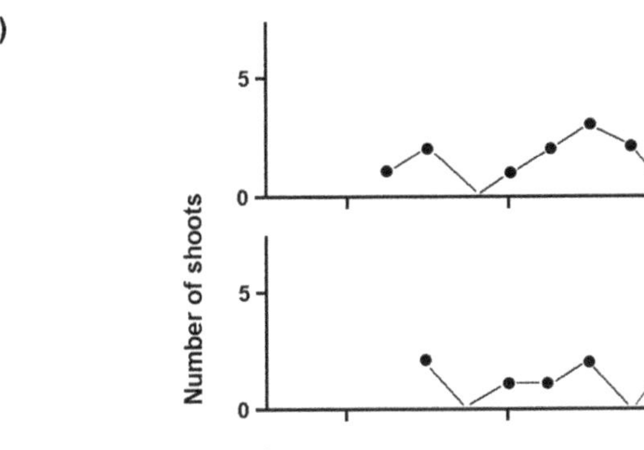


**Figure 4.4 (a, b, c)**

Growth curves in different populations (Haslam, 1970a)

English fens. Note that, as the reeds grow, the modal height moves to the right, and does so the more, the better the population.

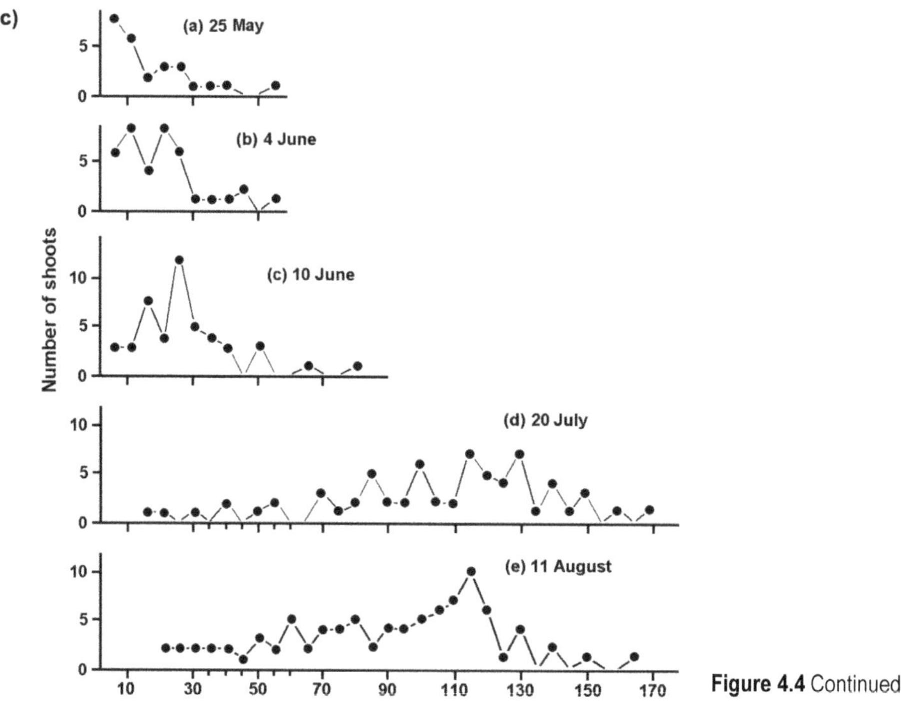


**Figure 4.4** Continued

however, large shoots emerging in June can flower in late autumn, when it is too cold in England for that last growth, meristematic tissue being killed. In general in England, June shoots will mature, and later July ones will not, but this also depends on habitat factors like temperature and water regimes. Severe June frosts may mean few shoots flower, as can heavy pest infection (see below). Internal competition and very late emergence also lead to reeds immature in autumn.

## Growth

In early emergence, necessarily, most shoots are very short, and only a few, rather short. The peak of the growth curve is therefore to the left. Figure 4.4 shows how the peak of the curve, the modal height, the height of the typical shoots of the stand, necessarily grows, i.e., moves to the right, as the shoots grow. However, most good stands are like this one in having, at full height, quite a few shoots taller than the peak height, and more, shorter.

There may be no main peak at all, the reeds varying evenly in height (Figures 4.5d and 4.4b). One such example is a good dominant stand, but in Malta, where variation is great; the other is a very poor (calcium-dominated) stand. (So, once more *Phragmites* is variable!)

Basal width (earlier, bud width) is the best determinant of final height (Figure 3.8), accidents excepted, though even this varies with larger habitat factors like temperature (Figure 4.6). That warm temperatures mean greater height does not mean warmer latitudes have taller reed.

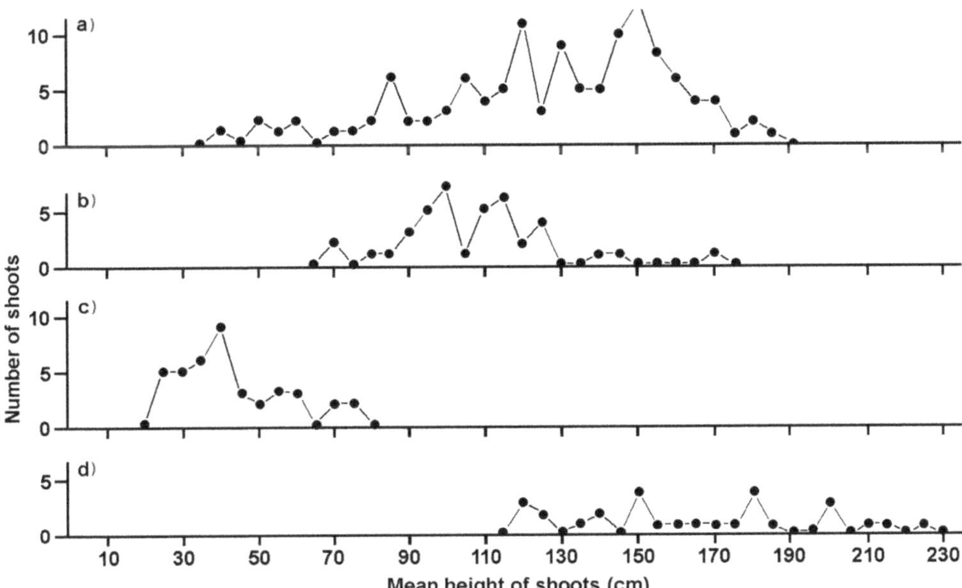


**Figure 4.5** Height distributions in different population types (Haslam, 1970a)

English fens, final height.

a) Good, optimum population. Height peak is at right.

b) Less good, sub-optimal population. Height peak central.

c) Depauperate, short stand, probably grazed. Height peak to left.

d) Bud development restricted by habitat. No peak.

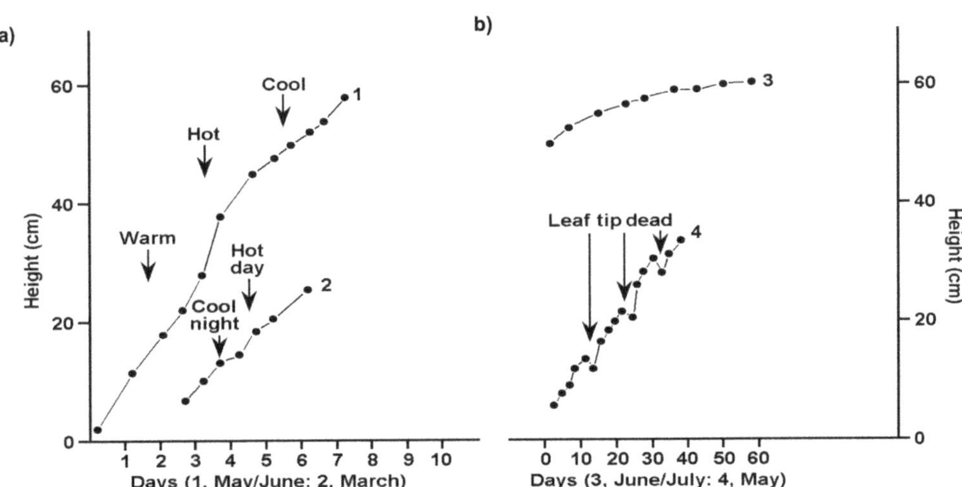


**Figure 4.6** Effect of daily temperature on growth (Haslam, 1969c).

Selected periods of growth. High day temperatures increase growth, the more the faster the growth.

a) England. Significant variation with day temperature.

b) Malta. Noticeable, but less such variation, growth rates are slower, so individual days' temperatures have less effect on the final height.

On the one hand giant reed is not found further north, it occurs in, for example, Romania, Turkey, Iraq and Egypt. Pakistan and Malta reed reaches greater heights than British. On the other hand not just Dutch but Swedish reed grows even taller than Maltese.

During growth, node number obviously increases and node number is proportional to height (Figure 3.8). Leaf width (modal leaf blade width of the widest leaf of the shoot) varies more than would be supposed (2–>80 mm, usually 10–40 mm in Britain), there being a strong genetic component also. It is said it is partly controlled by factors determined within the bud. (Leaf tips are often lost during growth, so leaf length is not reliable for study.)

In theory, all fully developed *Phragmites* shoots flower, and it can be thought that non-flowering ones have been hindered, by developmental or habitat reasons. The proportion of flowering reeds varies from 0% to nearly 100%.

The primary influence is that in the bud at emergence. Wider buds (see Chapter 3 and above) lead to taller shoots which are more likely to flower (Table 4.5). A reed 1 m tall is more likely to flower as a tall shoot in a short stand (modal height 90 cm) than as a small one in a tall stand (modal height 190 cm). Bad weather during or before flowering decreases it (by maybe 10%).

The panicles emerge slowly, over 2–6 weeks, around late July in England. The larger ones come first, as they are on the larger reeds. In England, once the large ones are out, there are frequently small (1–3 cm) ones unemerged. These come out, and flower in warm weather and good other conditions. In warmer Malta, over 90% of shoots in a dominant stand flower. No panicles have been found in British stands of under nine nodes (modal number). Those with over 12 always have panicles, and most individual flowering shoots have 14+.

## Early death and damage

Early death in some populations is shown in Figure 4.7 and Table 4.4 (and see Mook & van der Toorn, 1982).

Young shoots need food from rhizomes, and when there are large and small ones, the large may take the food and the young die. This is more common in Malta than in England, as really large shoots are common there. Next, there are frosts, killing 0–90% of young shoots, up to several times. Then, stem borers arrive, ‘reedbugs’ being a local and convenient term (see below). Internal competition between reeds potentially self-sufficient in food starts in July in England, and in May in Malta: earlier in Malta because wide shoots emerge earlier. Most dense reedbeds lose some small shoots. If these vulnerable, smaller shoots are in a different, e.g., Tall Herb, community, however, the rhizome system arranges they be supported, not have their food withdrawn or die. While most food goes back to the rhizome, the dead reeds remain, and in East Anglia are estimated to contain, in kg/ha, calcium 2.5–6.5, phosphorus 0.8–3, nitrogen 16–42, potassium 4–15, sodium 7–19, magnesium 2–5 (Haslam, 1969c). This returns to the bed when the reeds drop, but not when the bed is harvested; then this must, long-term, be replaced, for maintenance. Nutrient-rich water from farmland enters the beds. Death is mainly in small or summer shoots. Shading decreases vigour, but unless constant and heavy, does not kill. It presumably aggravates damage from the (more-severe) harm from food withdrawal.

No population shows massive deaths from all causes, and most have only minor damage, from one or more (Figure 4.8). Severe reedbug infection is local, and low in cold weather. The absence of litter increases frost damage but lowers pests. A replacement crop matures late, so has little damage from internal competition.

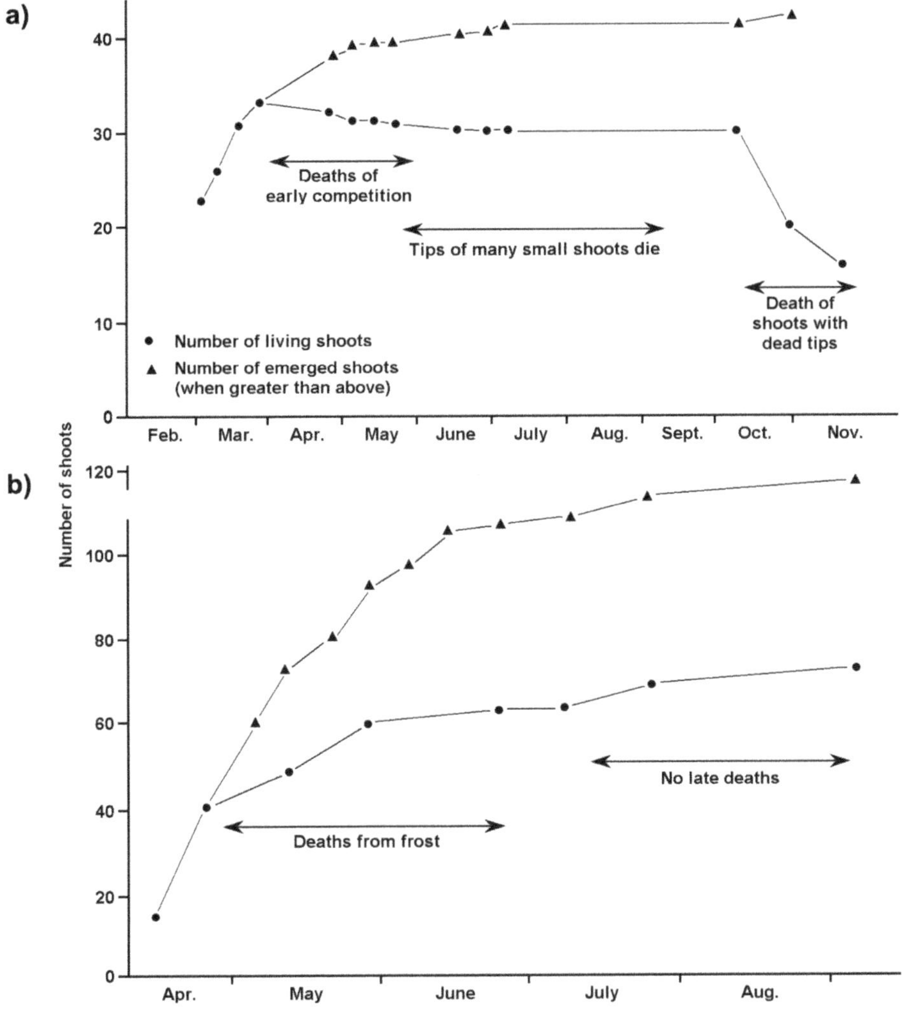


**Figure 4.7** Early deaths of shoots in dominant reedbeds.

a) Optimal stand, with a very wide range of stem widths encouraging internal competition. Malta.

b) Many frost deaths, sub-optimal stand, Cavenham, England.

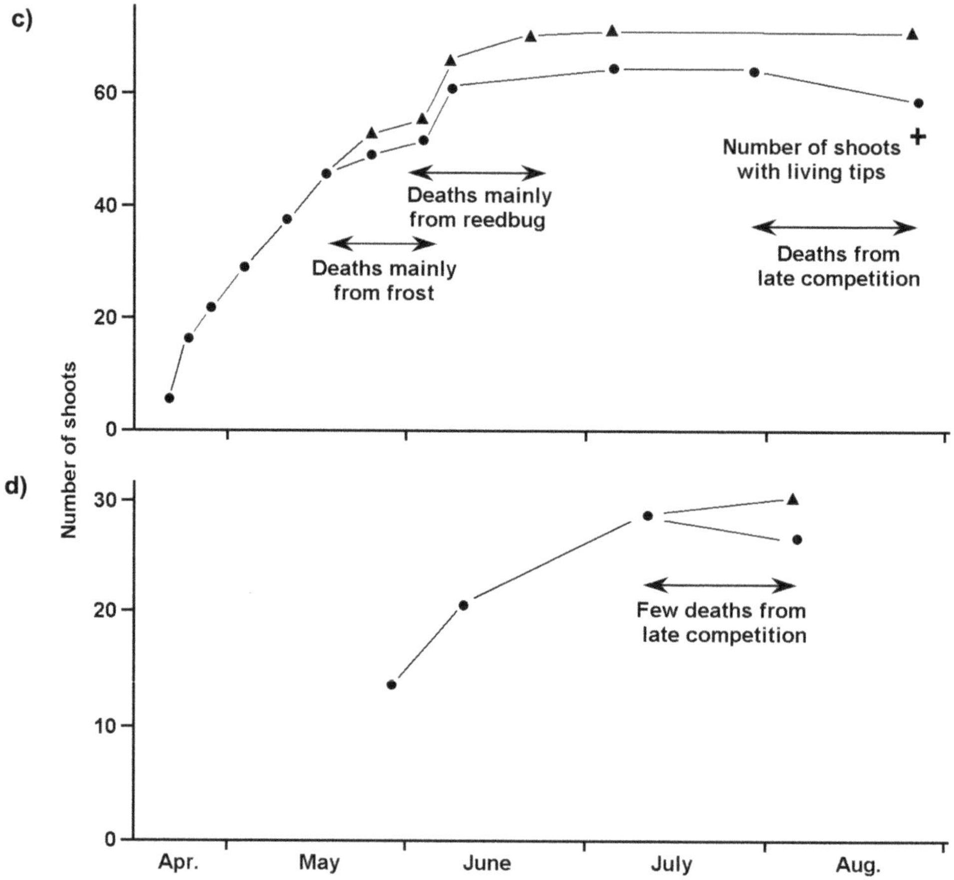

**Figure 4.7** Continued

c) Some frost, and litter mat encouraging reedbug, optimal stand, Cavenham, England.

d) Sparser. though dominant stand. Little frost, little litter, few deaths, Woodwalton, England.

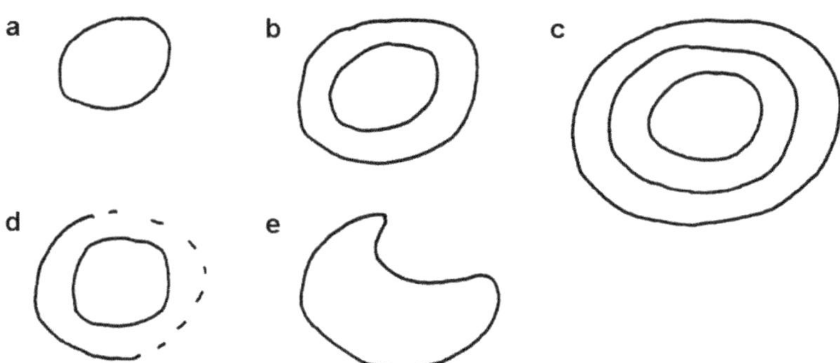


**Figure 4.8** Growth of young established clones, horizontal plan, Rosyth, Scotland (Haslam, 2003).

a) First. This is last year's size. b) Second year ring around first. c) Third year ring around both.

d) Clone has met unsatisfactory conditions, and second year shoots to the right are sparse, not dominant.

e) Piece 'bitten out' of the dominant clone by it meeting unsatisfactory conditions on one side.

## ADVANCING MARGINS

*Phragmites* spreads where it can, when possible forming dominant reedbeds. In a sense it is constantly spreading, as new horizontal rhizomes develop each year, and these spread the reeds as well as the rhizomes of each particular plant piece. Advancing stands, though, are usually considered as those spreading the reeds into places formerly without them, whether the wave advance of the reedbed or the sparse advance of sparse reed (Figure 4.9).

The reeds in two sides of an advancing young clone are shown in Figure 4.9. In (a) there was (near) bare soil, and there is dominant wave advance. In (d) grasses were already present, and root competition means that the reeds are sparse, and cannot form a shading canopy, so cannot keep out even short (effective) competition. The clone is the same. On one side a reedbed is increasing rapidly. On the other, there are merely a few odd reeds in grass. In both, the rhizomes are all travelling in the same direction, out from the clone centre. It is this which distinguishes the advancing stand: all new horizontal rhizomes travel forward from the margin.

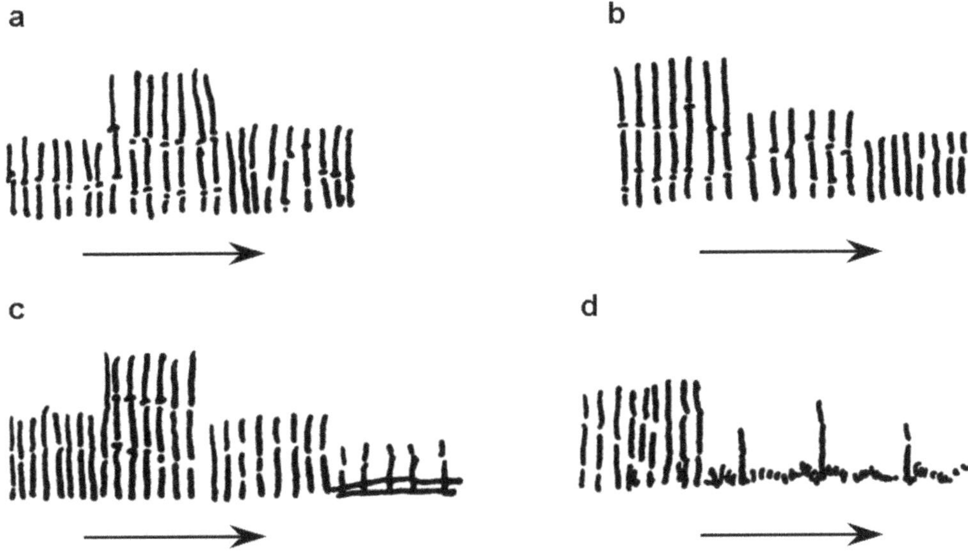


**Figure 4.9** Clones with advancing margin, vertical plan (diagrammatic), Rosyth, Scotland (Haslam, 2003)

Arrows indicate direction of growth.

a) A single ring this year. Last year's shoots are the tallest here, but other height patterns occur.

b) and c) show a second and shorter ring of shoots, grown in late summer. In b) this second ring is from new rhizomes, in c) it is from legehalme shoots, usually shorter and sparser.

d) This years' and last years' shoots were restricted by root competition with the other species, and are very sparse.

Figure 4.10 also shows a few other variants of advancing margins: variation of height where dominance occurs, the presence of *legehalme* (a clonal characteristic but needing good habitat), and dominant advance blocked by competition. To have just the one annual margin is the most frequent.

In (a) the clumps each contain the King reeds, and also later-emerging rather smaller lateral reeds. These all formed the same summer, the King first, then other smaller ones. Referring back to Figure 3.1, in the advancing state the vertical rhizome here can develop side shoots in its first year which, in an established stand, would take three years to grow. There is an internal difference. Inhibition of bud growth is relaxed, and food from behind powers the production of enough, and tall enough reeds to form a shading, killing wave. This may have up to 200 shoots per square metre.

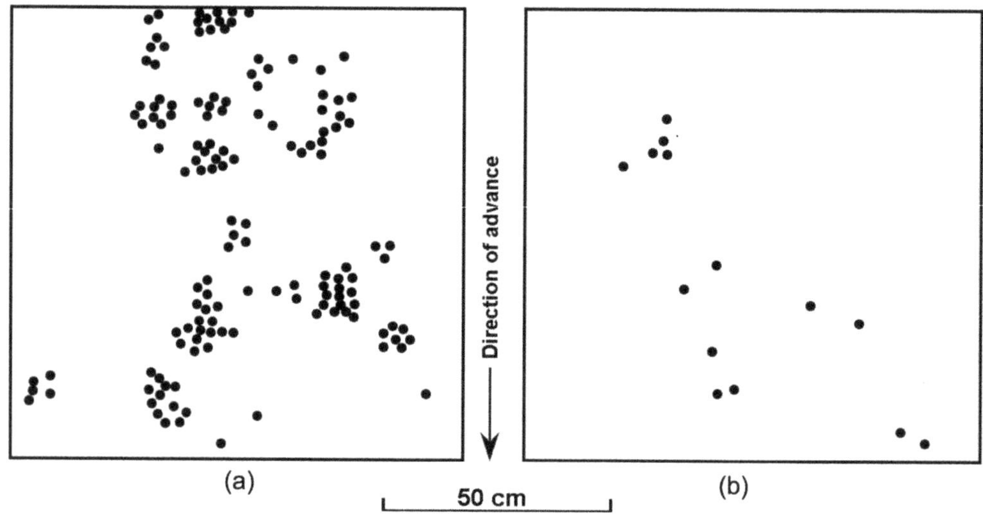


**Figure 4.10** Shoot distribution in advancing clones, Rosyth, Scotland (Haslam, 1970c)

a) and b) margin of clones.

a) Good conditions.

b) Opposite side of same clone. Affected by competition, few lateral shoots, reeds sparse.

c) Rapid spread, first year rhizomes, with up to 5 lateral shoots on new King reeds, becoming dominant.

Another pattern (Figure 4.9d) is more than that in Figure 3.1. The rhizomes are there, but there are few or no side laterals in the first year. The shoots are wider and taller than in the first year of Figure 3.1, but the only essential difference is that all the horizontal rhizomes are growing outwards in this first year.

The distance advanced in a year varies with habitat and genotype. If *legehalme* are not just present (as in 3 of 20 advancing clones in the Figure 4.10 marsh, but dense as in only one of the clones) advance can be extremely rapid. Here it is 5 m in a year, but up to 50 m has been recorded (van der Toorn, 1972). Shoots from the *legehalme* are necessarily late-growing, as the *legehalme* has to grow first itself, before bearing new shoots. So there is an outer fringe to the clone of short shoots, then the very tall shoots of the wave advance.

Even without *legehalme*, the wave advance can be double (or more?) the ordinary year's growth. In some clones there is the ordinary horizontal rhizome, growing in autumn, and then a second one, growing later. Both have their aerial reeds, but in the outer zone, the outer advancing zone, they are a little shorter. Again, as mentioned above with lateral rhizomes, there is a metabolic difference to established stands. Inhibition of bud growth is relaxed, energy is abundant, and two years' rhizome growth or more occur in a single year.

By the third year, in the hinterland, the population is like any other established, dominant reedbed: except it still has the ability to push food forward for the wave advance.

The first advance in the year, that from autumn-formed rhizomes, has the tallest shoots: food is ample. Later advances, by rhizome or *legehalme*, if present, are shorter, being shorter the later the shoots arise (see Chapter 3). This is partly because they have not had time (before winter) to grow to that height, partly because food supply is being cut back (Chapter 3). Sometimes, without the latter, the outer margin does reach the height of the inner (clonal differences).

## Dense tall stands, short stands, sparse stands

Established *Phragmites* can be any of these: tall or short, dense or sparse. On size, there is an overall genotypic variation and control (Chapters 1 and 5), but any reedbed with monodominant reed over 1 m high can be termed 'tall and dense' (not only those 8 m high!).

Figure 4.11 shows the distribution of shoots in such a reedbed in different years. The reeds are fairly evenly spread. As seen in Chapter 3, large aerial shoots tend to rise from autumn-formed buds developing low on the parent rhizome. These are found on rhizomes of all ages in dense stands, but typically only on young ones in sparse stands. Small shoots tend to arise near the ground surface, the buds often developing in winter and spring. They also occur on rhizomes of all ages in dense stands, but typically only on old ones in sparse stands.

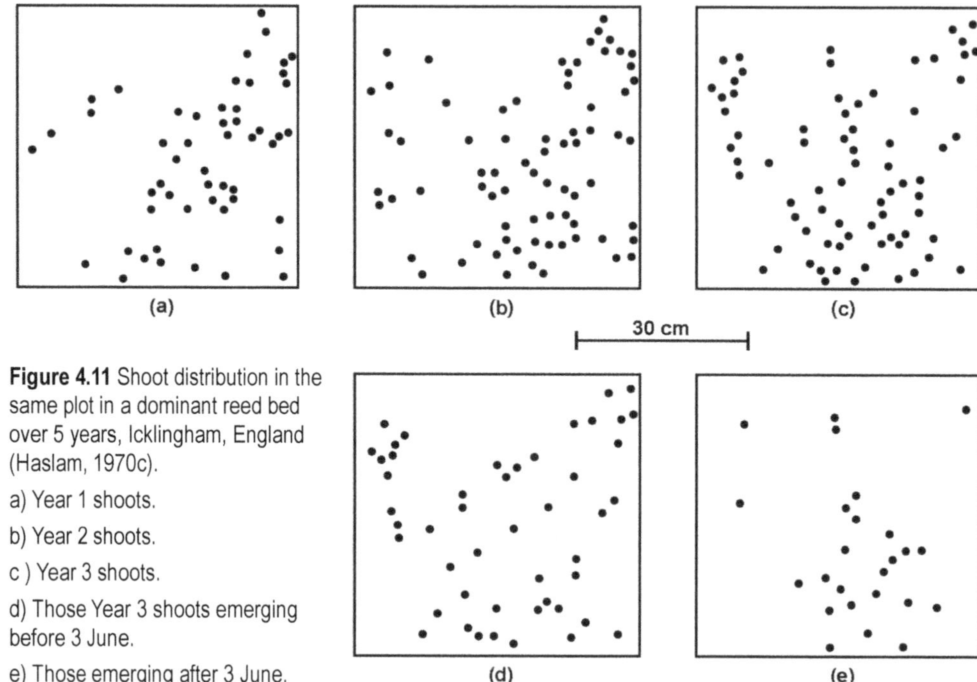


**Figure 4.11** Shoot distribution in the same plot in a dominant reed bed over 5 years, Icklingham, England (Haslam, 1970c).

a) Year 1 shoots.

b) Year 2 shoots.

c ) Year 3 shoots.

d) Those Year 3 shoots emerging before 3 June.

e) Those emerging after 3 June.

Shoots from one vertical system are likely to be close together in a sparse stand (e.g., within 0.25 $m^2$: Figures 3.1 and 4.12), because the rhizome branches are short, but in a dense one, although the central shoots are normally clumped, the others are likely to extend over a larger area (e.g., 1 $m^2$) and different rhizome systems may overlap. Any King reed is likely to have adjacent small shoots in subsequent years, irrespective of any longer branches. Once a shoot occurs in one place, therefore, others are likely to be there for several years, although they may be absent in any one year.

In Figure 4.11, over 5 years, pattern changes slowly. There are dense patches from three causes. In the top left, new vertical rhizomes invaded in one year, shoots were sparse in that year (b), dense in the next (c). In the centre of (a) the dense patch is from frosting (Chapter 3). In the top right, density is shown increasing, then decreasing (Figure 3.1).

Within these good, optimal stands, both height (by over 0.5 m) and density (by over 50/$m^2$) can vary from year to year (Table 4.6). The height curve (Figure 4.5) peaks to the right (negatively skew), most shoots being among the tallest of the reedbed. Variation is due partly to conditions in the previous summer, which is when the food to power this years' buds was made. It is also partly due to the current conditions, since up to half the shoot buds are made in late winter and spring. Suitable water and fluctuating temperatures increase bud formation (Chapters 3 and 6). Litter mat depth can change density between 30 and 260 reeds per square metre. Variation is more in the number of small, than of larger shoots since, generally speaking, it is the small ones that come from spring-formed buds, so are affected by the current litter. Table 4.7 shows the effect of nutrients within a dominant bed. When roots are able to utilise soil freely, shoots are dense, taller and heavier.

Poor stands may be short or sparse or, of course, both. Figures 4.5d, 4.10b and 4.12 show the height patterns of a typical restricted stand, but with even more restriction on buds than Figure 3.1. These stands usually occur where competition (see above) or nutrient deficiency (Chapter 6, Table 6.4) force the rhizomes to stop producing growing buds. These are restricted to, at least, single annual horizontal rhizomes, their terminal shoots, and 1–3 small laterals on these for the next two years. While emergence, the upthrusting of buds, is occurring, cutting or light grazing has much less effect than later in the summer when food is instead moving downwards (Table 4.8). A good replacement crop grows.

Short stands are characteristic of medium-level grazing (Figure 4.5c). (Over-grazing for some years means death, very low grazing means just a little decrease.) Sometimes they may occur elsewhere, as in ploughed fields (rhizomes cut annually), with summer cutting, after drying (degenerating stands), and occasionally with effective competition. Here, with few or no tall shoots, inflorescences are few or absent. Ordinarily the shoots resemble small shoots of good populations in their measurements, as indeed is to be expected, most being borne on short lateral rhizomes (grazing, cutting of rhizomes or shoot).

In a poor, short stand most rhizomes are the same width as in other populations, so the King reeds are wide at the base and are potentially tall. This potential can be stopped by lack of food in the rhizome, leading to a narrower, shorter reed, or by it being removed, and smaller laterals replacing it.

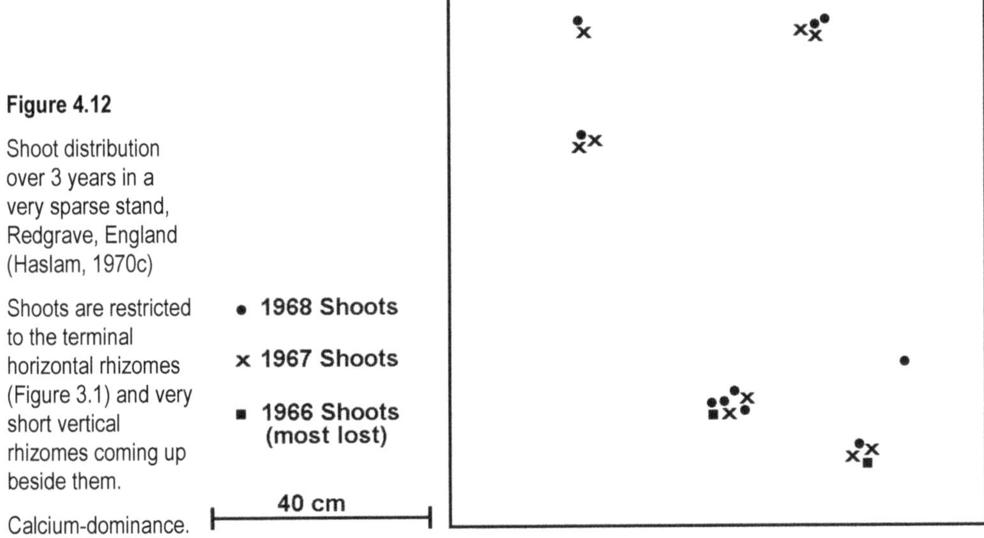


**Figure 4.12**

Shoot distribution over 3 years in a very sparse stand, Redgrave, England (Haslam, 1970c)

Shoots are restricted to the terminal horizontal rhizomes (Figure 3.1) and very short vertical rhizomes coming up beside them.

Calcium-dominance.

Stands deteriorate, in these days most often by drying, though also by increasing any other damage factor (grazing, mowing, shading, competition). In Figure 4.5a and c, the growth curve, starting at the optimum (a), moves through (b), and depending on conditions, passes through (c) or (d) and finally disappears.

## Protection of the population

There is the reedbed. There are plenty of other species willing and able to invade and colonise all this space. So, to remain in being, the reedbed must protect itself against such invasion. It does so very effectively. The reedbed's natural place in shallow water is generally impregnable, as is shown by the peat record: reedbeds remain for thousands of years, and only vanish when conditions change, either water rising (increase of sea level); or peat formation or water loss leading to wood, or bog forming instead (or, in recent millennia, human impact!).

A thick shading canopy deters most species from entering: or, rather, kills them on entering. Light on ground surface may be 0–1%.

A litter mat, something which only forms in drier places where, naturally, carr could invade, is an even stronger protection, as it shades all year round.

Evidence of *Phragmites* having root toxins in the soil which deter or kill other species is very doubtful. (The reverse, Chapter 6, does happen.) It is unlikely *Phragmites* litter kills *Phragmites*, (though see Armstrong & Armstrong, 2001) but a thick mat decreases biomass (e.g., Granéli, 1989). After a litter mat too thick for dense *Phragmites* is removed, new shoots emerge rapidly, indicating just a physical cause, e.g., removal in late May led to 64 new shoots in 3 weeks in a square metre. (Ten shoots came up in the control plot.) Also, of course, large dominant reedbeds grow on reed peat.

In shallow water, most plant species cannot survive. Indeed even many wetland species (e.g., *Juncus subnodulosus*, *Calamagrostis* spp.) are killed by flooding. At the water-land edge in reedbeds other species may start to invade in dry years, being killed off in wet years. Just a little drier in the habitat, and invasion can be successful, and the switch-over to drier communities begins.

The protection of the stand therefore comes from the external factor of water regime, preventing land, and many wetland plants growing in its typical habitat. It also comes from the plant itself. In all dense stands the shading canopy prevents many or all species growing, and in drier ones litter mats add to this lack of light.

Reedbeds are invaded, of course. This may be because disaster (unexpected drying or deep flooding; human impact) has weakened the reedbeds. 'Upsetting' *Phragmites* can leave a bed with late-emerging, sparse, short reeds for up to four years (the time it takes for most reeds to grow from new rhizomes, unaffected – directly – by the disaster). In this time, there is less or no canopy or litter mat, and invasion is easy. Whether it is successful depends on the species and habitat when *Phragmites* recovers. Since all reedbeds eventually end, those ended by bog or wood are successfully invaded, as is discussed in Chapter 11.

## Pests and diseases

Dvorak & Imhof (1998) reviewed the invertebrates of *Phragmites* in central and western Europe. The worst infestations are when the attack is in mid-May to Mid-June, lowering final height by 35–60%. July infestations are less important, as the reed is already at 80% of its final height. Infestation varies from 1–90% of shoots, and in the Czech Republic can lower above-ground net production by up to 13–14% (Table 4.9). (Pests of leaves are of little importance.) Tables 4.10–4.11 list the principal pests, and their effects. The pests can be divided into:

1. Species damaging the growing points without forming galls: reedbugs. *Archanara geminipunctata* (the earliest developer), *A. dissoluta* and *Platycephala planifrons* are the principal species, though *Arenostola* spp. are locally common. Both infestation and damage vary with circumstances. *A. dissoluta* may be sufficiently common in Poland to make shoots unsuitable for industrial use.
2. Species transforming the growing points of the reed into galls. The most serious are *Lipara lucens*, *L. similis, L. rufitaris* and *L. pullitarsis*. The galls are leaves bundling round stunted internodes. These are the most damaging type of pest, with losses of shoot biomass of *c.* 37% recorded. *Steneotarsonemus phragmitidis* is more locally damaging.
3. Species mining the stems. These occur in warmer regions, and include *Phragmataecia costoneae*, *Schoenobius gigantellus* and *Chilo phragmitellus*, *Rhizedra lutosa*, *Lasioptera arundis*, *L. hungarica* and *Giraudiella inclusa*, the last three of which are gall formers.
4. Leaf- and shoot-suckers. *Hyalopterus pruni* (aphid) and *Chaetococcus phragmitidis* (coccid) and others cause no substantial losses. (See, e.g., Durska, 1970; Mook, 1967, 1971; van der Toorn & Mook, 1982.)

Among fungi, *Ustilago grandis*, smut, infects young shoots, but is not obvious until July. Inflorescences do not develop, internodes are very short and swollen. Because infected shoots are brittle, if abundant, they render the crop useless. Attack may be most in reeds tall (over 2 m high) and dense (over 70/m$^2$). *Deightoniella arundinacea* leads to weak and non-maturing stems, with short upper internodes and inflorescences, if any, small and without ripe fruits. *Claviceps* spp. may infect half or even more of the inflorescences, the fungus replacing the fruit, so rendering the affected florets sterile. Other fungi also infect reeds, e.g., *Pucciniella phragmitidis*, but losses are low compared with invertebrate damage: and even that is very low compared with the size of the crop (and see Wazny & Wytwer, 1963).

Damage can be much reduced by studying the life cycles of the pest and disease organisms, and altering management accordingly, e.g., by drowning bottom-living overwintering organisms, removing litter mats that harbour others, burning, etc..

## Other animals

**Mammals** may live within the reedbed like some muskrats, or come down to the drier edges to graze, like moose or cattle. Table 4.12 lists the common mammals of central Europe, Table 4.13 includes those of the English Broadland (e.g., Grosch, 1978). (And see Carter, 2003; Henson, 2001, for water voles.)

**Fish** live in flooded reedbeds. There are many species, with a wide variety of diet, both herbivores (aquatic and wetland plants), and carnivores.

**Amphibia** also occur in reedbeds, mostly feeding on small invertebrates.

**Reptiles**, unlike the previous two groups, are not intrinsically dependent on a water environment, but may be frequent in reedbeds.

**Birds** are characteristic and functionally important components of all wetland ecosystems, and they are the most conspicuous animals (see Table 13). They are mostly adapted to life on water, mud flats and littoral stands of wetland plants. These offer numerous ecological niches (there are many patterns of water, substrate, and vegetation). Reedbeds are used for shelter, roosting, breeding and food supply (Tscharntke, 1999). Some birds live permanently in or near reedbeds; others visit from time to time, and many are migrants (see also Colenutt, 2003; Ward, 1999; and Gratton, 2005 for the recovery of arthropods after removing *Phragmites*).

**Invertebrate animals** are extremely numerous and often distinctive. Table 4.14 lists macrolepidoptera of different Broadland marshes, spread over that small area. This shows both the richness and the difference between marshes with – in the whole range of reedbeds – only slightly different history (see Chapter 11). (Mainly from Dvorak & Imhof, 1998. Also see Rudescu *et al.*, 1965; van der Toorn, 1972; Durska, 1970; Haslam, 1972 and 1973a; Westlake *et al.*, 1998.)

The heavy **phytoplankton** studied in the Hungarian Lake Balaton show this increases with nutrients (e.g., Lakatós, 1983, 1989; Lakatós & Biró, 1991; Lakatós *et al.*, 1982; Lakatós *et al.*, 1991).

## Water economy

Water economy has been reviewed by Królikowska *et al.* (1998). Reedbeds, like other wetlands, have all three phases of air, water and soil, and the relative importance of these varies in both space and time. This means complexity, and diversity of micro-environments (see Westlake *et al.*, 1998; see also Květ, 1973).

In *Phragmites* stands, temperature and humidity change with the amount and proportions of biomass and dead plant material (litter mat, standing dead stems) and vary in both space and time. A stand is a porous medium, which reduces air movement, e.g., by 90%, 15 m into a stand (Rychnovská & Smíd, 1973).

This means evaporation from the water surface is low, the shelter reducing the turbulent heat flux into the atmosphere. A refuge from the cold is thus created for animals in winter. Ice production within the stand is likewise reduced. Litter mats, of course, are also insulating. Dry soil heats up quicker than flooded soil (also see Chapter 6). Conversely, in summer it is colder within the reedbed. Harvesting, leaving bare soil also alters such conditions (see also Květ, 1973).

There is some, not much, evaporation from reedbeds with or without dead reeds. Transpiration starts as the new buds emerge, and is greatest as the shoots approach and maintain their maximum height (e.g., June to August, Table 4.15). Naturally enough, for such a variable plant, transpiration varies greatly with stand and habitat. *Phragmites* is one of the highest-transpiring species in what is generally the highest-transpiring vegetation (except, e.g., *Eucalyptus* in habitats with ample water). This is, of course, why reedbeds can be used to dry wet places (also see van der Toorn, 1972; Rudescu *et al.*, 1965; Schieferstein, 1999; Lippert *et al.*, 2001; Zhu *et al.*, 2003; Zeidler *et al.*, 1994; Rolletschek *et al.*, 1999a and b; Kühl & Neuhaus, 1993; Kühl *et al.*, 1999; Neuhaus *et al.*, 1993; Lissner *et al.*, 1999a and b).

## Ecophysiology

More recently, ecophysiology and other metabolic processes have been attracting attention, and collections of useful papers are in the *Phragmites* numbers in *Aquatic Botany* in 1989, 1999, 2001, 2004 and Květ & Westlake (1998) or photosynthesis, see Květ & Westlake (1998) and Lessman *et al.* (2001).

Papers on genetics, chromosomes, diversity, and various relations to behaviour include Clevering & Lissner (1999), Koppitz (1999), Pauca-Comanescu *et al.* (1999), Kühl *et al.* (1999), Rolletschek *et al.* (1999a and b), and Hanganu *et al.* (1999). Also see Rudescu *et al.* (1965) and van der Toorn (1972), Lissner *et al.* (1999a and b), Koppitz *et al.* (2004), Fürtig *et al.* (1996 and 1999) and Hartsendorf & Rolletschek (2001).

Oxygen and other gas studies include Armstrong *et al.* (1992), Armstrong & Armstrong (2001), Květ & Westlake (1998), Rolletschek *et al.* (1999), Weissner & Strand (1996), Lissner *et al.* (1999b) and Adams & Bate (1999).

For physiology, see the works of, e.g., Dykyjová, Fiala, Květ, Ondok, Szczepanski, and Westlake, and for productivity, the latter also.

Salt and other nutrition studies include Lissner *et al.* (1999a and b), Čížková *et al.* (1999), Hanganu *et al.* (1999), Adams & Bate (1999), Romero *et al.* (1999), Mendelssohn *et al.* (1999), Rodewald-Rudescu (1974), Verhoeven (1992), and Prach *et al.* (1998). Also see Rudescu *et al.* (1965), and van der Toorn (1972).

During the lifetime of this book it is likely that these studies will come to fruition.

For plants as food, also see Gaevskaya (1966).

**Table 4.1** Early-emerged buds are more likely to grow taller (accidents excluded) (Haslam, 1970a)

| | | Average final height of consecutive groups of shoots (cm) | | | | | | | | | |
|---|---|---|---|---|---|---|---|---|---|---|---|
| | *Site* | *1st* | *2nd* | *3rd* | *4th* | *5th* | *6th* | *7th* | *8th* | *9th* | *10th* |
| English fen | Cavenham, 1958 | 120 | 95 | 110 | 110 | 90 | 65 | 95 | 85 | 90 | 90 |
| English fen | Icklingham, 1958 | 160 | 140 | 150 | 135 | 135 | 130 | 95 | 90 | – | – |
| English fen | Icklingham, 1965 | 110 | 130 | 100 | 90 | 80 | – | – | – | – | – |
| English fen | Icklingham, 1965 | 160 | 170 | 70 | 60 | 50 | – | – | – | – | – |
| Malta | Salina, 1965 | 205 | 225 | 85 | 205 | 105 | 115 | 85 | 60 | 10 | 35 |
| Malta | Salina, 1966 | 165 | 185 | 110 | 79 | 125 | 100 | 85 | 35 | 65 | – |

**Table 4.2** Effect of exposure of *Phragmites communis* to frost in plots* in two Breck fens (Suffolk)

| Year | Treatment | Site | Number of stems | Modal height of stems (cm) | Dry weight (g) | Number inflorescences | Leaf dry wt/ stem dry wt ratio |
|---|---|---|---|---|---|---|---|
| 1957 | Spring frost | Icklingham | 391 | 120 | 1260 | 100 | 0.6 |
| | Control | Icklingham | 132 | 135 | 676 | 36 | 0.55 |
| | Spring frost | Cavenham | 208 | 85 | 332 | 20 | 0.7 |
| | Control | Cavenham | 88 | 85 | 220 | 4 | 0.7 |
| 1958 | Spring frost | Icklingham | 204 | 110 | 672 | 28 | 0.5 |
| | Winter frost | Icklingham | 154 | 160 | 579 | 7 | 0.5 |
| | Control | Icklingham | 184 | 150 | 784 | 28 | 0.4 |
| | Spring frost | Cavenham | 336 | 105 | 632 | 28 | – |
| | Winter frost | Cavenham | 124 | 80 | 12 | 9 | 1.0 |
| | Control | Cavenham | 164 | 120 | 480 | 4 | 0.7 |
| 1959 | Spring frost | Icklingham | 181(107†) | 165(200†) | 965 | 19 | 0.5 |
| | Winter frost | Icklingham | 154(120†) | 160(190†) | 579 | 7 | 0.5 |
| | Control | Icklingham | 152(138†) | 180(190†) | 1240 | 36 | 0.4 |
| | Spring frost | Cavenham | 151(91†) | 70(115†) | 208 | 5 | 0.9 |
| | Winter frost | Cavenham | 124(80†) | 80(80†) | 173 | 9 | 1.0 |
| | Control | Cavenham | 69(75†) | 85(75†) | 18 | 4 | 0.9 |

* The 'Cavenham' site was exposed, while 'Icklingham' was flooded and so more sheltered.

† Plots cleared in early April are referred to as exposed to 'spring frost', those in November as to 'winter frost' (i.e. winter plus spring frosts). Data refer to plots of 1 $m^2$.

**Table 4.3** Larger buds emerge earlier (Haslam, 1969b)

| Site | Date of emergence | Average width (mm) |
|---|---|---|
| Malta | before 9.2.66 | 6.6 |
| | 9.2.–24.2.66 | 10.0 |
| | 24.2.–5.3.66 | 7.2 |
| | 5.3.–12.3.66 | 5.0 |
| | 12.3.–26.3.66 | 4.0 |
| | 26.3.–30.4.66 | 5.4 |
| | 30.4.–3.5.66 | 3.0 |
| | 3.5.–16.6.66 | 6.0 |
| | before 1.2.67 | 8.0 |
| | 1.2.–11.2.67 | 19.0 |
| | 11.2.–21.2.67 | 7.0 |
| | 21.2.–9.3.67 | 5.0 |
| | 9.3.–22.3.67 | 4.0 |
| | 22.3.–5.4.67 | 4.5 |
| | 5.4.–12.5.67 | 4.5 |
| | 12.5.–24.5.67 | 2.0 |
| Icklingham | before 25.5.65 (2–3 weeks emergence) | 3.3 |
| | 25.5.–4.6.65 | 2.7 |
| | 4.6.–10.6.65 | 2.2 |
| | 10.6.–20.7.65 | 3.0* |
| | 20.7.–11.8.65 | 2.9* |

* Figures high because summer shoots are included.

**Table 4.4** Damage to emerged shoots

**a) Dense stand. Litter removed, frosts very light. All causes (England)**

| *Date of emergence* | *Number emerged* | *Cause of harm* | *Number* | *Date affected* | *Date dead* | *Date side shoots up* | *Notes* |
|---|---|---|---|---|---|---|---|
| Before 19.4. | 7 | Frost | 1? | 10.5. | 3.6. | 7.6. | Temporary recovery |
| | | Eaten at base | 1 | 15.7. | 15.7. | – | |
| 19.4.–23.4. | 11 | Frost | 1 | 24.5. | 3.6. | 7.6. | |
| | | Reedbug | 3 | 3.6. | – | 15.7.–26.8. | |
| | | Caterpillar on leaves | 1 | 3.6. | – | – | Damage minor |
| | | Late competition | 1 | 26.8. | – | | |
| 23.4.–27.4. | 6 | Reedbug | 1 | 3.6. | 26.8. | 21.6. | |
| | | Caterpillars on leaves | 1 | 3.6. | – | – | Damage minor |
| 27.4.–3.5. | 6 | Reedbug | 1 | 3.6. | 26.8. | 21.6. | |
| | | Reedbug | 1 | 14.6. | 9.7. | – | |
| | | Caterpillars on leaves | 2 | 3.6. | – | – | Damage minor |
| 3.5.–10.5. | 10 | Frost | 1 | 24.5. | 7.6. | 7.6. | |
| | | Reedbug | 1 | 14.6. | 21.6. | – | |
| | | Reedbug | 2 | 21.6. | 9.7. (one) | – | One living |
| | | Reedbug | 1? | 9.7. | – | – | |
| | | Broken | 1 | 22.7. | – | – | Broken by S.M.H. |
| 10.5.–17.5. | 9 | Caterpillars on leaves | 3 | 3.6. | – | – | Damage minor |
| | | Reedbug | 1? | 14.6. | 22.7. | – | |
| | | Late competition | 1 | 15.7. | – | – | |
| | | Late competition | 1 | – | 26.8. | – | |
| 17.5.–24.5. | | Early competition | 1? | 7.6. | 21.6. | – | |
| | | Late competition | 1 | 15.7. | 26.8. | – | |
| | | Late competition | 1 | 26.8. | – | – | |
| 24.5.–7.6. | 16 | Late competition | 1 | 4.7. | – | – | Shoots include side shoots |
| | | Late competition | 5 | 14.8. | – | – | |
| | | Late competition | 3 | – | 26.8. | – | |
| 7.6.–21.6. | 4 | Late competition | 1 | – | 26.8. | – | Shoots include side shoots |
| 21.5.–9.7. | 2 | Late competition | 1 | – | 26.8. | – | |
| 9.7.–29.7. | 2 | Late competition | 2 | 14.8. | 26.8. (one) | – | One living |

The queries indicate shoots which were harmed, but with the cause uncertain.

**b) Death or damage from late summer internal competition. Affected shots may have dead tips, be dying, or be dead (Haslam, 1970a)**

| | *Shoots <50 cm high* | | *Shoots 50–100 cm* | | *Shoots 100–150 cm* | | *Shoots >150 cm high* | |
|---|---|---|---|---|---|---|---|---|
| Site | Number of shoots | Number affected | Number of shoots | Number affected | Number of shoots | Number affected | Number of shoots | Number affected |
| Redgrave, 1968 | 12 | 8 | 23 | – | – | – | – | – |
| Horsey, 1968 | 7 | 5 | 19 | 6 | 40 | – | – | – |
| Ranworth, 1968 | 18 | 14 | 27 | 7 | 23 | – | – | – |
| Icklingham, 1968 | – | – | 26 | 1 | 44 | – | – | – |
| Icklingham, 1968 | 2* | – | 6* | – | 23 | 2 | 23 | 2 |
| Woodwalton, 1968 | – | – | – | – | 8 | 7 | 43 | 1 |

* Mostly recently emerged shoots.

**Table 4.5** To show that flowering is commoner on taller shoots of *Phragmites* (Haslam, 1970a)

| | Site | Shoots <50 cm high | | Shoots 50–100 cm | | Shoots 100 cm | | Shoots >150 cm high | |
|---|---|---|---|---|---|---|---|---|---|
| | | *Number of shoots* | *Number of inflorescence* | *Number of shoots* | *Number of inflorescence* | *Number of shoots* | *Number of inflorescence* | *Number of shoots* | *Number of inflorescence* |
| English fen | Icklingham, 1957 | – | – | 8 | – | 11 | – | 76 | 50 |
| | Icklingham, 1957 | 12 | – | 70 | – | 79 | 8 | – | – |
| | Icklingham, 1957 | 8 | – | 39 | – | 35 | 13 | – | – |
| | Eriswell, 1958 | 3 | – | 25 | – | 31 | 10 | 8 | 8 |
| | Cavenham, 1959 | 13 | – | 56 | 5 | 1 | 1 | – | – |
| | New Forest, 1966 | 51 | – | – | – | – | – | – | – |

**Table 4.6** Comparison of performance in the same stand in different years (Breck fens, England 1 $m^2$), aerial stems (Haslam, 1972)

| | Weight (g) | | | Modal height (cm) | | | Number of stems | | | Number of inflorescences | | |
|---|---|---|---|---|---|---|---|---|---|---|---|---|
| | *1957* | *1958* | *1959* | *1957* | *1958* | *1959* | *1957* | *1958* | *1959* | *1957* | *1958* | *1959* |
| Icklingham* | 700 | 1000 | 1240 | 190 | 200 | 205 | 90 | 125 | 150 | 45 | 35 | 40 |
| Icklingham* | 440 | 1000 | 1240 | 105 | 155 | 170 | 135 | 165 | 160 | 30 | 35 | 44 |
| Cavenham** | 245 | 365 | 150 | 105 | 110 | 85 | 100 | 125 | 70 | 3 | 5 | 5 |
| Eriswell | 240 | 380 | 625 | 95 | 1110 | 120 | 45 | 65 | 110 | 5 | 15 | 50 |
| Eriswell | – | 130 | 310 | – | 90 | 85 | – | 85 | 100 | – | 0 | 3 |

* Stands in which a high water level persisted until late summer in 1959 (favourable conditions for *Phragmites communis*).

** Stand in which water level dropped early in 1959 (unfavourable conditions).

**Table 4.7** Competition for soil nutrients, Cavenham, England (Haslam, 1970a)

| | Year before treatment | | Year of treatment | | |
|---|---|---|---|---|---|
| *Plot size* | *Number* | *Modal height (cm)* | *Number* | *Modal height (cm)* | *Weight (oven dry, g $m^{-2}$)* |
| a) With free soil | | | | | |
| Control | 97 | 105 | 110 | 125 | 364 |
| 1 x 1 m | 89 | 100 | 134 | 105 | 306 |
| 0.5 x 0.5 m | 84 | 90 | 172 | 90 | 328 |
| 0.25 x 0.25 m | 128 | 100 | 352 | 105 | 816 |
| b) No free soil | | | | | |
| Control | 110 | 125 | 69 | 85 | 148 |
| 1 x 1 m | 81 | 105 | 78 | 75 | 169 |
| 0.5 x 0.5 m | 120 | 120 | 104 | 70 | 142 |
| 0.25 x 0.25 m (av.) | 200 | 120 | 233 | 70 | 312 |

**Table 4.8** Effect of cutting during and after emergence (Haslam, 1969b)

Living shoots were cut, leaving the litter dead standing aerial stems. Plots 1x1 m. Woodwalton, England, 1967.

| | **Final crop** | | **Cut crop** | | **Crop of previous year** | |
|---|---|---|---|---|---|---|
| *Date of cut* | *Number of shoots* | *Modal height (cm)* | *Number of shoots* | *Modal height (cm)* | *Number of shoots* | *Modal height (cm)* |
| 19.4.67 | 74 | 130–80 | 9 | 10 | 36* | 165 |
| 28.4.67 | 59 | 120 | 15 | 15 | 27* | 150 |
| 5.5.67 | 57 | 120 | 26 | 15 | 24* | 160 |
| 1.6.67 | 42 | 120 | 13 | None | 34 | 155 |
| 9.6.67 | 53 | 120, 155 | 33 | 80 | 64 | 175 |
| 23.6.67 | 33 | 80 | 81 | 21, 85 | 48 | 165 |
| 30.6.67 | 23 | 65 | 71 | 35, 80 | 41 | 175 |
| 7.7.67 | 4 | – | 83 | 50, 80 | 50 | None |

* Recorded after some stems were broken and lost.

**Table 4.9** Influence of invertebrates on the production of *Phragmites australis* in Czechoslovakia (Dvorak & Imhoff, in Květ *et al.*, 1998)

| **Species** | **Decrease of dry matter production by attacked stems** | **Average number of damaged stems (% total)** | **Total decrease of dry matter production in Czechoslovakia (% normal)** | **Damage** | **Scale of importance of insect pests** |
|---|---|---|---|---|---|
| *Archanara geminipuncta* | 51 | 3.0 | 1.56 | Destroy the growing point | 4 |
| *Giraudiella inclusa* | 36 | 3.0 | 1.08 | Drooping of stems | 5 |
| *Hypalopterus pruni* | minimum | 0–100 | – | Damage to leaves | 10 |
| *Lasioptera arundinis* | – | 1 | – | Galls on lateral shoots | 11 |
| *L. hungarica* | 50 | 1.5 | 0.75 | Drooping of stems | 8 |
| *Lipara lucens* | 60 | 1.5 | 0.9 | Gall on growing point | 7 |
| *L. pullitarsis* | 45 | 7.0 | 3.15 | Gall on growing point | 1 |
| *L. rufitarsis* | 47 | 4.5 | 2.12 | Gall on growing point | 3 |
| *L. similis* | 47 | 2.0 | 0.94 | Gall on growing point | 6 |
| *Platycephala planifrons* | 46 | 0.03 | 0.01 | Destroy the growing point | 9 |
| *Steneotarosenemus phragmitidis* | 48 | 6.5 | 3.1 | Gall on growing point | 2 |

All stem species (weighted mean).

**Table 4.10** Effects of *Lipara* spp. galls (Dvorak & Imhoff, 1998)

a) Influence of gall formation by *Lipara* species on the length of *Phragmites* stems at Bohemian localities

| Species | Length of infested stems/length of healthy stems |
|---|---|
| *Lipara lucens* | 0.43 |
| *L. pullitarisis* | 0.56 |
| *L. rufitarsis* | 0.58 |
| *L. similis* | 0.73 |

b) Changes in dimensions of internodes of *Phragmites australis* caused by *Lipara* spp. infestation (percentage of the dimensions of unaffected stems; internodes numbered from top)

| Internodes | Length of internodes | | | Diameter of internodes | | | Strength of internodes | | | |
|---|---|---|---|---|---|---|---|---|---|---|
| | *First* | *Fourth* | *Ninth* | *First* | *Fourth* | *Ninth* | *First* | *Fourth* | *Ninth* | *Total stem* |
| *Lipara lucens* | 111 | 88 | – | 76 | 72 | – | 42 | 33 | – | 57 |
| *L. rufitarsis* | 103 | 102 | 80 | 95 | 94 | 96 | 86 | 68 | 50 | 74 |

**Table 4.11** Infestation of *Phragmites australis* by some stem miners and some characteristics of the affected stand at Neusiedlersee (Dvorak & Imhoff, 1998)

| | **Short mixed *Phragmites*** | **Mostly pure *Phragmites*** | |
|---|---|---|---|
| | *Mostly in terrestrial ecophase* | *Flooded with terrestrial ecophase during the summer* | *Permanently flooded* |
| Mean stem length (cm) | 166 | 218 | 215 |
| Number (stems $m^{-2}$) | 65 | 75 | – |
| Water depth (cm) | 0–25 | 25–50 | 50–110 |
| Infested stems (% total) | | | |
| by Phragmataecia | 8.2 | 12.2 | 10.1 |
| by Pyralidae | 7.8 | 1.0 | 2.7 |
| by other miners | 0.8 | 0.5 | 0.8 |

**Table 4.12** Mammals in Central Europe

Species diversity and some semi-quantitative data on mammals in Central European littoral reed belts (Dvorak & Imhof, 1998).

| SPECIES | Occurrence in ecophases[a] | | | Food composition (% of volume) | | Population density (No. $ha^{-1}$) | |
|---|---|---|---|---|---|---|---|
| | *lit.* | *lim.* | *terr.* | *plants* | *animals* | *averages* | *maximum* |
| **Herbivorous** | | | | | | | |
| *Ondatra zibethicus* | ++ | + | + | 100.0 | | 4.0–40.0 | 55.0 |
| *Arvicola terrestris* | + | ++ | | 100.0 | | 6.1–8.1 | 15.1 |
| *Micromys minutus* | ++ | ++ | ++ | 64.2 | 35.8 | – | – |
| *Microtus oeconomus* | | + | + | 100.0 | | – | – |
| *Microtus arvalis* | | + | ++ | 100.0 | | 3.0–212.9 | 447.6 |
| *Clethrionomys glareolus* | | + | ++ | 88.1 | 11.9 | 9.4–42.2 | 49.8 |
| *Rattus norvegicus* | | ++ | + | + | + | – | – |
| *Mus musculus* | | | + | 74.6 | 25.4 | 0.6–2.8 | 11.1 |
| *Apodemus agrarius* | | + | ++ | 72.5 | 27.5 | – | – |
| *Apodemus sylvaticus* | | | + | 74.9 | 25.1 | 0.8–7.9 | 15.9 |
| *Apodemus flavicollis* | | | + | 37.2 | 67.8 | 0.3–1.4 | 3.2 |
| *Apodemus microps* | | | + | 100.0 | | 0.1–1.3 | 6.6 |
| *Lepus europaeus* | | | + | + | | – | – |
| *Capreolus capriolus* | | + | + | + | | – | – |
| *Cervus elaphus* | | | + | + | | – | – |
| *Sus scrofa* | | | + | + | + | – | – |
| **Insectivorous** | | | | | | | |
| *Sorex araneus* | | ++ | ++ | | + | 0.9–31.6 | 70.0 |
| *Sorex minutus* | | ++ | | | + | 0.9–32.9 | 70.6 |
| *Neomys anomalus* | (+) | (+) | (+) | | + | – | – |
| *Neomys fodiens* | ++ | ++ | | | + | – | – |
| *Crocidura suaveolens* | | | (+) | | + | – | – |
| *Erinaceus europaeus* | | | + | | + | – | – |
| *Talpa europaea* | | | (+) | | + | – | – |
| **Carnivorous** | | | | | | | |
| *Mustela nivalis* | | + | + | | + | – | – |
| *Mustela erminea* | | + | + | | + | – | – |
| *Putorius putorius* | | + | + | | + | – | – |
| *Vulpes vulpes* | | + | + | | + | – | – |

Scale of occurrence: ++ frequently present; + present; (+) accidental occurrence; – no data.

[a] Abbreviations for ecophases: lit. – littoral; lim. – limosal (swampy soil); terr. – terrestrial.

**Table 4.13** Mammals and birds of Broadland (characteristic species) (George, 1992)

| Mammals | Birds | | |
|---|---|---|---|
| | *Alder carr* | *Sallow carr* | *Open fens* |
| Chinese water deer | blackcap | cetti's warbler | bearded tit |
| coypu (probably extinct) | chiff chaff | | bittern |
| fox | great spotted woodpecker | | cuckoo |
| muntjac | lesser spotted woodpecker | | grasshopper warbler |
| otter | long-tailed tit | | heron |
| rabbit | marsh tit | | little tern |
| water vole | redpoll | | marsh harrier |
| | sparrowhawk | | oyster catcher |
| | willow tit | | redshank |
| | wren | | reed warbler |
| | | | ringed plover |
| | | | savi's warbler |
| | | | sedge warbler |

**Table 4.14** The 'macrolepidoptera' recorded from Broadland over the 1960s to 1990s, and their status (George, 1992)

| | | Bure fens and marshes | Ant fens and marshes | Thurne fens and marshes | Yare fens and marshes | Status |
|---|---|---|---|---|---|---|
| * | Reed Leopard (*Phragmataecia castaneae*) | x | x | x | – | 2 |
| | Goat Moth (*Cossus cossus*) | – | – | x | x | Nb |
| | The Forester (*Adscita statices*) | – | – | x | – | Nb |
| | White-barred Clearwing (*Synathedon spheciformis*) | x | – | – | – | Na |
| | Barred Hook-tip (*Drepana cultraria*) | x | – | – | – | Nb |
| | Blotched Emerald (*Comibaena bajularia*) | – | x | – | – | L |
| * | Rosy Wave (*Scopula emutaria*) | – | – | x | – | Nb |
| | Purple-bordered Gold (*Idaea muricata*) | – | x | – | x | Na |
| * | Single-dotted Wave (*I. dimidiata*) | x | x | x | x | – |
| * | Small Scallop (*I. emarginata*) | x | x | x | x | – |
| * | Oblique Carpet (*Orthonoma vittata*) | x | x | x | x | L |
| | Red-green Carpet (*Choroclysta siterata*) | x | x | – | – | L |
| | Scarce Tissue (*Rheumaptera cervinalis*) | x | – | – | – | Nb |
| | Brown Scallop (*Philereme vetulata*) | – | x | – | – | Nb |
| | Dark Umber (*P. transversata britannica*) | x | x | x | x | L |
| | Marsh Carpet (*Perizoma sagittata*) | – | – | – | x | 2 |
| | Valerian Pug (*Eupithecia valerianata*) | – | – | x | x | Nb |
| | Dentated Pug (*Anticollix sparsata*) | x | x | x | x | Na |
| | Dingy Shell (*Euchoeca nebulata*) | x | x | – | – | L |
| | Small Seraphim (*Pterapherapteryx sexalata*) | x | x | x | x | L |
| * | Bordered Beauty (*Epione repandaria*) | x | x | x | x | – |
| | Dark Bordered Beauty (*E. paralellaria*) | – | – | x | – | 3 |
| | Satin Beauty (*Deileptenia ribeata*) | x | – | – | – | Nb |
| * | Common Wave (*Cabera exanthemata*) | – | x | x | x | – |
| | Broad-bordered Bee Hawk-moth (*Hemaris fuciformis*) | – | – | x | – | Na |
| | Alder Kitten (*Furcula bicupsis*) | x | x | x | x | Nb |
| | Poplar Kitten (*F. bifida*) | x | x | – | x | L |
| | Maple Prominent (*Ptilodontella cucullina*) | x | x | x | x | Nb |
| | Small Chocolate-tip (*Clostera pigra*) | – | x | x | – | Nb |
| | Scarce Vapourer (*Orgyia recens*) | x | x | – | x | 2 |
| * | Round- winged Muslin (*Thumatha senex*) | x | x | x | x | L |
| | Muslin Footman (*Nudaria mundana*) | – | x | – | – | L |
| * | Dotted Footman (*Pelosia muscerda*) | x | x | x | – | 3 |
| * | Small Dotted Footman (*P. obtusa*) | – | x | x | – | 1 |
| | Orange Footman (*Eilema sorocula*) | – | – | – | x | Nb |
| * | Water Ermine (*Spilosoma urticae*) | x | x | x | – | Nb |
| | Dotted Rustic (*Ryacia simulans*) | – | – | – | x | Nb |
| | Lunar Yellow Underwing (*Noctua orbona*) | – | x | – | – | Nb |
| * | Fen Square-spot (*Diarsia florida*) | – | x | x | – | L |
| | Square-spotted Clay (*Xestia rhomboidea*) | x | – | – | – | Nb |
| | Heath Rustic (*X. agathina agathina*) | – | x | – | x | L |
| | Silvery Arches (*Polia hepatica*) | – | – | – | x | Nb |
| | Bordered Gothic (*Heliophobus reticulata marginosa*) | x | x | – | – | Nb |
| | Light Brocade (*Lacanobia w-latinum*) | – | x | x | – | L |

Continued overleaf

**Table 4.14** Continued

| | | Bure fens and marshes | Ant fens and marshes | Thurne fens and marshes | Yare fens and marshes | Status |
|---|---|---|---|---|---|---|
| | Dog's Tooth (*L. suasa*) | x | – | x | x | L |
| | Marbled Coronet (*Hadena confusa*) | – | – | — | x | L |
| * | Striped Wainscot (*Mythimna pudorina*) | x | x | x | x | L |
| * | Southern Wainscot (*M. straminea*) | x | x | x | x | L |
| * | Obscure Wainscot (*M. obsoleta*) | x | – | x | x | Nb |
| * | Flame Wainscot (*Senta flammea*) | x | x | x | x | 3 |
| | Star-wort (*Cucullia asteris*) | – | – | x | – | Nb |
| | The Suspected (*Parastichtis suspecta*) | x | x | x | – | L |
| * | Pink-barred Sallow (*Xanthia togata*) | – | x | x | x | – |
| | Dusky-lemon Sallow (*X. gilvago*) | x | – | – | x | L |
| * | Reed Dagger (*Simyra albovenosa*) | x | x | x | x | Na |
| * | Double Kidney (*Ipimorpha retusa*) | – | – | – | x | Nb |
| | Angle-striped Sallow (*Enargia palacea*) | – | – | x | – | Nb |
| * | Crescent Striped (*Apamea oblonga*) | – | – | x | – | Nb |
| * | Small Clouded Brindle (*A. unanimis*) | x | x | x | x | – |
| * | Double Lobed (*A. ophiogramma*) | x | x | x | x | L |
| * | Middle-barred Minor (*Oligia fasciuncula*) | x | x | x | x | – |
| | Lyme Grass (*Photedes elymi*) | – | x | x | – | Na |
| * | Mere Wainscot (*Photedes fluxa*) | – | x | x | x | Na |
| * | Small Wainscot (*P. pygmina*) | x | x | x | x | – |
| * | Fenn's Wainscot (*P. brevilinea*) | x | x | x | x | 3 |
| * | Ear Moth (*Amphipoea oculea*) | x | x | x | – | – |
| * | The Butterbur (*Hydraecia petasitis*) | x | x | – | x | Nb |
| * | Haworth's Minor (*Celaena haworthii*) | – | x | x | x | L |
| * | The Crescent (*C. leucostigma leucostigma*) | x | x | x | x | L |
| * | Bulrush Wainscot (*Nonagria typhae*) | x | x | x | x | – |
| * | Twin-spotted Wainscot (*Archanara geminipuncta*) | – | x | x | x | Nb |
| * | Brown- veined Wainscot (*A. dissoluta*) | x | x | x | x | Nb |
| * | Webb's Wainscot (*A. sparganii*) | – | x | x | x | Nb |
| * | Rush Wainscot (*A. algae*) | – | – | x | x | 3 |
| * | Large Wainscot (*Rhizedra lutosa*) | x | x | x | x | – |
| * | Fen Wainscot (*Arenostola phragmitidis*) | x | x | x | x | L |
| * | Small Rufous (*Coenobia rufa*) | x | x | x | – | – |
| * | Silky Wainscot (*Chilodes maritimus*) | x | x | x | x | Nb |
| | Marbled Clover (*Heliothis viriplaca*) | x | – | – | – | 3 |
| * | Silver Hook (*Eustrotia uncula*) | x | x | x | x | Nb |
| * | Silver Barred (*Deltote bankiana*) | – | – | x | – | 2 |
| * | Cream-bordered Green Pea (*Earias clorana*) | x | x | x | x | Nb |
| * | Gold Spot (*Plusia festucae*) | x | x | x | x | – |
| * | Lempke's Gold Spot (*P. putnami gracilis*) | – | – | x | x | L |
| | Scarce Silver Y (*Syngrapha interrogationis*) | x | – | – | – | Nb |
| * | Straw Dot (*Rivula sericealis*) | x | x | x | x | – |
| * | Pinion-streaked Snout (*Schrankia costaestrigalis*) | x | x | x | x | L |
| * | Marsh Oblique-barred (*Hypenodes humidalis*) | – | x | – | – | Nb |
| * | Dotted Fan-foot (*Macrochilo cribrumalis*) | x | x | x | x | Na |

* Species confined to fens, marshes or wet woodland.

**Table 4.15** Transpiration and evaporation during summer from open water by reed in the Danube Delta (Królikowska *et al.*, 1998)

| Period | Evaporation ($E_o$) (mm) | Transpiration ($E_c$) (mm) | $E_c/E_o$ |
|---|---|---|---|
| **1959** | | | |
| 1 June–23 June | 164 | 534 | 3.25 |
| 24 June–22 July | 60 | 120 | 2.00 |
| 23 July–22 August | 208 | 417 | 2.10 |
| 23 August–19 September | 142 | 485 | 3.42 |
| **1960** | | | |
| 14 April–12 May | 68 | 74 | 1.15 |
| 1 June–30 June | 62 | 60 | 0.98 |

# Chapter 5
# Variation: genetic and clonal

## Introduction

Reeds are variable, reedbeds are variable. This is illustrated by the photographs in this book, Table 1.1, Figures 3.10, 6.2–6.10, 8.1–8.16, 11.1, and by anyone looking at living reed!

Reeds vary greatly, and do so for several causes:

a) genetic, the subject of this chapter;

b) environment, habitat and climate;

c) history of the reedswamp, more clearly called 'Bed Variation'.

Much too little is known about genetic variation. It can be detected only by:

- genetic studies, including isozymes and ploidy;
- in a uniform habitat, seeing clear differences in characters (Table 1.1) such as emergence and leaf death dates, panicle shape and colour, patches that are dog-legged or substantially taller or shorter than the rest. This is not completely certain, as there could be local unseen variation in past management, water (e.g., water below peat) or nutrients;
- transplants to a uniform habitat and waiting at least four years for the transplants to settle and have new growth, and finding them different;
- growing young plants from seed (necessary here to have both parents from the chosen reed type) and finding them different from those from another type.

If the habitat is thoroughly checked for all easily seen habitat variables, genetic variation is probable.

## Infra-red colour photography

Infra-red colour photographs are taken in the 520–900 nm wave band. While ground level, or oblique aerial photographs can be used, much the best are vertical aerial photographs, looking down on easily measured patches of colour.

*Phragmites*, seen in this way, may be red, the typical colour of inland reedbeds, though also found seawards; pink, found both in inland and coastal reedbeds; green, only found in brackish marshes, and blue, for dead, not full-grown and very depauperate stands. (Blue can be ignored.) Red, pink and green in fact intergrade. There are many different shades of colour, found in parts of Britain, Europe and the USA: so probably occurring throughout the range. However, the seeded Dutch polders are red (van der Toorn, 1972). Inland East Anglian reedbeds are red, with some pink, usually with diffuse, well-integrated pattern. Picking out separate clones is, as one would expect, difficult or impossible. The reedbeds are old or ancient and clones could be intergrading, or the bed becoming monoclonal, with the passage of centuries.

Coastal beds are different. Figure 5.1 of the Delaware Estuary, USA, shows one of the coastal marshes invaded by *Phragmites*. Table 5.1 gives some details of the clones in relation to the main variable, salt, as assessed by distance from the sea and the land. (There is no evidence of aggressive clones in this region, Chambers *et al*., 1999.)

Green clones are seawards, and in the 'green zone' green clones can usually grow into those of other colours (shown as 'taking bites' out of them). In the middle, predominantly pink zone, the pink clones are usually dominant and can compete successfully against green and red ones. Finally, landwards, red clones are dominant, and up on the land at the back, the typical large, perhaps mixed areas of inland marshes, are starting to appear.

More detail is shown in Figure 5.2 where the clone edges, and successful competition patterns are clear, also pools or ditches may, or may not, be outlined to perhaps demarcate clones.

A more complex pattern is shown in Figure 5.3, from an older bed: the tidal River Waveney in East Anglia. Inland of the sea bank, where brackish water rarely penetrates, the reed is in fairly large patches, mainly red and pink. The amount of pink and green can be attributed to the minor (and mostly old, pre-sea wall construction) salt in the marsh.

On the riverwards side of the sea wall there are variations in level. There are many small compartments, the centre being low, with much brackish water, and the reeds are green. On the higher ground around, usually well above water, the reed is mostly red, with some pink.

By the dykes, the clones are different to that in the middle. This may be of more general significance. Frequently bands of taller reed occur by narrow dykes through reedbeds. It has usually been assumed that nutrients in the dyke water are responsible for the bigger reeds. Now, therefore, it is clear different clones do occur in this habitat, and nutrient causes can only be postulated if there is no clonal difference.

**Figure 5.1** Infra-red colour photographs of the Delaware marshes, USA (Courtesy of the Earth Satellite Corp. MDA Federal Inc.)
Above, seaward; next page, landward. Seawards, *Phragmites* (most) invading and replacing *Spartina* (the hollow-looking vegetation). Red, pink and green *Phragmites* patches, clones, occur, red being mainly landwards, green, seawards (see Table 5.1).

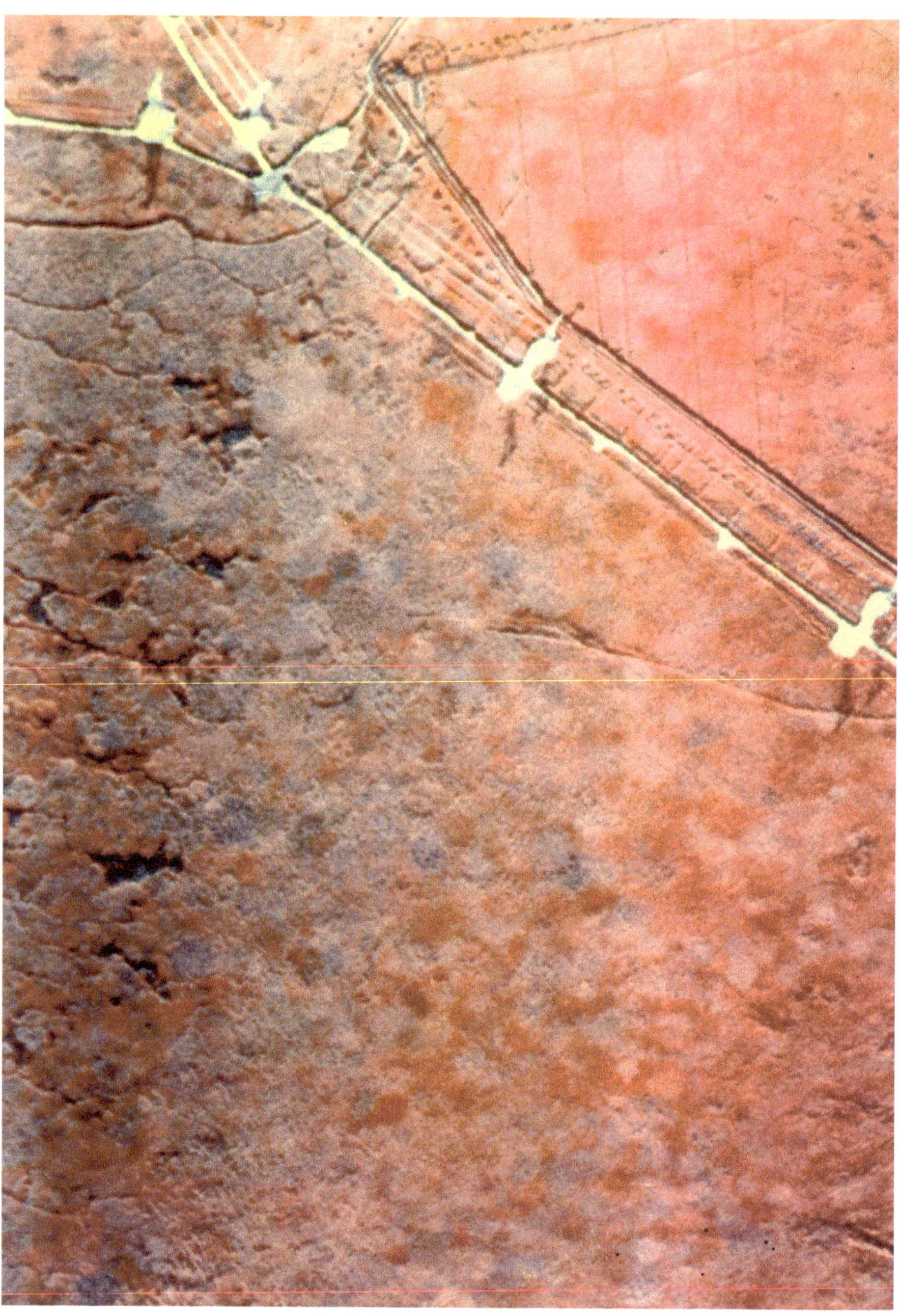


**Figure 5.1** Continued

**Figure 5.2** Infra-red colour photographs. New Jersey marshes, USA (Courtesy of the Earth Satellite Corp. MDA Federal Inc.).
a) Small-scale pattern. b) Larger scale pattern.
The shape of the clones are clear, sometimes bounded by a dyke or creek, sometimes not. Clone movement will be visible. Crescent patterns mean one clone is advancing into another. In (b) clones clearly mingle at some edges, but have sharp separation at others. One clone occurs actually inside another (advancing or retreating?).

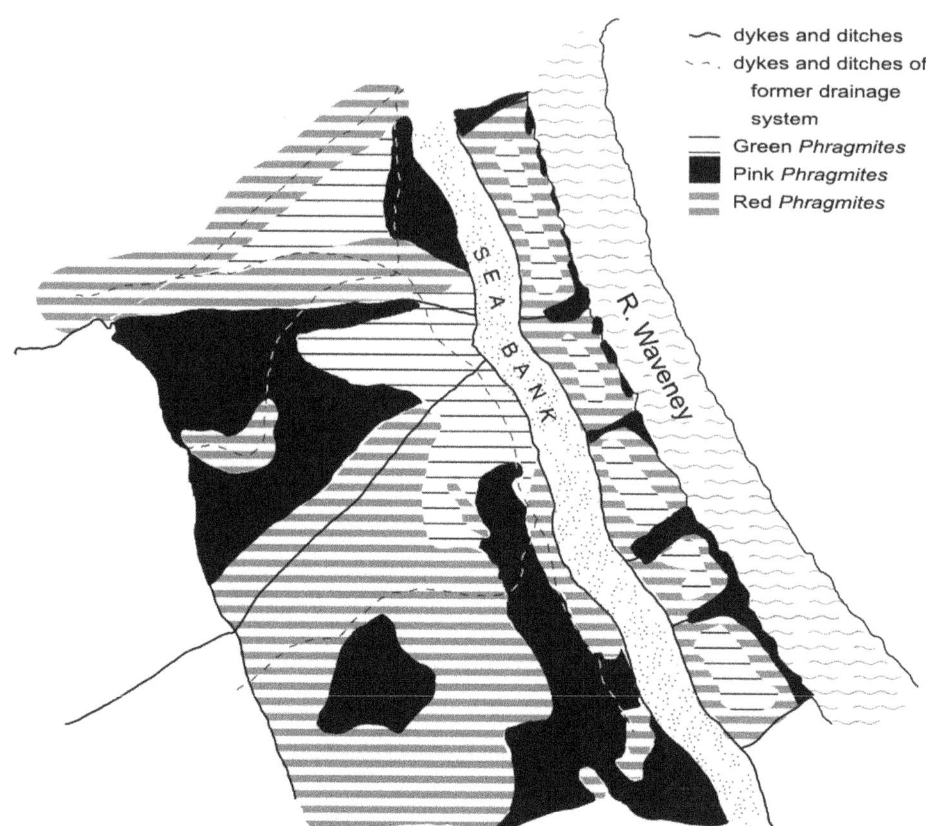


**Figure 5.3** Re-drawn from infra-red colour photographs. River Waveney estuary, England (taken from photograph by Committee for Aerial Photography, University of Cambridge).

Unlike Figures 5.1 and 5.2, these are managed reed marshes. On the river side of the sea wall, there are compartments, sunk with green clones centrally; raised round the edge, with red clones. Inside the sea wall, sea flooding is rare. Patterns in pink and red are partly governed by habitat, partly by plant factors (see text).

Infra-red colour photographs can elucidate reed patterns especially in coastal clones. What it is that causes the difference is, unfortunately, still unknown (Table 5.2). The clonal colours differ genetically, and the overall colour patterns are related to salt. Genotypes differ in their tolerance to salt (also see Chapter 6). It therefore seems likely that there are also ranges of genotypes with different tolerances to other habitat factors, perhaps to water regime, nutrient regime, light, perhaps even to grazing or mowing. More likely, *Phragmites* is able to make the best use of its genotypes to grow well in varying habitat. Indeed, Rolletschek *et al.* (1999a and b) does record this in relation to phosphorus, nitrogen and amino acids.

In a reedbed with many clones it is easy to wander, photograph in hand, and find each clone on the ground. It is better, though, to pick out clones when shoots are young, when emergence date, shape of shoot tips, shades of green, etc., differ between clones, or in autumn, when fruiting, time of leaf death, colour of dying leaves, and pattern of abscission vary. Late summer is the best time to photograph, but a waving 2 m (+) reedbed can look too uniform then to walk round and find clones! These clones are so clearly different in the single feature of infra-red colour. However, those of a single colour differ in many morphological and behavioural characters (Table 5.2). These – height, emergence-date, inflorescence type, and so on – are *not* correlated with, let alone causally part of, the infra-red colour. Pink clones can be tall or short, early or late: just like green and red clones. Presumably, plenty of genes are involved.

In one marsh (Cley Mill, Norfolk) with many small and variable clones, it was easy to link photograph and clone on the ground. The marsh was examined over more than a decade, and change in clone outline was not much at all. The idea that many clones means recent colonisation, and much competition and rapid change, is not true, at least in this marsh.

Yet movement there is. Figure 5.1 shows this, though the rate is not known. Figure 5.3 would seem a much more stable pattern, clones, growing in the habitats best suited to their colour, being without the mix of colours and uneven clone edge found in Figure 5.1.

Despite all the variation between the clones in Cley Mill reedbed, the dead reeds themselves are all obviously from this bed. The reed type only varies within limits 'set by' the bed (Chapters 6 and 11).

The cause within the reed which leads to its infra-red colour is none of those in Table 5.2, but may be electrophoretic.

Cytotype, ecotype and ploidy are described by, e.g., Koppitz (1999), Pauca-Comanescu (1999) and Kühl *et al.* (1999).

One last point. Reed clones compete with each other, not just with other species. The clone best adapted to the incident habitat is successful. *Phragmites* can be ruthless.

## Ploidy

This is the genetic character which has been most widely investigated, and over the longest period of time. Ploidy shows the number of sets of chromosomes in the cells. The basic number is 12 ($x$=12). The ordinary number is double this, (2n=24), which is the ordinary diploid number. Most clones tested are diploid or tetraploid (2n=48), in Europe, and higher in Asia and North Africa, where many are octoploid. Both higher and lower (haploid, 2n=12) than this are recorded (triploid, heptaploid, hexaploid) and clones with non-standard numbers, e.g., 2n=50. (See Clevering & Lissner, 1999; van der Toorn, 1972; Koppitz, 1999; Björk, 1963 and 1967; Löve & Löve, 1961; Pauca-Comanescu *et al.*, 1999; Roberts, 2000, and standard floras of various countries.) Pazourková (1973) considers higher ploidy may be optimum. Both monoclonal and polyclonal marshes have been recorded, for ploidy (e.g., Clevering & Lissner, 1999, and Koppitz, 1999, record both).

Linking ploidy to the morphology and behaviour of *Phragmites*, however, is another matter. Table 5.3 shows Dutch chromosome number (tertraploid and variants) with not much linkage to shoot height or habitat. The haploid *Phragmites* found and studied by Dr D. Dykyová was frail and small, more like a shaded stand than a normal one. Since haploid is not a normal ploidy, though, this is more interesting as a curiosity than as an adaptive or behaviour-linked feature. Little phenotypic differences were found by Pauca-Comanescu *et al.* (1999), in polyclonal marshes, *cf.*, infrared colour. It is generally an independent character. As indicated in Chapter 6, clones may be adapted to salt concentrations (Clevering & Lissner, 1999; Krzakova, 1996; Krzakova & Drapikowska, 2000). Caryology is described in Pazourková (1973).

There is a suggestion that both ploidy and height increase southwards through Europe to North Africa, but the two have not been causally linked.

## Isozymes, etc.

These comprise the recent tests, when genetic components could be separated and identified. However, again it is unusual to link genetics to performance. Hauber *et al.* (1991) and Pellegrin & Hauber (1999), in the Mississippi Delta and Gulf Coast, found great differences in isozymes. Also, looking at recent invasion, they found circular clones in a non-patterned background, as also occurred in clones of the Gulf Coast. The same variation was found in the mid-west of the USA and (in Bahrman & Gorenflot, 1983) in Europe, North Africa and central Asia. The Gulf patterns were correlated, patches being different also in morphology, and presumed invasive populations were dissimilar to stable old ones. This is a direct link between isozymes and performance. Twenty-one multilocus isozymic phenotypes were found; 140 isozymes from 16 enzymes systems were examined. Koppitz (1999) finds both higher diversity on land than in water, and that geographic distances between stands are matched by genetic diversities. Schieferstein (1999) recorded genetic diversity in 80 German beds, where some were more alike than others. (Also see Djerbrouni, 1992; Koppitz *et al.*, 1997; Kühl *et al.*, 1999; Kühl & Neuhaus, 1993; McKee & Richards, 1996; Neuhaus *et al.*, 1993; Schröder, 1979; Zeidler *et al.*, 1994).

This subject clearly needs far more investigation. Unfortunately, *Phragmites* does not attract the funding due to, say, ancient cereals or mice, let alone *Homo sapiens*. A start has been made; performance in a few clones on the Gulf Coast of the USA differs between clones, and these clones differ in isozymes. This is, though, a long way from testing isozymes of clones in East Anglia to determine those best to transplant for thatching reed or water purification! Or to determine which will best protect a river bank from scour.

## STABLE ECOTYPES

Different clones, stable after transplant, have been studied for performance characters, in the days before isozyme analyses. Genetic differences, are, however, clear. The main publications are van der Toorn (1972), van der Toorn & Mook (1982) for The Netherlands, and Dykyjová & Hradecká (1976) in the Czech Republic, and Björk (1967) for Sweden. (See also Květ, 1973, and Soetaert *et al.*, 2004). Also see Table 5.4.

A Dutch transplant experiment (Figure 5.4) showed the different ecotypes remained stable and different after 15 growing seasons. The ploidy is irrelevant to these differences. Figure 5.5 shows, diagrammatically, the differences between these types, in shoot size, height, leaf position, density; and in rhizome depth and pattern, root depth and pattern, and associated flora and substrate. They also show habitat variation within the clones. It seems that, as would be expected, the clones colonised and developed on the habitats most suitable for them, that the habitat selected (out of the available genotypes) those with the best character to grow there. (Many more characters and detail are given in van der Toorn, 1972.)

**Figure 5.4** Dutch transplant experiment (van der Toorn, 1972).
Transplanted material remained different for over 14 years.
Left: peat ecotype (Kalenberg).
Right: riverine ecotype (Kampen provenance).

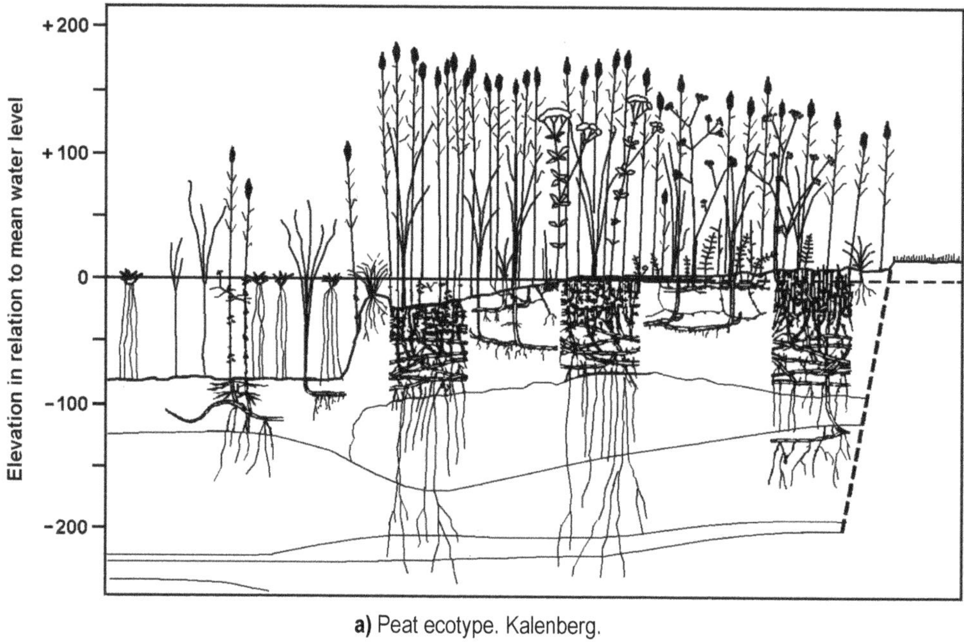


**a)** Peat ecotype. Kalenberg.

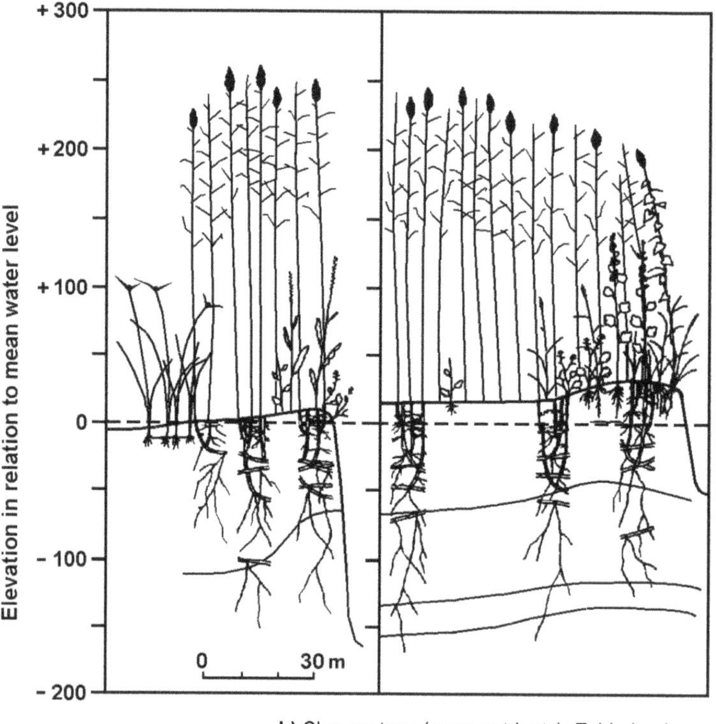


**b)** Clay ecotype (more nutrients). Zuiderland.

**Figure 5.5** Vegetation and soil profiles in different habitats, with different clones in The Netherlands (after van der Toorn, 1972). Much detail is shown: size and density of aerial shoots, position of their leaves, pattern and depth of rhizomes, amount of water roots (most on the peat), and the other species present. The shortest stand does not form a fully shading canopy.

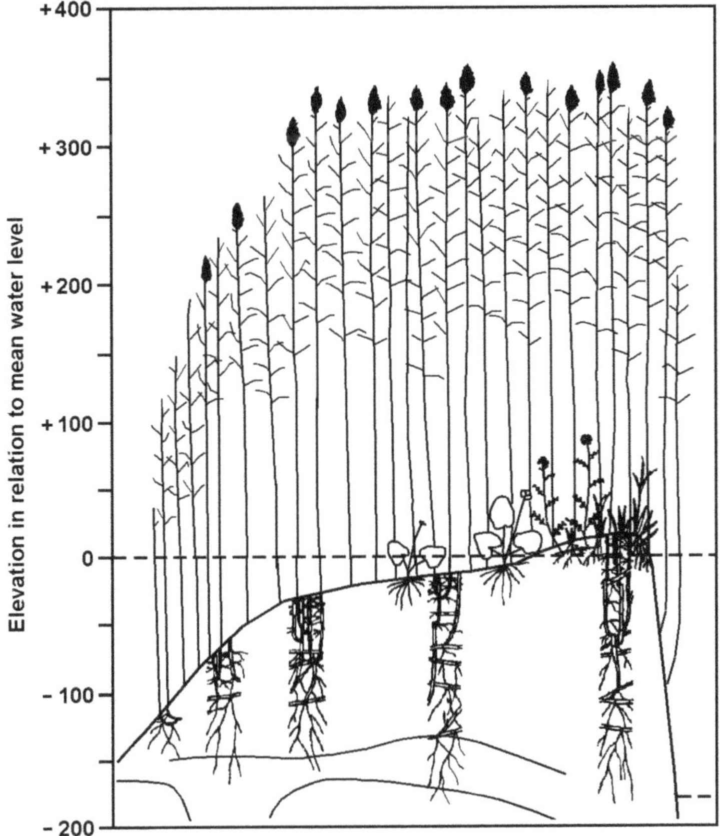


**Figure 5.5** Continued.
**c)** Clay/sand ecotype with tidal movement. The Biesbosch.

The Czech study showed the same result (Figure 5.6). The two types investigated indeed differed in habitat, in the waters of a Czech fish pond (a large lake-sized pond), and out on the shore. Like those in The Netherlands, the differences were only partly due to habitat, mostly, to genotype. Figures 5.6 and 5.7 show a difference in leaf and inflorescence pattern, as well as in rhizome. The water type has deeper and thinner rhizomes, and this depth may be partly or mainly due to the deeper water: but other differences remain. A German study also showed transplants maintained their differences (Koppitz, 1999), where, though, similar genotypes have diverged because they grew in different habitats, transplant will become similar after the usual four years.

A different type of transplant test is described by Bastlová *et al.* (2006), who cultivated six clones from each of six countries (Sweden, The Netherlands, Czech Republic, Hungary, Romania and Spain). The southern transplants grew taller and more vigorously (and flowered later, see Chapter 6) than the northern ones. Other transplants, using Europe-wide, and a few Australian, African, Maltese and Israeli clones, run into the difficulty that mid-Europe climates kill southern shoots in autumn, before they reach full height, so the height of the transplant does not reflect the height of the original. This is avoided when the transplants are to warmer climes, e.g., Malta. In general, the more extreme characters lessen on transplant, they become closer to the 'typical'.

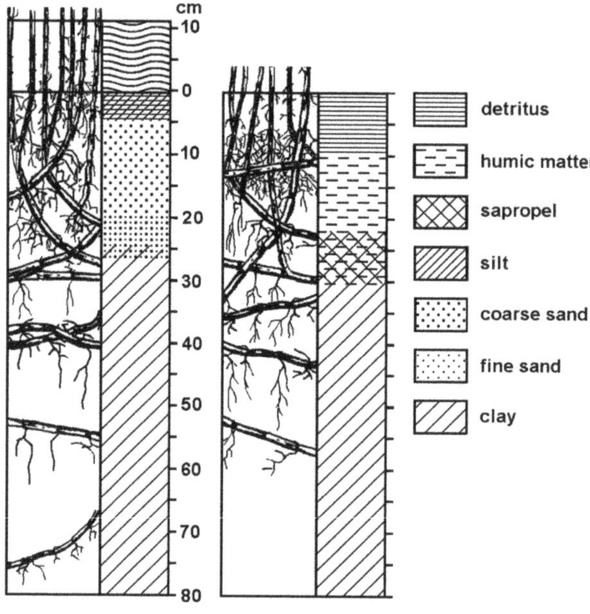


**Figure 5.6** Rhizomes in two ecotypes in a fish pond in the Czech Republic (after Dykyjová & Hradecká, 1976). Left: underwater; Right, marshy. The marsh ecotype forms dominant, highly-productive stands, more so than those of the shallow water. They are also wider, with more roots.

**Figure 5.7** Inflorescences (panicles) of two ecotypes in a fish pond in the Czech Republic (after Dykyjová & Hradecká,1976) The same two ecotypes as in Figure 5.5b. The marsh plant is on the right, with the larger and more upright leaf blades and denser panicles.

## Conclusion

*Phragmites* has very many different genotypes, and many of these vary in either or both of morphological characters (height, leaf size, etc.) and in adaptation to habitat characters (water, substrate, salt). Consequently many visibly different clones may occur in a (near-) uniform marsh, and many physiologically different ones in a marsh with a habitat gradient, and an apparently uniform reedbed. Geographic variation certainly exists (Chapter 4), clones being genetically adapted to the climatic growing season as well as varying with the annual weather.

*Phragmites*, once more, is complex!

**Table 5.1** Infra-red patterns in the Delaware marsh (see text)

The records are of the number of clones of one infra-red colour invading those of another colour in each of the main habitats (Haslam, 1979).

| | **Green dominant** | | **Pink dominant** | | **Red dominant** | |
|---|---|---|---|---|---|---|
| *Habitat* | *over pink (and pink-green)* | *over red (and red-green)* | *over green (and pink-green)* | *over red (and pink-red and red-green)* | *over green (and pink-green and red-green)* | *over pink (and red-pink)* |
| Seawards green band | 20 | 2 | 4 | – | 1 | – |
| Inner mixed band with much green | 12 | 4 | 1 | 3 | 5 | 2 |
| Outer mixed band with much pink | 9 | 1 | 5 | 6 | 4 | 4 |
| Landwards dry red-pink band | – | – | – | 8 | 1 | 14 |
| Landwards dry red-pink area, embanked from the sea | – | – | – | 12 | – | 8 |

**Table 5.2** Characters independent of clone infra-red colour (from Haslam, 1979)

| Character | |
|---|---|
| Shoot density | |
| Stem height and width | |
| Leaf size, shape, orientation, density | |
| Leaf colour | |
| Terrestrial and littoral forms (see below and Dykyjová, 1970) | |
| Relative maturation of clone, relative time of shoot death, and of leaf abscission in autumn, pigmentation during autumn | |
| Inflorescence size and density | |
| Standing dead shoots, litter, bare ground (the colours of these can be seen as different to that of standing living shoots) | |
| Cutting, burning, grazing and other management practices | |
| Rust (*Puccinia*) and other pests and diseases (present but not severe in the habitats studied) | |
| Water regime, from dry land to standing water over 1 m deep (including stands on dry land during mild droughts) | |
| Soil nutrient regime, from mesotrophic fen peat to eutrophic mineral alluvium | |
| Salt regime, from fresh water to permanent salt causing dwarf populations, or damaging sea water floods | |
| Leaf nutrient content (nitrogen, phosphorus, potassium, calcium, magnesium, sodium) | |
| Leaf surface (as seen with a scanning electron microscope) | |
| Leaf anatomy (and as inflorescences and leaves are the same colour, all anatomical characters differing between inflorescences and leaves) | |
| Chlorophylls, phaeophytins | |
| Carotenoids | |
| Flavonoids | |
| Anthocyanidins, leucanthocyanidins | |
| Early products of photosynthesis | |
| Water vapour or carbon dioxide patterns immediately above the leaves | |
| Wilting<br>Chilling<br>Membrane alteration with ethylene | tested on other species |

**Table 5.3** Chromosome number, shoot length, and habitat *of Phragmites australis* provenances in The Netherlands (van der Toorn, 1972)

| No. | Provenance | Chromosome number (2n) *mean* | Chromosome number (2n) *range* | Shoot length[1] | Habitat |
|---|---|---|---|---|---|
| 1 | Scherpenzeel | 48.8 | 47–55 | I | peat |
| 2 | Wapserveen | 48.1 | 45–51 | I | |
| 3 | Kalenberg-E | 46.8 | 43–50 | I | |
| 4 | Kalenberg-W | 47.5 | 44–50 | I | |
| 5 | Tjeukemeer | – | – | II | |
| 6 | Nieuwkoop | 47.4 | 45–49 | I | |
| 7 | Ameide | 46.5 | 43–50 | III | riverine[2] |
| 8 | The Biesbosch | 48.0 | 45–51 | II | |
| 9 | Kampen | 48.7 | 46–53 | II | |
| 10 | Zuidland | 46.8 | 45–48 | I | estuarine |
| 11 | Ossendrecht-S | – | – | I | |
| 12 | Ossendrecht-H | 47.3 | 45–50 | III | |
| 13 | Elburg | 46.7 | 43–52 | II | IJsselmeer |

1 I = short form
II = intermediate form
III = long form

2 The location Kampen belonged to a fresh water tidal area up to 1932.

**Table 5.4** Ecotypes of *Phragmites australis* (Květ & Westlake in Květ *et al.*, 1998)

**The Netherlands** (van der Toorn, 1972)

| | *Shoot density (no. m⁻²)* | *Shoot height (cm)* | *Above-ground biomass (g dry wt m⁻²)* | *Tolerance of submersion* | *Tolerance of frosts* | *Tolerance of salt* | *Habitat* |
|---|---|---|---|---|---|---|---|
| | | | | | *(% 1° shoots killed)* | | |
| 'Peat' | 220 | <250 | 990 | Low | 39 | Low | Lake margins |
| 'Riverine' | 90 | >300 | 2400 | High | 52 | High | Tidal freshwaters |

**Czechoslovakia** (Dykyjová & Hradecká, 1973)

| | *Shoot density (no. $m^{-2}$)* | *Shoot height (cm)* | *Above-ground biomass (g dry wt $m^{-2}$)* | *Underground biomass (g dry wt $m^{-2}$)* | *Nitrogen biomass load (g N $m^{-2}$)* | *Habitat* |
|---|---|---|---|---|---|---|
| 'Littoral' | 103 | 225 | 1500 | 5890 | 28 | Fish pond; marginal water |
| 'Limosal' | 101 | 235 | 2070 | 3360 | 47 | Fish pond; marginal land |

# Chapter 6
# Variation: environmental

## Introduction

The previous chapter described genetic and clonal (presumed genetic) variation, and showed heredity is responsible for much variation in performance, in height, density, competitive vigour, and being best suited to a particular habitat, etc.. In fact, from all that, it could be supposed all variation is genetic. This is not true at all. There is quite as much, indeed, actually, more variation due to environmental factors. Looking (again) at Table 1.1 shows some part of this.

A book could be written just on the environmental variation! Space only permits selecting a few environmental variables, and describing, very briefly, the variation in the reed because of it. Since there are very many causes of variation, the actual *Phragmites* shoots, reeds and underground parts are influenced by all of these, giving extreme complexity. There is variation between reeds in different reedbeds, especially if reeds are strong, when differences are greater (Chapter 10).

## History: past management

This is the effect now of one or more habitat changes in the past. (It could be the simple act of removing or planting a population.)

The most striking example is of 'cultivated reed'. In Britain, this describes a harvested, or recently harvested bed. The bed may have been in existence for only decades, or for long enough to accumulate a metre or more of peat. This is irrelevant. The cultivated reed (Figures 4.4a and 10.8) is as uniform as is possible. Cutting and spring-burning for 1–4 years turns a non-uniform population, with many very tall and very short reeds (Figure 4.6), a long period of emergence, and usually a generally untidy look caused by sheaths being loose on the stem, by slight bending at the nodes and by dead stems dropping. The cultivated reed has the reverse. Cultivated reed is usually straight, though occasionally a very dog-legged clone stays slightly dog-legged.

So cultivated reed is good for thatching (Chapter 10), easy to harvest, making up into uniform bundles. It is fortunate that wild reed can be so easily transformed to cultivated reed!

If harvesting is abandoned, the reed may revert to wild in less than five years. On the other hand it may remain in the cultivated form for at least 90 years, so probably for a century and more. The most extreme example seen is a fen where drainage halted harvesting probably around 100 years before. After 50 years the dry area was Tall Herb mix, with sparse *Phragmites*. Fifty years on again, and reed had taken over: and in the cultivated form. This form was presumably the earlier one, which had been present in the lower levels of the fen, where harvesting stopped around the 1930s.

When reed is gradually forced out from ex-harvested beds by drying, the cultivated form may be lost. In one fen the reed remained straight and 'clean', but became more varied in height. This was because the rhizomes became poorly, and their reeds became very unduly small (Figure 6.1).

Historic management can therefore be most important in determining characters of the population. Mis-interpretation can occur when history is not known and considered.

Wicken Fen (Cambridgeshire) is on calcium-rich, other nutrient-medium, sedge peat, organic mud, etc., but no *Phragmites* peat. That occurs in more silted, nutrient-richer, places. The fen beside Wicken was converted to arable in the Second World War, which meant fertilising, and nutrient release from disturbance (ploughing, etc.). So when this was returned to fen afterwards, it was the more nutrient-rich *Phragmites*, not the nutrient-medium *Cladium mariscus*, fen sedge, which colonised and did very

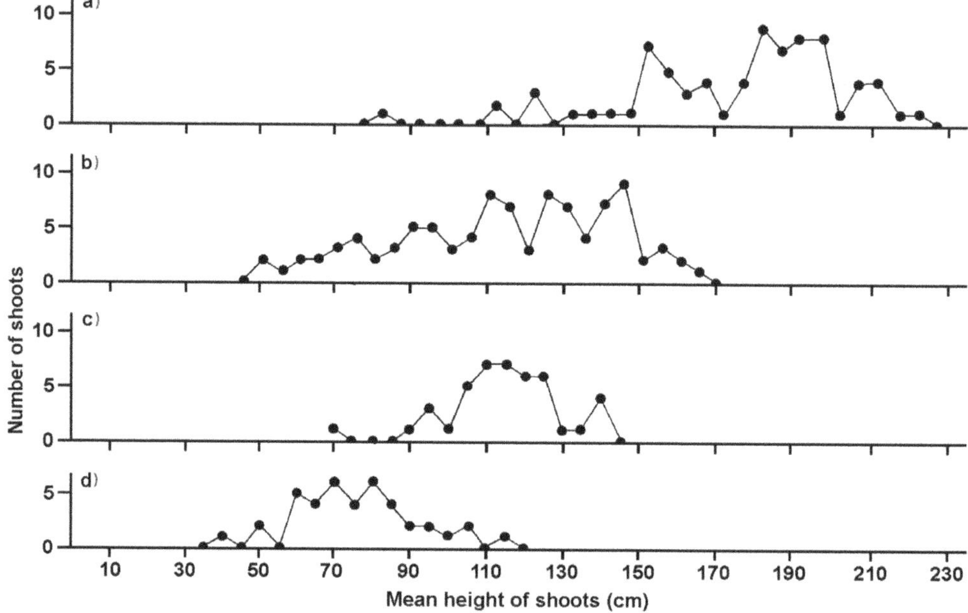


**Figure 6.1** Stands deteriorating over 8 years (Haslam, 1970)

These dry Valley fens are being much further dried. In both fens modal height became less, and the mode moved to the left, (a) and (b) Icklingham, England; (c) and (d) Cavenham, England.

well. These events happened 60 and 45+ years ago, respectively. In another 50 years will the memory be lost, and the reedbed mis-interpreted? Will reedbed be thought to be native to the area? (The reed near-doubled in size in the new reedbed, compared with the main Fen, reaching 3 m high.)

In Malta, there is at least one patch of Egyptian reed, imported into a wildfowl decoy pond, by a farmer shooting wildfowl. If the stand and pond remain in existence for another 50 years, will botanists record the immense dispersal distance of *Phragmites* fruits?

The principal feature in this section, however, is one where the causes have indeed been forgotten.

Looking at reeds of one bed, there may be considerable clonal variation, but all reeds usually have characters in common, which probably make them different to reeds of a different bed. Exactly what past management?

*Phragmites* transplants are increasingly common, for constructed reedbeds for water purification, and for new beds, e.g., for the EU bittern project, where those concerned do not want to wait for spread from the nearest ditches, or the nuisance of collection from a few kilometres away.

The trouble is that most such transplants are alien, often Dutch. This means alteration of the gene pool of the country, and, for a wild plant in semi-wild conditions, is very wrong. Once more, in a century will their provenance be forgotten, and the *Phragmites* investigated as native?

## SALT

After the difficulties involved in bed variation, salt – though also partly unknown – is at least simpler! More is known about the effect of sea water (salt: also found near old salt mines, etc.) than of any other chemical factor. It influences the plant in various directions.

In different habitats, at different seasons and in different concentrations, salt can alter:

- **internode length**, so shoot height. Sudden exposure to salt in the growing season (or new shoots growing into such a habitat) slows growth;
- **bud width**, and so potential shoot height (Chapters 3 and 4). Salt narrows buds. Mildly toxic conditions;
- **bud density**, and so reed density. Salt, and the saltiest possible conditions, inhibit bud development;
- **emergence date**, and so probability of maturation;
- **adverse habitat factors** (e.g., sea water incursion) may delay emergence, the longer (up to a month), the more adverse the factor (e.g., toxicity);

- **leaf size**, thickness, toughness and hardness. Sudden toxic conditions weaken leaves, chronic salt toughens them;
- ***legehalme* production**. In clones with this potentiality, these develop more often in (chronically) slightly salty conditions;
- **health and death**. Sea water in excess quantities kills. High quality reedbeds grow, e.g., at the back of saltmarshes;
- **reed hardness**. Chronic mildly brackish water produces harder reed; sudden sea incursions, softer reed;
- **reed brittleness**, as the last. When hard, reed also becomes brittle, easily breaking;
- **reed fibrousness**. This is less affected than the above components of strength (see Chapter 10) but is likewise weakened by sea water incursions.

Lethal limits for salt vary slightly geographically and genetically and as most analyses were done before it was known that different genotypes differ in salt tolerance, the following figures are difficult to interpret. Chlorinity at 1.2% in soil water 10 cm below ground level (Ranwell *et al.*, 1964) is recorded for Britain, in 10–20% sea water in The Netherlands, where the optimum growth of saline populations is 1.75–3.5% chlorine, and freshwater ones, 0–1.75% chlorine (see previous chapter for variation in salt tolerance). Hartsendorf & Rolletschek (2001) found best growth in 0–1.5%. Salt tolerance may increase when transpiration is high (Lissner *et al.*, 1999a) and warmer climates – with different genotypes – may have more salt tolerance (Lissner *et al.*, 1999b). In salt-sensitive populations, *Phragmites* decreases in saltier water (Mendelssohn *et al.*, 1999), in Denmark, as elsewhere. In Romania, giant reed is less tolerant to sea water: despite growing well in mildly brackish areas (Hanganu *et al.*, 1999). In Southampton estuary, Bartlett (1999) found reed growing taller towards the freshwater side (see also Burdick & Konisky, 2003). Soil cellulose decomposition is less the more salty the water (Mendelssohn *et al.*, 1999): the usual disinfectant effect?

Lissner *et al.* (1999a and b) conclude that salt tolerance is a cation adjustment to water loss. Soetaert *et al.* (2004) record above-ground biomass nearly double in the oligohaline than the mesohaline site, and reed also emerged earlier, with lower shoot density, and more, but smaller leaves, and more new rhizomes. While principles still hold, actual details may vary with salt tolerance. It is probably significant (Adams & Bate, 1999) that in estuaries the deep underground rhizomes are in less salty soil than is at the surface.

The dwarf *Phragmites* is one of the more extraordinary forms (Figure 8.6c; see Chapters 1 and 3 also). Shoots average 5 cm long, in prostrate rosettes. The King reeds, those terminating long horizontal rhizomes, are barely 30 cm high. The stems are very tough, strong, highly fibrous and not brittle. The tough, small leaves can scratch skin. This form occurs in habitats brackish, dry and trampled. In a Malta site, decreasing salt, adding fertiliser and shading gave, 30 years later, shoots 10–40 cm high and less tough. Without the salt, *Phragmites* can grow equally small, in dry and trampled places, but is neither tough nor strong.

In dykes, *Phragmites* tolerates slightly brackish waters, which harm less salt-tolerant species. *Phragmites* and *Potamogeton pectinatus* are frequent, together with saline species. One flood, though, raised dyke chlorinity to 2.5%, where it was usually under 1%, and did lead to slow death of *Phragmites* shoots. The next year shoots were sparse (bud development hindered), but of normal height (bud width and internode length both unharmed).

Whether as storm spray or as flood, salt can hinder the development of growing internodes, so leading to short reeds. A sea flood on a basically freshwater marsh usually leads to a late-emerging, and so late-maturing sparse, short population. Reed beds receiving little salt may recover in one year, those with more, often four years (the time, see Chapter 3, it takes for the rhizomes to be nearly a new crop), or even longer. In this last, the recovery has to wait for the salt concentrations in the soil to be reduced to non-harmful levels.

Pallis (1958) records small clusters of vertical stems on wet salty sand, but when there is more water and less salt, a horizontal rhizome develops. Once more, habitat controls growth.

After one small sea water flood the reeds became harder. They also became less fibrous, so the reed, overall, was less satisfactory for thatching (see Hanganu *et al.*, 1999; Godwin & Turner, 1933; Peake, 1960). Hard, brackish-water reed is often very brittle, so less satisfactory for thatching than it first appears.

In permanently (slightly) brackish areas of commercial reedbeds, reeds are hard. A brackish part of Hickling marsh was for years known for being harvested triple wale. That is, to be harvested once in three years, the dead reeds remaining hard and upstanding for longer than elsewhere.

Salt, therefore, affects *Phragmites*, and sufficient concentrations kill. Milder ones have a mix of effects, varying with the incoming salt pattern, chronic or sudden flood. The results also depend on the salt-tolerance of the affected clones. That which may kill salt-susceptible clones may be entirely tolerable for salt-tolerant ones. Both morphology and seasonal behaviour are altered.

Germination seems unaffected in one test by low salt levels, e.g., 2%. However, establishment is different. None became established in 2%. At 1% salt, the same pattern applied as with nutrients: plants are more tolerant in wetter conditions, and did become established, though not in the drier soil. At 0.5% salt, more established, though again more in wet than in dry water regimes. Older plants were more tolerant, remaining alive, though dwarf, in 2% salt in dry conditions (see also Soetaert *et al.*, 2004).

## Nutrients

While mineral regimes are generally more important than the more localised salt patterns, even less is, so far, known about nutrients. *Phragmites* grows in a wide range of substrate and nutrient types, but low nutrients lead to poor performance (Table 6.1). High nutrients may come from substrate, by accreting substrate or from water. Since far too many low-nutrient stands are in heath, moor, bog or equivalent, which are usually (lightly) grazed, there is unfortunately this second complicating factor influencing performance (Figure 6.2).

**Figure 6.2** Valley bog with *Phragmites* in flush, New Forest. Reeds, short and chewed, grow in small valleys flushed with storm water.

That being said, populations from a wide range of stands in the same year had, in nutrient-poor places, modal height not exceeding 0.75 cm and stem densities under 40 per m$^2$. (In flushed, boggy stands, modal heights were under 40 cm, densities under 16 per m$^2$.) Higher nutrient regimes had stands of at least 1 m modal height, and densities of at least 70 per m$^2$. These may be monodominant stands, usually in shallow sheltered bays, or be in mixed stands. (Low-nutrient vegetation types are usually short.)

Plants of course, like animals, require a range of minerals, and *Phragmites* suffers if any nutrient is over-low. In fens, with calcium-dominance or calcium-influence restricting P, also N and K availability, it is these (in, e.g., East Anglia and Anglesey) which are limiting, and additions greatly improve the stand (Table 6.2). It takes time, the second year of adding nutrients has normal size shoots, and only in the fourth year do some horizontal rhizomes have enough nutrients to carry these outside the fertilised 1 m$^2$.

The extreme P-deficiency (also with N and K lesser deficiency) leads to shoots sparse and short, emerging late, with tough, ribbed leaf blades, and early leaf death, with bright yellow colouring from July. Where this is dried or disturbed, the *Phragmites* becomes more normal: more nutrients (see also Harding, 1992).

As well as fens, there are bogs. And in north west Scotland, *Phragmites* patches occur where there are calcium patches, either from local limestone outcrops, or lime-rich run-off into the bog. In wet heath, very depauperate stands (Chapter 4) may have browny leaf discolouration from late summer. No doubt there are also stands restricted by deficiency in other nutrients.

Nutrient addition experiments elucidate further. Calcium dominance, as indicated above, is cancelled by added other nutrients. Elsewhere, standard fertiliser treatment barely alters habit, but has a substantial effect on smaller shoots, and replacement shoots. The primary (King) reeds (from the largest rhizomes) are little affected, except where they are already kept short. The small shoots from the back of the rhizome, those which are started with just the food the horizontal rhizomes have over (they put most food into front stems), these may be substantially taller – and, if sparse, denser – with extra nutrients (Ca, N, P, K). This can also happen after cutting or grazing. Ordinarily shoots cut after June in commercial reedbeds are barely replaced (Chapter 4), but with extra nutrients, a good crop emerges (Table 6.2): though it is doubtful whether they mature before winter kills them. The same happens in low-nutrient habitat (Figure 6.3). Light grazing reduces *Phragmites* largely because the shoots do not get a chance to photosynthesise properly before they are grazed. Then, in this nutrient-low stand, food in the horizontal rhizome is too low to produce a good crop of replacement buds. If, though, extra nutrients are added, there is a good crop. (Björk, 1967, and A. Szczepanski, 1970 pers. comm. both record reed growing well in pure sewage.)

Nutrients, in appropriate places, can make up for an over-low water table. In flooded places, where water regime is optimal, extra nutrients are seldom needed. Where water is inadequate, they may be.

Tables 6.3–6.5, of shoots and rhizomes, shows that those in really deficient stands have lower levels of nutrients, but that once nutrients are adequate there is little difference in concentrations in middle- and high-nutrient habitats. Plant calcium may or may not reflect habitat calcium. Poor bog stands in north-west Scotland are *not* limited by (major) nutrient deficiency.

There is no real correlation between strength of dead reeds and their nutrient content. Strong reeds are not low – or high – in nutrients (Tables 6.6–6.9). This applies both to the overall reed, to the butt of the reed (most exposed to weathering on the roof), and to the elements tested separately. Leaf sheaths have the highest content, and nodes (where buds arise) have more than internodes. There is no evidence that thatching reed has increased in nutrient content with increased land fertiliser. (In Table 6.9. there is a suggestion of decrease.)

Nutrient translocation along rhizomes is less than would be expected, as the fen example above shows. It was four years before rhizomes moved fertilisers from where it was added. Also, when *legehalme* cross a dyke, the new shoots growing in the dyke (Figure 8.8) are short. Tall ones are only where roots can grow down in the soil at the far side. Food is not moved far horizontally (except in advancing margins).

a)

b)

**Figure 6.3** Riverside reeds (infrequent), Hungary.

a) Small stream.

b) Large river (River Danube).

In the young plant, nutrients (particularly phosphorus) may be crucial (Chapter 2).

Nutrient status therefore affects bud stimulation and bud size. Variation due to nutrients affects habit (tough yellow leaves, etc.), total density and size of shorter shoots, in behaviour after summer culling and, overall, in height, deficient stands being, generally shorter, though with exceptions. (Romero *et al.*,1999, also find *Phragmites* can grow well in a range of nutrient regimes, and find interactive effects.)

An interesting paper by Votrubová *et al.* (1997) describes anatomical abnormalities in eutrophic habitats. These included increased lignification, blockages of air channels and vascular tissues in roots and rhizomes, mostly in damaged organs. High nutrients do not make the abnormalities, they allow damaged parts to do so. Romero, *et al.* (1999) also find *Phragmites* growing well in varying nutrient regimes, with interactive effects.

All of these defend the plant against flooding, phytotoxins entering, and spread of pathogens. However, replacement of damage does take energy. This is thought-provoking. How much of *Phragmites* is so adapted to its functions? (See also Květ, 1973; Bayly & O'Neill, 1972; Dykyjová, 1979; Planter, 1970.) Trémolières *et al.* (1994), demonstrate effects of changing groundwater nutrients.

(Study of organic acids is in the early stages, e.g., Cižková *et al.*, 1996, 1999a and 2001, find high levels, and Rolletschek *et al.*,1999, find amino acid metabolism affected by phenotype. Kickin & Meltzer, 1998, have carbohydrate differences.)

## LIGHT

Shade alters habit and density. Shaded shoots (Figure 6.4) tend to be around and up to *c.* 1 m, reaching up to maybe 1.5 m in glades. They are therefore shorter than in full light. The leaves are more horizontal, bent to get most light (unlike dominant stands, which get sufficient light while droopy). The leaves may also be paler and frailer. A shading canopy is rare: not enough is grown. There is an interesting complication with light. Short shoots may be shaded in woods, where trees are tall. *Phragmites* in Tall Herb is much the height of the shorter to medium *Phragmites* shoots in the reedbed. Reedbeds themselves have shoots of varying heights. In the first two habitats, they live. In the third, many short shoots within a reedbed, may die (Chapter 4). Because Tall Herb can cast as much shade as a reedbed, this death is the choice of the horizontal rhizomes, withdrawing food from the short shoots. In Tall Herb even a little food made in shade is worth having. In the reedbed, it is more efficient to concentrate on the tall reeds, which can have much food removed from them without dying, to be stored below (Figure 6.5).

There is also the 'bush effect'. When reeds grow by and in a bush, so that tall shoots come into light, the taller shoots grow taller than others in the full light nearby. Their internodes grow longer. At the (sunlit) upper part, leaves are like others in the light, not like those in shade.

**Figure 6.4** *Phragmites* in wood, England.

Short, narrow, non-flowering reeds, with leaves adjusted to horizontal to catch the most light.

a)

b)

**Figure 6.5** *Phragmites* in Tall Herb vegetation, England.

(a) is of rather lower nutrient status than

(b) approaching a rich mire. Leaves darker green, shoots look 'healthier'.

## Water regime

This is, after all, the primary control and influence on *Phragmites*. Reeds are limited at the lower end by over-deep water. Most photosynthesis is above water, so short shoots extend less far into deep water than tall ones. Shoots >3 m high can still photosynthesise well in 1–1.5 m of water. Indeed, in Poland, reed has been seen in 4 m water. At the dry end, shaded *Phragmites* may extend into wet woodland, or may be lost as bog develops above the fen. It is, at the least, likely that different clones are found in different water regimes, as with salt (see above and Chapter 5).

*Phragmites* has a narrower range of water regime in acid than in alkaline soils, in mineral-poor than in mineral-rich ones. How far this is differently-adapted genotypes, how far alkaline nutrient-rich soils are preferable for the species, is not known. There may well be genetic variation (Chapter 5), and undoubtedly some chemical regimes can extend the water level range (see above, and below). In the centre of the range it is factors other than water table which determine performance (see also Spence, 1964).

The same applies to autumn flooding or drought. In a normal year a dry stand next to a normally-flooded one yellows first. In a drought autumn, it is the other way round.

Wetter dominant reeds are usually sparser and taller. Adaptation to new conditions is faster in wet stands (even those only wetted half the time). The influence of water level is greatest between +5 cm and -15 cm of ground level.

In dry Malta, admittedly with the aid of brackish conditions, *Phragmites* now occurs in generally drier habitats than further north. *Phragmites* is here being stressed at the far limit of its water regime. It is able to grow because nutrients are ample, and salt is adequate and, presumably, genotype is satisfactory. Underground, rhizome pattern alters with water as well as other factors, but because the influences are so many, water regime cannot be quantified, and statements do not apply to all stands. First, and clearly, horizontal rhizomes in wet conditions are oval (wider horizontally), they also 'look' in better health; not quantifiable! Water is above the ground. Because of wet and difficulty, only a small number of rhizome patterns have been excavated. On these, however, it seems likely, and logical, that wet places have horizontal rhizomes from fairly high in the soil down to around 2 m (seen at the side of dykes being bulldozed). Dry sites, however, are likely to have fewer both high and low. Also, the complex sloping branched vertical rhizomes (Figure 3.3) are more typical of wet sites, and more nearly vertical rhizomes with close, vertical (short) branches, of dry places.

Weisner & Strand (1996) in Sweden, found horizontal rhizomes shorter and higher in the soil in deep water. This affects nutrient and oxygen processes.

As a general tendency, drier reedbeds are not over 1.5(–2) m modal height, and may be shorter. Tall reed, whether 3 m or 10 m (in modal height) is usually flooded for much of the year. Without genetic data, and knowing there are exceptions, it is not possible to say whether reeds cannot grow tall in dry places, or whether the genotypes best adapted to dry places happen not to have tall reed.

*Phragmites* is more susceptible to competition at the dry end of its range (Chapters 1 and 7). This is largely because the competitors here are more effective. But is it, in part, also because *Phragmites* has less vigour?

Rooting also varies. Water roots (Chapter 3) only flourish in stands flooded at least in the growing season. The root felt towards ground level is also thick and dense only when this level is wet in the growing season and most of the rest of the year. (It can be 'squishy' rather than flooded.)

The total range of water level is wide, from water well over 4 m, perhaps 6 m in deep lakes, to perhaps >6 m below ground level. The latter occur in, e.g., Malta seepage areas and Hungarian railway embankments. At the ends of the range there must be very favourable factors which do not just cancel, but reverse the normal water regime checks. The driest Malta site was on a former spring, now dried but still with water down below. There were no competitors and the perseverance of Maltese *Phragmites* under drying compels admiration (Chapter 8). Even in England *Phragmites* can grow 2–3 m up railway banks above the wet level it is otherwise confined to, below. In Hungary, the same is seen in plains, but up in the hills reeds occur in places actually dry by mid-summer (see below), and it grows >3 m above that. This time also, some chemical factor associated with steam trains overrides the dryness (Figure 6.6)

**Figure 6.6** Reeds on railway embankment, Hungary.
Dry stream in foreground. Railway line several metres up, at arrow.

Water regime is not only water in the growing season, it is water integrated over the year, and it is also whether the water is (nearly) still, or flowing, or fluctuating greatly up and down. Generally, as long as the regime is regular from year to year, it is well tolerated. A stable spring can be a good habitat. So can water present after snow melt in late spring, but dry for most of the year, as in Hungarian mountain streamside willow-reed bands. In this last, though, it is necessary to have plentiful, so not rapidly dispersed, snow melt. Reed was 0.5 m shorter in a year when snow had been inadequate.

*Phragmites*, though, does like a *regular* water regime. Unexpected drought in spring can mean emergence a month or even two months late, and with lower density than usual, so reeds do not mature in autumn and, in summer, may leave a habitat open to invasion. Over-low nutrients may lead to short populations in only 75 cm water, while more nutrients may encourage reeds to grow in deeper water (e.g., Björk, 1967; Misra, 1938; Stålberg, 1939). Zonneveld (1960) in a Dutch estuary found the optimum fluctuation was -1–0 m, but -1.5 m–>25 cm, possible. The same occurs if the habitat is unexpectedly flooded. Disruption causes damage (Figure 6.7; see also McCartney, 1999).

Fluctuating water levels may increase bud density (see also Pallis, 1916), though a high stable level may produce the best-performing stand. Transplants show in those young plants that growth of rhizomes as well as shoots is better around water level than in drier places, although all those near water level did well. (See Lessmann *et al.*, 2001, for photosynthetic variations.)

**Figure 6.7**

Floodplain reed, in habitat usually drained in Europe, Hungary

Finally, drained reedbeds gradually lose their reeds (see also Chapter 11). One fen had commercial reedbeds. Two decades of abandonment and drying led to most being reed–*Arrhenatherum elatior–Urtica dioica*, and other Tall Herbs. *Phragmites* was locally dominant, but not tall and not excluding all competitors (the water table, except after heavy rain, was 0.75–1 m down). Elsewhere on peat cuttings (where ground level was lower) reeds were sparser, and in the highest level, they were absent. The fen was being drained. Even over a few years the bed performance deteriorated (Figure 6.1). Another two decades and the little reed left, obviously stressed by drought, was eliminated by grazing.

Loss of water by human impact, for drainage, abstraction or both is a primary cause of loss of wetlands. The implications of hydrological balance in new reedbeds are discussed by McCartney (1999).

Water regimes not tolerated include scour. *Phragmites* is rare along (non-brackish) rivers (Haslam, 1982), and is away from scour, or ephemeral where present. Also, the extreme *and irregular* variations of water level on reservoir shores are rarely tolerated.

In Australia, habitats can be classified as, broadly: surface water, ground water, and estuaries with brackish water. The first are in or by watercourses, pools, flood plains, 'chains of ponds', or a mosaic of wetlands and channels, where droughts may occur. Groundwater wetlands are few and small, such as 'mound springs' with continuous artesian wetlands. The habitat is newly increased by boreholes with ponds. Estuarine *Phragmites* is in large shallow coastal lagoons, mainly in the southeast. Salt and nutrients vary greatly with river flow and drought. Over-salty conditions kill the *Phragmites* (Roberts, 2000).

## Temperature

In its response to temperature, *Phragmites* has many genotypes adapted to the various climates occurring in its cosmopolitan distribution. A few of these are shown in Table 3.3. Emergence, length of growing season, fruiting and shoot death are all fitted to the climate. When winter is little and late, leaf death is late, and shoots may survive the winter. Where there is none, they may remain green through the winter. The larger Great Reed, *Arundo donax*, in Malta remains green through a warm winter, but the leaves die, in a cold one. Next year's growth is good, in both. Reed stands from non-frosted countries are, in the milder frosts, more tolerant than those from frosty ones. Severe frosts, of course, kill all, but the lower tolerance of the native genotype means the reeds can die back gradually.

The first source of temperature variation is thus the climate in which the reed genotype grows. As far as seasonal behaviour goes, northern genotypes could grow in the south, but not the reverse. Southern ones are killed by the north's winter cold. And with a longer available growing season, the southern types avail themselves of it, filling up the seasons, so growing slower for the same final height.

Then there is the response to frost, which, of course, is only found in the north. Whether there are, in a given climate, genotypes adapted to different frost regimes is not known. To make studies more difficult, in the last half-century in Britain climate change has led to far fewer frosts. Earlier, killing frosts occurred most years in September and June. Now they may be few even between November and March.

Buds just above soil surface are killed by moderate frosts, growing shoots, by mild ones. The apical meristem, and perhaps 2–3 upper intercalary meristems are killed. Affected shoots die. Shoots over 0.5 m high usually survive because their apices are above the level of ground frosts. In the past, 85% of young shoots could be killed, in frost-hollows (and see Figure 4.7).

Any soil not protected by water or a thick insulating litter mat is warmer by day, cooler by night, and fluctuating temperatures greatly increase bud production. So does killing the first-emerging shoots, as they are usually the largest (Table 6.1) and are replaced by several smaller ones. Frosting and temperature fluctuations therefore make beds more uniform, denser and shorter. Harvesting reed, leaving bare soil behind, does this, so good thatching reed is encouraged (Chapter 10).

Autumn frosting, provided food has moved to the rhizome, encourages quicker final maturation, and does no harm.

Frosting is native, the plant is adapted. Burning is another matter, as Europe – and large reedswamps anywhere – have no tradition of being burnt by lightning. Burning is, though, a form of temperature fluctuation, which has the same effect on emergence as frosting, though in autumn and winter it just removes all dying and dead stems.

Burning breaks intrinsic dormancy (Chapter 3) so, except in cold-imposed dormancy, leads to a new emergence in a few weeks (within a year, growth rhythms again become adapted to the seasons, whether or not there is a cold winter). Burning produces (even more than frosting) a dense, uniform, shorter stand. Severe burning leads to maybe six times the density, though if the upper soil itself is affected, this is too much, and emergence is delayed, and poor.

Suitable burning is therefore the treatment of choice for restoring commercial reedbeds to good performance. Dense, straight, uniform reed is produced.

The growth-rate of *Phragmites* is influenced by the growing temperature (Figure 4.6). Warm days have faster growth.

## Stand edges, sparse stands

At the edges of monodominant stands, unless there is a barrier, like a wall or sudden access to grazing, there is a band, narrow or wide, of decreasing size and density of shoots. This accounts for a lot of variation!

Sparse stands abound (see above and Chapters 4 and 7). Mostly, the shoots are typical, more often restricted than depauperate in population type. The food supply to small shoots, and the habit of shaded shoots is described above.

A few variations are worthy of comment: the pink in tall sparse shoots (not on the associated short ones), is a sign of high performance in some clones, and missing in years of poor weather.

Strongly ribbed leaves are usually but not always in sparse stands, and seldom with high performance (see Chapter 7 for *Phragmites*' relations with other species). There are the sparse, but monodominant stands in nutrient-poor lake bays, which can be very sparse: but remain dominant as no other species invades. *Phragmites* varies!

## Conclusion

Only a few of the habitat features influencing *Phragmites* are presented here. (Grazing and cutting, mowing and harvesting are described more in Chapters 9, 10 and 11).

*Phragmites* varies greatly with habitat, even more than it varies with genotype! Reeds are immensely variable, and interpretation needs those who have properly studied the plant. Van der Toorn (1972) supplies a similar but complementary account. Other excellent studies include Rudescu *et al.* (1965), Björk (1967), the collected works from Třebon (Květ, Dykyjová, Prach, Ondok, etc., see bibliography list), Bittmann (1953) and Hürlimann (1951).

*Phragmites* is not stable and fixed, it has plasticity and is versatile: part of the secret of its success.

**Table 6.1** *Phragmites* performance in habitats of different nutrient status (1950s, 1960s data)

| a) Site | Number of shoots/m² | Modal height (cm) | Population type | A1 | A2 | B | C1 | C2 | D | E |
|---|---|---|---|---|---|---|---|---|---|---|
| | | | | (see below) | | | | | | |
| Eriswell | 4 | (100) | restricted (see Figure 3.1. Sparse, but reeds tall and short) | + | | | | | | |
| Eriswell | 2 | nil | restricted | + | | | | | | |
| Redgrave | 2 | (15) | restricted | + | | | | | | |
| Garboldisham | 31 | (100) | restricted | + | | | | | | |
| Inverkirkaig | 22 | nil | restricted | | + | | | | | |
| Kylesku | 5 | (150) | restricted | | + | | | | | |
| Maiden | 36 | nil | restricted | | + | | | | | |
| Boarlan | 10 | (40) | restricted–depauperate (partly grazed) | | + | | | | | |
| Glen Canisp | – | 175 | optimal (modal height high, on right of graph) | | | + | | | | |
| Eddrachillis | 122 | 40 | sub-optimal (lightly grazed, modal height central) | | | + | | | | |
| Gairloch | 74 | 125 | optimal | | | + | | | | |
| Assynt | 7 | 20 | depauperate (short, no shading canopy) | | | | + | | | |
| Clashnessie | 2.5 | 40 | depauperate | | | | + | | | |
| Inverkirkaig | 3 | 50 | depauperate | | | | + | | | |
| Cranesmoor | 25 | 10 | depauperate | | | | + | | | |
| Matley | 4 | 15 | depauperate | | | | + | | | |
| Burley | 16 | 30 | depauperate–sub-optimal | | | | + | | | |
| Lochluichart | 53 | 70 | sub-optimal | | | | | + | | |
| Burley | 68 | 50 | sub-optimal | | | | | + | | |
| Cranesmoor | 55 | 125 | sub-optimal | | | | | + | | |
| Eriswell | 108 | 120 | nearly optimal | | | | | | + | |
| Foulden | 72 | 135 | optimal | | | | | | + | |
| Garboldisham | 200 | 170 | optimal | | | | | | + | |
| Icklingham | 115 | 155 | optimal | | | | | | | + |
| Santon Downham | 74 | 160 | optimal | | | | | | | + |
| Ranworth | 196 | 145 | optimal | | | | | | | + |
| Cley | 71 | 210 | optimal | | | | | | | + |
| Rosyth | 30 | 250 | optimal (different clones in 'uniform' undisturbed marsh) | | | | | | | + |
| Rosyth | 144 | 160 | optimal (different clones in 'uniform' undisturbed marsh) | | | | | | | + |
| Rosyth | *c.* 100 | 100 | optimal (different clones in 'uniform' undisturbed marsh) | | | | | | | + |

**b) The habitat types. Communities, nutrient status and *Phragmites* performance in the above list**
There are of course many more communities than those shown in this table (see Chapter 12).

**A** *Phragmites* stands restricted by nutrient deficiency.
- **1** Nutrient-poor fens, with *Schoenus nigricans* and one or more of: *Betula* spp., *Drosera* spp., *Pinguicula vulgaris*, and *Sphagnum* spp.
- **2** Oligotrophic lochs.

**B** Ungrazed or hardly grazed bog stands, with more than one of: *Drosera* spp., *Erica* spp., *Eriophorum* spp., *Menyanthes trifoliata*, *Molinia caerulea*, *Pinguicula vulgaris*, *Sphagnum* spp.

**C** Lightly-grazed habitats often with light grazing making population type poor, or towards this.
- **1** With several of the species in B above, boggy.
- **2** Flushed bogs or poor fens, with better *Phragmites* performance (*Drosera* etc. absent, and *Molinia* etc. more abundant).

**D** Nutrient-medium fen stands, with one or several of the following common in the habitat, though not necessarily so in the monodominant *Phragmites* stand: *Calamagrostis canescens*, *Cladium mariscus*, *Cirsium palustre*, *Galium uliginosum*, *Juncus articulatus*, *J. subnodulosus*, *Lotus uliginosus*, *Molinia caerulea*.

**E** Nutrient-rich fen and marsh stands with one or several of the following common in the habitat, though not necessarily so in the monodominant *Phragmites* stand: *Agrostis stolonifera*, *Carex acutiformis*, *C. riparia*, *Cirsium arvense*, *Epilobium hirsutum*, *Galium aparine*, *Glyceria maxima*, *Lythrum salicaria*, *Salix* spp., *Urtica dioica*.

**Table 6.2** The effect of added nutrients (N, K and P) on *Phragmites* (Haslam, 1995)

**a)** Effect of fertiliser on a calcium-rich, East Anglian fen, deficient in phosphorus and other available nutrients. Records after fertilisers were added every three months for eighteen months.

| | *Deficient (control)* | *Fertilised plot* |
|---|---|---|
| Number of shoots | 31 | 43 |
| Number over 1 m high | 6 | 25 |
| Number of inflorescences | 4 | 15 |
| Number very yellow (P-deficient) | 30 | 4 |

**In calcium-dominant fens, *Phragmites* is severely restricted by lack of available nutrients.**

**b)** Effect of fertiliser on replacement crops growing after summer cutting in East Anglian nutrient rich fens (one treatment).

| *Month of cut* | *Fen* | *Year* | *Number of original shoots* | | *Number of replacement shoots* | |
|---|---|---|---|---|---|---|
| | | | *Fertilised plot* | *Control plot* | *Fertilised plot* | *Control plot* |
| June | Woodwalton | 1965 | 33 | 35 | 63 | 48 |
| July | Woodwalton | 1966 | 63 | 60 | 9 | 2 |
| June | Woodwalton | 1967 | 24 | 24 | 108 | 52 |
| July | Woodwalton | 1967 | 47 | 83 | 8 | 4 |
| June | Icklingham | 1967 | 42 | 65 | 74 | 66 |
| July | Icklingham | 1967 | 78 | 54 | 28 | 15 |

July cuts are too late for a full replacement crop; density, however, is higher in the fertilised plots.

**Table 6.3** Nutrient content of aerial shoots in June

Results as per cent dry weight. Analyses kindly done by the Ministry of Agriculture, Fisheries and Food, Cambridge, 1960s.

| Site | *Phragmites* type | Habitat | P | K | N | Ca | Na | Cl | Total CaNPK |
|---|---|---|---|---|---|---|---|---|---|
| *a) Breck fens* | | | | | | | | | |
| Icklingham | dominant | nutrient-rich wet fen | 0.06 | 1.86 | 1.91 | 0.08 | 0.18 | 0.82 | 3.91 |
| Icklingham Mill | restricted (competition) | nutrient-rich dry fen | 0.02 | 0.86 | 2.17 | 0.19 | 0.05 | 0.74 | 3.24 |
| Cavenham | sub-optimal (competition) | nutrient-rich dry fen | 0.05 | 1.54 | 2.11 | 0.07 | 0.17 | 0.78 | 3.77 |
| Foulden | dominant | edge of nutrient-poor area | 0.04 | 1.00 | 1.39 | 0.02 | 0.04 | 0.60 | 2.45 |
| Redgrave | sub-optimal | ditch bank, nutrient status fair | 0.006 | 0.81 | 1.80 | 0.12 | 0.04 | 0.38 | 2.74 |
| Redgrave | restricted | nutrient-poor (*Cladium* etc.) | 0.03 | 1.70 | 1.45 | 0.11 | 0.17 | 0.54 | 3.29 |
| Redgrave | restricted | nutrient-deficient (*Schoenus* etc.) | 0.02 | 1.37 | 1.02 | 0.12 | 0.04 | 0.45 | 2.53 |
| *b) North-west Scotland* | | | | | | | | | |
| Gairloch | dominant | nutrient-rich swamp | 0.09 | 1.88 | 2.13 | 0.08 | 0.04 | 1.02 | 4.18 |
| Clachtoll | dominant | edge of nutrient-rich loch | 0.21 | 2.51 | 2.03 | 0.10 | 0.04 | 1.40 | 4.85 |
| Eddrachillis | sub-optimal, grazed | flushed wet valley bog | 0.11 | 1.88 | 3.32 | 0.06 | 0.19 | 1.29 | 5.37 |
| Applecross | sub-optimal | nutrient-rich marsh | 0.05 | 1.96 | 2.28 | 0.09 | 0.08 | 1.23 | 4.38 |
| Oldany | sub-optimal | flushed valley bog | 0.11 | 1.67 | 1.36 | 0.04 | 0.03 | 0.60 | 3.18 |
| Oldany | sub-optimal (short) | grassy slope | 0.20 | 1.96 | 2.01 | 0.08 | 0.04 | 1.02 | 4.25 |
| Oldany | depauperate, dwarf | saltmarsh | 0.09 | 1.96 | 1.80 | 0.07 | 0.20 | 1.18 | 3.92 |
| Assynt | depauperate, grazed | bog | 0.05 | 1.96 | 2.28 | 0.09 | 0.79 | 1.23 | 4.38 |
| Clashnessie | depauperate, grazed | bog | 0.04 | 0.91 | 1.77 | 0.09 | 0.11 | 0.76 | 2.81 |
| Inverkirkaig | depauperate, grazed | bog | 0.05 | 0.96 | 2.47 | 0.08 | 0.13 | 0.89 | 3.56 |
| Inverkirkaig | depauperate, grazed | nutrient-poor loch | 0.08 | 1.44 | 2.33 | 0.07 | 0.09 | 0.89 | 3.92 |
| Lochluichart | large spring shoots | semi-flushed bog | 0.03 | 1.78 | 1.80 | 0.04 | 0.001 | 0.69 | 3.65 |
| Lochluichart | small spring shoots | semi-flushed bog | 0.07 | 1.86 | 1.87 | 0.05 | 0.06 | 0.82 | 3.85 |
| Lochluichart | large summer shoots | semi-flushed bog | 0.05 | 2.66 | 1.93 | 0.00 | 0.06 | 0.72 | 4.64 |

June is typically when new shoots no longer need food from the rhizomes. Calcium is higher in lime-rich sites, as expected. High nitrogen values are (except Icklingham Mill) associated with livestock. K and P show little correlation. The plant can in some part 'choose' solutes to take up, it is not just a passive receptor of the habitat. All these values come from rapidly-growing shoots, so are sufficient for this.

**Table 6.4** Nutrient content of rhizomes in January

Analyses as Table 6.1. (Data comparable with other 1960s analyses, not necessarily with those of the 1980s, Haslam, 1995).

| Site | Rhizome type | N | P | K | Ca |
|---|---|---|---|---|---|
| *Nutrient-rich fen* | | | | | |
| Icklingham | bud | 0.93 | 0.11 | 0.78 | 0.08 |
| | vertical | 0.81 | 0.08 | 0.72 | 0.20 |
| | horizontal | 0.53 | 0.08 | 0.39 | 0.12 |
| Woodwalton | bud | 2.04 | 0.21 | 2.56 | 0.07 |
| | vertical | 1.17 | 0.09 | 1.69 | 0.05 |
| | horizontal | 1.02 | 0.12 | 0.43 | 0.19 |
| Icklingham Mill | bud | 1.76 | 0.19 | 1.75 | 0.06 |
| | vertical | 0.91 | 0.11 | 0.36 | 0.08 |
| | horizontal | 1.28 | 0.13 | 0.46 | 0.10 |
| *Nutrient-poor fen* | | | | | |
| Redgrave | bud | 1.63 | 0.09 | 1.26 | 0.10 |
| | vertical | 1.60 | 0.05 | 0.75 | 0.04 |
| | horizontal | 1.00 | 0.11 | 0.98 | 0.95 |
| Redgrave | vertical | 1.02 | 0.02 | 1.57 | 0.08 |
| | horizontal | 1.21 | 0.04 | 1.69 | 0.09 |
| *Valley bog* | | | | | |
| Cranesmoor | bud | 0.73 | 0.08 | 1.81 | 0.05 |
| | vertical | 0.62 | 0.05 | 1.88 | 0.03 |
| | horizontal | 0.53 | 0.05 | 1.45 | 0.05 |
| Cranesmoor | bud | 1.28 | 0.10 | 2.36 | 0.14 |
| | vertical | 1.21 | 0.03 | 1.31 | 0.04 |
| | horizontal | 0.65 | 0.12 | 0.31 | – |

Reasonably consistently, the highest nutrient levels are in the buds, as the plant is getting ready for the great flow from rhizome to aerial shoot at emergence.

**Table 6.5** Nutrient content of horizontal rhizomes of different ages (Haslam, 1995)

Norfolk samples from high-nutrient peat dykes; tank samples in mineral soil, Cambridge; Boarlan samples from nutrient-poor loch, north-west Scotland. Analyses as Table 6.1 (Haslam, 1965).

| Site | Month of sampling | Age (years) | N | P | K | Ca | Mg | Na | Mn |
|---|---|---|---|---|---|---|---|---|---|
| Norfolk (1) | April | 1 | 2.03 | 0.28 | 2.50 | 0.03 | 0.049 | 0.41 | 0.002 |
| | | 2 | 1.66 | 0.24 | 1.88 | 0.08 | 0.078 | 1.16 | 0.002 |
| | | 3 | 1.66 | 0.32 | 1.88 | 0.07 | 0.077 | 1.14 | 0.006 |
| | | 3 or 4 | 1.42 | 0.37 | 2.19 | 0.10 | 0.099 | 0.76 | 0.004 |
| | | 4 or 5 | 1.09 | 0.22 | 2.13 | 0.07 | 0.093 | 1.18 | 0.004 |
| | | 5 or 6 | 1.16 | 0.19 | 3.44 | 0.07 | 0.089 | 0.91 | 0.002 |
| | July | 1 | 0.76 | 0.14 | 2.06 | 0.09 | 0.055 | 0.41 | 0.004 |
| | | 2 | 1.04 | 0.22 | 1.75 | 0.10 | 0.058 | 0.67 | 0.002 |
| | | 3 | 1.03 | 0.26 | 2.56 | 0.10 | 0.054 | 0.91 | 0.002 |
| Norfolk (2) | April | 1 | 1.44 | 0.28 | 2.25 | 0.03 | 0.056 | 0.41 | 0.006 |
| | | 2 | 1.75 | 0.27 | 3.25 | 0.03 | 0.056 | 0.24 | 0.004 |
| | | 3 | 1.20 | 0.22 | 1.69 | 0.09 | 0.008 | 0.66 | 0.008 |
| Tank, wet | March | 1 | 1.55 | 0.24 | 2.19 | 0.18 | 0.038 | 0.18 | 0.004 |
| | | 2 | 1.05 | 0.18 | 1.63 | 0.47 | 0.046 | 0.18 | 0.009 |
| Tank, dry | March | 1 | 1.46 | 0.26 | 1.69 | 0.16 | 0.063 | 0.16 | 0.002 |
| | | 2 | 1.37 | 0.21 | 1.19 | 0.23 | 0.063 | 0.68 | 0.004 |
| Boarlan | August | new | 0.84 | 0.15 | 3.13 | 0.08 | 0.071 | 0.15 | 0.006 |
| | | 1 | 0.34 | 0.081 | 2.50 | 0.04 | 0.053 | 0.14 | 0.009 |
| | | 2 | 0.30 | 0.027 | 1.94 | 0.07 | 0.063 | 0.28 | 0.008 |
| | | 3 or 4 | 0.35 | 0.040 | 2.38 | 0.06 | 0.081 | 0.37 | 0.011 |
| | | 4 or 5 | 0.29 | 0.055 | 2,75 | 0.06 | 0.080 | 0.35 | 0.006 |
| | | 5 or 6 | 0.31 | 0.048 | 2.19 | 0.08 | 0.075 | 0.30 | 0.004 |

The pattern is inconsistent. Higher levels could be expected in newer rhizomes, which only sometimes occur.

**Table 6.6** Variation in nutrient levels and reed strength (Haslam, 1995)

From marshes of apparently uniform habitat with conclusive evidence of clonal variation. Nutrient results expressed as per cent (oven) dry weight in internodes. Silica skeleton remaining after ashing: varies from powder to patterned tube recorded as 0–15 (high). Strength measured as fibrousness, lack of brittleness and hardness, from Poor to Good.

| Reed strength | N | P | K | Ca | Mg | C | Ash | Si skeleton |
|---|---|---|---|---|---|---|---|---|
| *Cley Mill, Norfolk* | | | | | | | | |
| Poor–Fair | 0.212 | 0.115 | 0.114 | 0.042 | 0.028 | 53.9 | 2.91 | 13.0 |
| Fair | 0.251 | 0.046 | 0.156 | 0.027 | 0.022 | 53.0 | 3.84 | 15.0 |
| Fair | 0,218 | 0.062 | 0.251 | 0.026 | 0.025 | 53.8 | 3.10 | 14.0 |
| Fair | 0.222 | 0.012 | 0.070 | 0.025 | 0.020 | 53.7 | 3.40 | 13.0 |
| Fair–Good | 0.227 | 0.040 | 0.081 | 0.020 | 0.014 | 53.9 | 2.95 | 14.0 |
| Good | 0.258 | 0.055 | 0.180 | 0.020 | 0.018 | 53.8 | 3.14 | 14.0 |
| *Rosyth, Scotland* | | | | | | | | |
| Poor | 0.379 | 0.078 | 0.096 | 0.047 | 0.041 | 53.3 | 4.08 | 1.0 |
| Fair–Good | 0.305 | 0.096 | 0.151 | 0.040 | 0.022 | 53.4 | 3.86 | 8.0 |

Strength is not correlated with nutrient content.

**Table 6.7** Reed strength and nutrient content (of internodes at the butt, the part most exposed to weathering on the roof) (Haslam, 1995)

| Strength | Number of samples | N | P | K | Ca | Mg | C | Ash | Si skeleton |
|---|---|---|---|---|---|---|---|---|---|
| Good | 10 | 0.239 | 0.059 | 0.101 | 0.054 | 0.023 | 53.98 | 2.83 | 9.4 |
| | | (0.091) | (0.021) | (0.067) | (0.038) | (0.009) | (0.61) | (1.10) | (3.9) |
| Fair–Good | 21 | 0.238 | 0.063 | 0.095 | 0.045 | 0.022 | 53.99 | 2.69 | 8.6 |
| | | (0.111) | (0.039) | (0.058) | (0.045) | (0.011) | (0.57) | (1.06) | (3.5) |
| Fair | 23 | 0.223 | 0.055 | 0.089 | 0.030 | 0.018 | 54.01 | 2.52 | 7.1 |
| | | (0.069) | (0.048) | (0.053) | (0.012) | (0.009) | (0.64) | (1.06) | (4.2) |
| Poor–Fair | 23 | 0.191 | 0.044 | 0.049 | 0.042 | 0.024 | 54.14 | 2.42 | 6.7 |
| | | (0.064) | (0.025) | (0.028) | (0.029) | (0.011) | (0.53) | (0.98) | (4.0) |
| Poor | 52 | 0.268 | 0.069 | 0.076 | 0.041 | 0.023 | 53.97 | 3.17 | 5.7 |
| | | (0.113) | (0.038) | (0.102) | (0.026) | (0.011) | (0.76) | (1.44) | (3.1) |

The correlation coefficients between strength and N, P, K, Ca, Mg, C and Ash are well below any relevant level of significance. Strength and nutrient content are not correlated. That between strength and Silica skeleton is much better than 0.001. This strong correlation, however, merely means the Silica skeleton is one factor determining strength.

**Table 6.8** Comparison of nutrient contents of internodes, nodes and sheaths of dead reeds (Haslam, 1995)

Nutrient results expressed as per cent (oven) dry weight, standard deviations in brackets. Internode and node levels are all correlated to a significance better than 0.001. Internode and sheath levels of K, Ca, Mg and Ash are correlated to a significance better than 0.001; those of N and P to 0.05.

*a) Combined samples (33)*

| | N | P | K | Ca | Mg | Ash |
|---|---|---|---|---|---|---|
| Internode | 0.243 (0.083) | 0.087 (0.096) | 0.130 (0.186) | 0.060 (0.042) | 0.031 (0.027) | 5.00 (2.48) |
| Node | 0.476 (0.107) | 0.135 (0.093) | 0.148 (0.151) | 0.109 (0.170) | 0.041 (0.022) | 6.09 (2.80) |
| Sheath | 0.792 (0.268) | 0.215 (0.084) | 0.094 (0.054) | 0.255 (0.389) | 0.044 (0.035) | 6.95 (2.11) |

**Table 6.8** Continued

*b) Individual samples*

| Strength | | N | P | K | Ca | Mg |
|---|---|---|---|---|---|---|
| **100-year thatch, Norfolk** | | | | | | |
| Fair | Internode | 0.460 | 0.088 | 0.028 | 0.112 | 0.018 |
| | Node | 0.596 | 0.102 | 0.030 | 0.125 | 0.018 |
| | Sheath | 1.300 | 0.368 | 0.090 | 0.224 | 0.026 |
| **Hungary** | | | | | | |
| Good | Internode | 0.231 | 0.083 | 0.245 | 0.045 | 0.033 |
| | Node | 0.425 | 0.141 | 0.261 | 0.081 | 0.053 |
| | Sheath | 0.712 | 0.227 | 0.117 | 0.151 | 0.023 |
| **Norfolk** | | | | | | |
| Poor | Internode | 0.178 | 0.047 | 0.042 | 0.040 | 0.012 |
| | Node | 0.389 | 0.041 | 0.051 | 0.045 | 0.019 |
| | Sheath | 0.670 | 0.181 | 0.096 | 0.159 | 0.021 |
| **Malta, larger** | | | | | | |
| Good | Internode | 0.147 | 0.037 | 0.063 | 0.144 | 0.044 |
| | Node | 0.372 | 0.190 | 0.102 | 1.018 | 0.092 |
| | Sheath | 0.708 | 0.212 | 0.066 | 2.318 | 0.146 |
| **Czechoslovakia** | | | | | | |
| Poor | Internode | 0.408 | 0.043 | 0.054 | 0.041 | 0.032 |
| | Node | 0.639 | 0.116 | 0.064 | 0.058 | 0.043 |
| | Sheath | 1.178 | 0.268 | 0.143 | 0.123 | 0.031 |
| **Ohio, USA** | | | | | | |
| Good | Internode | 0.138 | 0.026 | 0.015 | 0.046 | 0.011 |
| | Node | 0.332 | 0.063 | 0.025 | 0.061 | 0.017 |

Leaf sheaths have the highest nutrient content. Nodes, where the leaves arise, where there are buds, and where there is more meristem tissue, generally have more nutrients than the internodes.

**Table 6.9** Nutrient content and strength of older reed (internodes) (Haslam, 1995)

Nutrient results expressed as per cent (oven) dry weight, standard deviation in brackets. Silica skeleton as number on scale (15 is high).

| Site | Number of samples | N | P | K | Ca | Mg | C | Ash | Si skeleton |
|---|---|---|---|---|---|---|---|---|---|
| *a) Harvested and analysed 1960s. Internode plus node. No data on reed strength.* | | | | | | | | | |
| Commercial beds | 9 | 0.300<br>(0.057) | 0.016<br>(0.003) | 0.117<br>(0.057) | 0.056<br>(0.027) | 0.040<br>(0.009) | – | – | – |
| All samples | 16 | 0.388<br>(0.183) | 0.019<br>(0.013) | 0.110<br>(0.079) | 0.075<br>(0.084) | 0.069<br>(0.101) | – | – | – |
| Sample A26 | 14 | 0.380 | 0.020 | 0.070 | 0.090 | 0.047 | – | – | – |
| *b) Harvested 1960s and stored. Analysed 1987/8. Internodes only. Mean reed strength: Fair.* | | | | | | | | | |
| Reedbeds | 7 | 0.309<br>(0.057) | 0.668<br>(0.018) | 0.087<br>(0.034) | 0.040<br>(0.014) | 0.027<br>(0.009) | 53.46<br>(0.46) | 3.76<br>(0.85) | 7.6<br>(4.8) |
| Sample A26 | 1 | 0.350 | 0.057 | 0.072 | 0.043 | 0.023 | 52.73 | 5.10 | 15.0 |
| *c) From thatched roofs, aged 5 to 100 years, analysed 1987/8. Internodes only.* | | | | | | | | | |
| Combined | 6 | 0.321<br>(0.086) | 0.086<br>(0.022) | 0.034<br>(0.010) | 0.056<br>(0.028) | 0.021<br>(0.007) | 53.81<br>(1.11) | 3.16<br>(1.96) | 4.5<br>(1.26) |
| Breaking down after 5 years | | | | | | | | | |
| Poor | 2 | 0.302 | 0.085 | 0.042 | 0.039 | 0.031 | 52.67 | 2.66 | 5.5 |
| Good quality and condition after *c.* 100 years | | | | | | | | | |
| Strong | 1 | 0.466 | 0.088 | 0.028 | 0.112 | 0.018 | 54.99 | 2.80 | 4.0 |

Sample A26 was the one available for duplicate analyses in (a) and (b), and the values of the two are remarkably close. About a dozen more old thatch samples were analysed, in a different laboratory. As in these samples, old reeds contain more nitrogen and phosphorus.

1960s reeds contain more nutrients than 1980s, in general. There is no evidence that recent reed breaks down sooner because of 'all those nasty nitrates'.

# Chapter 7
# Mixed stands: competition, invasion, succession and decline

## INTRODUCTION

Reedstands are obvious. Looking out over the wide lake or river, there is the wide reedswamp, too (see Frontispiece 'The Changing Seasons'). How does it maintain itself? In lakes there are few potential competitors. Submerged or floating plants do not interfere, and are reduced in the shade of *Phragmites*. Reed size, together with the space taken up, makes invasion unusual.

With intermittent drying, reeds are dense enough to cast heavy shade. Becoming a little drier again, there is probably a thick litter mat, once more hindering invasion. Only gaps in the reedbed can be colonised. Over time, this is how carr invades and succeeds. Or, alternatively, and without the litter mat, bog. In East Anglia, the primary English wetland area, harvesting reed includes removing young woody plants or other competitors, so keeping the reedbeds in being where, without intervention, they would not be.

*Phragmites*, as known from the peat record (Chapter 1), is the primary European reedswamp peat-former. Therefore monodominant *Phragmites* is, historically, the primary type, which can exist for at least 8,000 years. That puts into perspective the, indeed fascinating, interactions in transient populations, often caused by human impact!

## WEAPONS OF COMPETITION

Competition between species is silent, but is as strong and ruthless as any war: the silent battlefield. There are, of course, very many ways in which one species can dominate another, but the principal ones here are:

### *A PLANT FACTORS*

1. **Shading from above**. Conversely, preventing a potentially shading plant from growing tall enough.
2. **Litter mat preventing new growth**. Conversely, ability to withstand the litter mat of other species.

3. **Production of toxins** (from the roots) which will hinder good growth of other species. Conversely, tolerance to such toxins from other species. (*Phragmites*' tolerance is low, and its production is probably low, see text, but also see Cížková *et al*., 1999.)
4. Development of **advancing margins** which pass through other vegetation like a tank. Conversely, ability to re-grow and successfully compete in the less vigorous growth behind the advancing margin.
5. **Occupy the space** completely, and do so early in the year before other species can.
6. **Pull down other species** (lodging). Also have the ability to stand upright, not be lodged.
7. **All factors hindering bud development**. To grow well, *Phragmites* must grow many buds from the rhizomes, and all factors, plant or otherwise, which restrict this development, also favour the competition.
8. Have **plant vigour**. Also have some other competitive factor (shade, toxins, quick growth, etc.).

### B External factors which favour one or other species in a mixed stand

1. **Water level**. Raising water level removes aquatic and many wetland species not adapted to (as relevant) deep-flooded, shallow-flooded or damp places. It may also alter nutrient status. Lowering water level acts in reverse. Water regime is the primary factor.
2. **Nutrients**. Nutrient status determines which species can, potentially, flourish, grow adequately, or pine and die. In general, species of higher nutrient status are more able to shade and suppress than those of lower nutrient status, when habitat allows them entry.
3. **Impact**. This has many effects, primarily to destroy reedbeds by drying and disturbance. However, winter harvesting favours reeds, heavy grazing removes it, etc..
4. **Disease** affecting one species but not another. (A leaf virus killed *Epilobium hirsutum* in a Tall Herb community. It was still absent after four years. *Glyceria maxima* died of rhizome rot in a mixed stand with *Phragmites*, which became mono-dominant.)
5. **Aeration**. Some species, e.g., alder, *Alnus glutinosa*, invades fens better where aeration (dryness, moving spring water, etc.) is good. Others, e.g., *Salix cinerea*, have a wider range. It is considered, without flow evidence, that in commercial beds *Cladium mariscus* is more favoured by through-flow than is *Phragmites*; and with evidence, *Glyceria maxima* is less so (Buttery, 1959; Buttery & Lambert, 1965; Buttery *et al*., 1965. Also see Kiendl, 1953).

6. **Weather**. Lack of sunshine and unexpected lack of water in the growing season lead to shorter shoots, perhaps by 0.5 m, perhaps even more. Competitors only a little shorter than *Phragmites* then have an advantage. Date of emergence/early growth/germination of the species concerned can be crucial in determining which has a shading canopy, or at least shading capacity, first.
7. **All factors hindering bud development**. To grow well, *Phragmites* must grow many buds from the rhizomes, and all factors, plant or otherwise, which restrict this development, also favour the competition.

All these, and more, interact and control the outcome. Competition is highly complex: or very simple. Some examples are described below. *Phragmites* of course grows as an associated species in a remarkably wide range of communities, which means there are a remarkably wide range of competitive relations! In each community, each plant and each species 'tries' to increase its control over the community. The fact that, in stable conditions, most vegetation remains stable shows not that all is quiet and calm, but that the state of tension reached by each in the combination is a sustainable balance.

The examples start with the most historically common patterns.

## Invasion of open water by reedswamp

The depth in which *Phragmites* can grow varies with the nutrient status, the substrate texture, the (lack of) scour, and, quite likely the genotype: *c.* 75 cm to well over 4 m are recorded. *Phragmites* cannot establish by seed under water (Chapter 2), so must grow out from the shore, or grow from rhizome, etc., pieces, also probably on the shore (Figure 3.4 j and k).

*Phragmites* will not invade far into substrate where it is difficult for its rhizomes to penetrate (rock, boulders, stony ground without much silt, etc.); or where wave or river scour will disturb or remove them. It will invade deeper if nutrients are in good supply, which may be in all, part or none. A reedswamp, reed patches, or reed absence may result.

There may be other species present. In good conditions a substantial flora of bottom-living (well-submerged mostly rooted), water-filling or floating plants is likely, such as water lilies, strap-shaped leaved, finely-divided leaved, and rosette species. *Phragmites*, however, is not hindered by these. If reed can grow there in the physico-chemical conditions, it will. It is above water level, so if dense enough, it shades all below. The aquatic plants do not hinder reed by toxins, or in any other way. To say 'competition' is a misnomer. *Phragmites* ignores them as much as it does the grains of sand.

Seldom, there are other species in the deeper reedswamp. With the shoots often very sparse, patches may develop which show little competition, as different layers are used (Figure 3.4i). Landwards, competition for rhizomes in denser swards, powered by shade and substrate type, may occur.

## Bog invasion

Bog is one of the two natural and traditional successors of *Phragmites* (Chapter 1). Reedbeds form peat, so bringing soil surface up to ground level (where further fen peat cannot form as it oxidises in dry conditions). Alternatively, wider changes in, e.g., sea level, may bring substrate to the surface. *Sphagnum* can invade:

- when its propagules can reach the site;
- when rainwater can accumulate on the ground surface;
- where nutrient-richer flooding is absent or vary rare;
- where the soil is bare beneath the reeds (no thick litter mat).

When the reedbed is flooded, regularly or constantly, the nutrient-rich water of the reedbed would both flood any *Sphagnum* and prevent its growth. *Sphagnum* invades only where there are rainwater pools. The soil is nutrient-rich, but the rainwater is not, and *Sphagnum* can grow. Without much incursion of richer water, *Sphagnum* continues to grow. Its leaves and tips absorb water, and hold it in, so water accumulates within and between the shoots (Figure 7.1), which is difficult (though of course possible) to dry out. Gradually, *Sphagnum* spreads, becomes thicker, and more bog-like species arrive to join, and supersede the fen species.

**Figure 7.1** *Sphagnum* Reedland. *Sphagnum* on fen surface (Haslam & Bone 2020).

*Sphagnum* can hold many times its weight in water. Once it is established on a surface, the water is held, just below the moss tips covering the surface. Rainwater lenses occur. Evapotranspiration is reduced because the main water is sub-surface. Downwards loss is reduced because the substrate below (peat, rock, etc.) is hardly pervious. Sideways loss is reduced because of the low slope and texture of the vegetation and peat. Where the conditions are met, the surface water is acid and nutrient deficient, and bog grows. (There are at least two *Phragmites* clones here.)

The Dutch, who have a good deal of this community, call it '*Sphagnum* reedland'. In England now, with all the drying, communities are small, though the peat record shows that, e.g., in the west Fenland, there used to be much *Sphagnum*, though colonising not *Phragmites* but other fen types.

*Sphagnum* and *Phragmites* can co-exist for a relatively long period. The *Phragmites* rhizomes are well down in the mineral-rich peat. Even the vertical rhizomes, at this *Sphagnum*-reedland stage, are almost entirely in that rich peat too. The *Sphagnum* has its nutrient-deficient rainwater. Both grow. The *Phragmites* is no longer forming peat but, if conditions stay stable, *Sphagnum* does so. The two form peat quite differently. *Phragmites* roots, leaves and stems accumulate at the bottom of the water and become humified, accumulating ever higher in the water. *Sphagnum* builds peat with its tips constantly growing upwards and spreading, and its old stems and leaves turning to peat. It stays wet because the top absorbs rainwater, and the rest keeps it. *Sphagnum* peat therefore grows upwards and, to keep pace, the *Phragmites* rhizomes and bud position grow upwards too, basing growth, as usual, on the ground surface, which is here the top of the *Sphagnum*. Therefore gradually first the upper part, then all of the *Phragmites* plant comes to be within *Sphagnum* peat, and the plant dies from nutrient deficiency. Nutrient-rich water may come in but in terms of millennia, a little nutrient is irrelevant since the bogs can grow. Unless there is human impact, the bogs will grow.

Competition is indirect. *Phragmites* could perfectly well shade out *Sphagnum*, especially if aided by a thick litter mat. If it does so, there is no *Sphagnum* mat. Similarly, if there is drying (as too often in England), there are neither rain-water pools for bog peat, nor river flooding for fen. Once the *Sphagnum* layer is there, it seems likely it can decrease *Phragmites* bud formation within it (i.e., the development of buds is hindered first at the top, then, as bog peat grows, also further down: rhizomes being now in bog peat).

## Sallow (*Salix cinerea*) carr invasion

Tree invaders of *Phragmites* beds are usually *Salix* spp., sallows and willows, or alder, *Alnus glutinosa*, though there are others.

Sallow carr is very common. It tolerates stagnant habitats, but not calcium-dominant fens (where it is replaced by birch, *Betula* spp.). Sallow only invades when the bed is unflooded for part of the year for a few years while the tree seeds first germinate, then start to grow and become established. (Figure 7.2 shows a similar pattern with tussock sedge.) To develop from mono-dominant reed to sallow carr with sparse *Phragmites* in the undergrowth, can take between a few decades to a century or more. Sallow seed production is profuse in spring, as is its germination in early summer. Young plants abound wherever they can grow and can flourish given good light, damp but not flooded land: preferably without the numerous invertebrates of Tall Herb. Good light can come from:

**Figure 7.2** *Carex paniculata* succeeding *Phragmites* (Y. Bower in Haslam, 2003)

*C. paniculata* grows in *Phragmites* shade and when large enough, pushes *Phragmites* away. By the time this tussock dies, many other tussocks are present.

- **animal damage**, e.g., grazing, walking, pheasants, pests;
- **severe late frosts** giving open patches in June;
- **human impact**.

Damp but unflooded habitats can come from:

- **unusual drought** in early summer;
- **raised parts**, e.g., new clods of soil, past peat cutting.

Invasion therefore depends on accident; on the 'window of opportunity' that allows sallow seeds to arrive where they can germinate, become established, and not be drowned or eaten until the saplings are large enough to withstand flooding and be only partly eaten, etc.. Sallow saplings may start in the open, but frequently the reedbed soon closes round them, and they are shaded. Sallow then grows very slowly, but does find enough light filtering through the canopy to stay alive. As the years pass, and the saplings look constantly unhealthy, even on the point of death, they grow taller. Finally, sallow being taller than reed, the bushes emerge above the canopy. Now they can, suddenly, use the light and grow and spread fast. Rapidly a reedbed hiding small sparse saplings becomes carr, the shaded reed declining into short, sparse, frail shoots (Chapter 4). Even if alder invades the sallow as it dries further, the reed will remain much the same: unless and until wet woodland becomes dry woodland.

Windows of opportunity may happen this year, or not for twenty-five years and then only for a few sallows. Predictions are impossible: except that if the fen is dried and unflooded and Tall Herb takes over, invasion seems much slower (seeds eaten by invertebrates? Too dry?) and if it is then turned to grass or arable, it will not happen.

## Competition with *Carex paniculata*

*Carex paniculata* is the great tussock sedge, with tussocks reaching over 1 m high. As a dominant plant it grows in nutrient-medium, usually calcium-rich places, which are shallow-flooded part or all of the year. This means it is a habitat in which *Phragmites* also can dominate, and *C. paniculata* is probably dominant because of past history, or because of the lower nutrients and (incoming or through) calcium influence. In the investigated fen, there was peat cutting and management of tussocks for domestic use, as well as incoming calcium-sandy spring water. Water level varied from *c.* 20 cm above to *c.* 10 cm below ground level.

A tussock species, *C. paniculata* relies on seeds for propagation, and without constant young tussocks the community is not sustainable. The leaves cast heavy shade beside the tussock, but as the tussocks are separate, there is light shade, or even open glades between them. *Phragmites*, of course, being rhizomatous, and when able to, forms a complete shading canopy. It is taller, and has a (thin) litter mat, which means invasion by seed is restricted to occasional breaks in the mat.

Logic, therefore, says that without human impact, *Phragmites* will, given decades, fill up the gaps between the tussocks, prevent *C. paniculata* establishment, and shade it out.

In fact, the reverse may occur (Figure 7.2).

*Carex paniculata* mostly germinates in late spring–early summer, in similar conditions to sallow (above), except the young plants can survive in less light. A well-grown, five-month old plant which has 3–5 shoots, 15–30 cm high, is able, therefore, to stand winter flooding 10–20 cm deep. Within dominant *Phragmites*, *C. paniculata*, once established in a window of opportunity, continues as spindly, weak young tussocks, looking as though shade will kill them by next year. In fact, after 15 years the tussocks had widened and grown up, the tough leaves were thrusting sideways and pushing the reeds away. *C. paniculata* was preparing to dominate over *Phragmites*.

Where *C. paniculata* can become established in quantity, it, therefore, can dominate over *Phragmites*. Where it cannot, the reedbed remains. As well as breaks in the litter mat and canopy, there are also the chemical and water factors. Dominant *C. paniculata* has a restricted nutrient range, indeed within the much greater range of *Phragmites*, but only occupying part of it. It is also not found in deep water or (sustainably) in unflooded areas. So, though locally *C. paniculata* can do better, in its whole range *Phragmites* dominates.

## Competition with *Schoenus nigricans*

*Schoenus nigricans* is also a tussock species, also growing in fens. Its smaller size, and ability to grow from its base, so spread after damage, do not alter the competitive relations. *Phragmites* towers over *S. nigricans* even more than over *C. paniculata*, so if dense and tall, completely shades it. *S. nigricans* is not, unlike *C. paniculata*, shade-tolerant. The outcome, this time, is surely certain? Well, it is, if *Phragmites* is dense and tall, which is only in a disturbed and dried *S. nigricans* community. For in fens, *S. nigricans* is the calcium-dominated community *par excellence*, the one where even nutrient-medium species cannot do well, or grow at all. *Phragmites* (Table 3.8) is made sparse and short, unable to dominate or shade anything (quite apart from its early-yellowing thick leaves, shorter growing season, etc.).

Tussocks, in due course, must die, and if not replaced, their community dies. (Unlike *C. paniculata*, where all shoot tips are on top, *S. nigricans* has them all over the tussocks. If the tussock is badly grazed or burnt, etc., so the tussock itself is gone, shoots from the base can form new tussocks. If, though, tussocks die in the ordinary way, they do not re-grow.) Rhizomes individually die at the back, but each can continue, and grow into the same gap or area time and again, which on the face of it, gives the rhizomatous species an advantage. A shading canopy, if nutrients permit, can be built up slowly.

Unlike the pattern with *C. paniculata*, that with *S. nigricans* is not plant-powered competition, it is the removing of *Phragmites* from any competition by the chemical nature of the substrate.

## Seasonal patterns in competition: Malta

In order to compete, both parties to the competition must be present. Obvious, but a not always remembered fact. In the Mediterranean climate of Malta, tall monocotyledons and trees are summer-green, in the same way as they are further north. However, most other plants are winter-green and die back, or if annuals, die, in summer.

It is not surprising that *Phragmites* can grow into drier habitats than in, say, England. Tall Herbs and similar competitors are just not there, in its growing season.

## Competition with *Arundo donax* (a southern species)

*Arundo donax*, the great reed, is larger than *Phragmites*, being, e.g., 3–4 m high and very thick, where adjoining *Phragmites* is, e.g., 1.5 m, and thinner in both reed and vegetation. *A. donax* is, basically, a damp species: *Phragmites*, a wet one. In the Camargue, for instance, *A. donax* commonly dominates on dyke banks, above *Phragmites*, which is dominating around water level. *A. donax* has higher-level and shorter rhizomes. In disturbed and rapidly drying Malta, the pattern has been more disrupted but is basically similar. The two can overlap (Chapter 8), but *A. donax* grows up the bank and up the hill (seepage areas), while *Phragmites* is down in the wet, or once-wet areas. It is obvious *A. donax* forms stands in which *Phragmites* cannot grow. So why is there *Phragmites* where both can occur? First, the drying of the past few decades has altered *Phragmites* habitats so *A. donax* can also grow, but there has not yet been time for the take-over. Secondly, disturbance is damaging and lowers competitive vigour: so competitors can grow alongside, while these conditions exist. Root toxins have not been investigated, but there is no need to postulate them.

Given a flooded habitat, however, *Phragmites* dominates, *A. donax* cannot tolerate long-term flooding.

## Competition with *Juncus subnodulosus*

*Juncus subnodulosus* is a mat-forming rush, forming a dense sward 0.5–1 m high, with a remarkably dense rhizome network a little below ground surface. Rhizomes, of course, mean roots.

There are no large amounts of *J. subnodulosus* peat in the peat record. It is not, therefore, a long-term historical peat-building community. While, obviously, all native species have (un-impacted) native communities, it seems that most East Anglian dominant *J. subnodulosus* was developed by management, by mowing (e.g., rushes for strewing) and in nutrient-medium, semi-dried fen.

*Phragmites* is often present, though sparse and short. Investigation shows that, as buds grow up through the *J. subnodulosus* mat, they narrow, not by just the usual 0–2 mm, but by 2–5 mm. This means narrow, and therefore short, shoots which have over-short internodes, as well. These are quite unable to form a shading canopy (rarely reaching 1 m high, while the reed alongside the *J. subnodulosus* community, with 15 rather than 10–11 nodes, was up to 190 cm).

Additionally, *Phragmites* emergence is two weeks later than in the reed population alongside, and it dies two weeks earlier, so giving a month less for food production.

If the rhizome mat is broken, as in clods being dug up, then both these hindering factors nearly vanish. Reeds turn green, grow taller, and live longer.

This is one of the strongest examples of toxin production. There can be no other explanation.

## Competition with *Calystegia sepia* and *Galium aparine*

*Calystegia sepia* and *Galium aparine* are unusual in that they can both bring down reeds to the ground. Both are annual scramblers. They overlap, but *C. sepia* is of rather wet, not very disturbed fens and marshes, and *G. aparine* of Tall Herb, rather dry, nutrient-rich fens. Probably, in the past and when these habitats were wetter, they were used for harvesting or mowing.

When conditions are good for the two species, which is far from every year, they are plentiful by mid-summer, scrambling over the reeds and pulling them down, so that over large swathes the reeds are flat on the ground by late summer.

Since, like this, they still cover the ground, the community has not been opened: it is no more likely to allow fresh invasion. Next year, it is unlikely *Phragmites* has been weakened, and the balance of dominance is not altered. It is just that when the weather (and perhaps other factors) favours the growth of these, large reeds are flattened.

## Competition with *Calamagrostis canescens*

*Calamagrostis canescens* is one of several short grasses very 'obviously' controlled by *Phragmites*, only growing where light comes through *Phragmites* beds. If the bed is flooded, *C. canescens* cannot tolerate that, and *Phragmites* dominates. If *Phragmites* can produce a shading canopy, it again dominates. If neither of these apply, *C. canescens* can keep *Phragmites* buds sparse, and itself dominate. This is an unstable relationship. Water regime is primary, plant factors secondary. (And see Godwin & Bharucha, 1932.)

## Commercial reedbeds

Commercial (and ex-commercial) reedbeds in East Anglia have been used and managed perhaps since the twelfth century (see Chapter 1 for early use). They may be in or near the (peat-excavated) Broads, their rivers, or near the coast.

These old beds may originally have been deeply flooded, but for a long time now have been partly shallow-flooded, partly damp. At this water level wet woodland can invade and dominate, reducing the reed to being sparse and of the shaded habit. The habitat could also bear wet meadows, various Tall Herb and other communities, such as do occur, frequently. To keep reed dominance, management is needed, and the particular management for reed is winter cutting (and perhaps burning, see Chapters 4 and 11).

Winter cutting removes young woody plants before they become large, shading, invading trees. It gives bare ground in spring, and this means greater temperature fluctuations which, in turn, means more *Phragmites* buds at emergence and these emerging earlier, so there is a shading canopy of *Phragmites* before its competitors (e.g., *Calamagrostis canescens*) develop.

In beds the richer side of nutrient-medium, and calcium-rich, *Cladium mariscus*, fen sedge, can also grow well. This species used to have many uses (fuel, thatch, litter, etc.), but is now mostly used for patterns on reed-thatched roofs.

*Cladium mariscus* has perennial leaves which store much food. Annual winter cutting of a mixed stand leads to *Phragmites* dominance. Conversely, summer cutting every 3–4 years leads to *C. mariscus* dominance (*Phragmites*, see Chapters 6 and 11, is much reduced by summer cutting which is too late for a replacement crop) (see Godwin, 1941).

In these instances, the outcome is powered by management. In the same part-flooded, part-damp habitat, *Phragmites* may be monodominant, or sparse amidst a number of different communities, depending entirely on the harvesting regime.

## Competition in bog

*Phragmites* is larger and more vigorous than the species of bogs, so can easily shade them out: in those few places where it can grow well and is not grazed. Such places are where nutrients have been raised, by e.g., flushes, mineral-rich springs, outcrops of calcium-rich rocks below, etc..

(As described in Chapters 6 and 9, with very low mineral nutrients, replacement shoots after grazing are few.)

Competition is ineffectual, but in order for *Phragmites* to develop and dominate, nutrients must be added from outside.

## Competition in Poor Fen, *Parvocaricetum*

Poor Fen has more nutrients than bog but, unlike bog, is often caused by, and due to, management, past and present. Light grazing is or was common and this, with the sub-adequate nutrients, mean that it is difficult for *Phragmites* to be more than a short, sparse associate. If (as happened in Wicken Fen), nutrients are increased, then *Phragmites* can grow taller and become more frequent: without management, this could lead to a reedbed, in the course of time (or, of course, carr and wood).

No doubt there are a few species with root toxins or other defences, but studies are little, and the primary response of *Phragmites* is that if nutrients and grazing or other impact allow it to grow tall and dense, it will shade out the shorter species of the *Parvocaricetum*.

## Competition in rich, Tall Herb fens

This community is usually found in dried, ex-managed fens and marshes, and most of the species are 1–1.5+ m high, much the same as the *Phragmites* living sparsely within it (or locally, by dykes: or indeed not at all).

This community can be stable over decades, but it is not a historically or presently viable vegetation type. Although the dry soil and generally numerous invertebrates militate against tree invasion, it happens (as for sallow, above), probably slowly and certainly surely, unless management changes. It is also possible for *Phragmites* to take over the area as a very dry reedbed.

## Discussion

Interactions between *Phragmites* and other species may be very complex as with *Carex paniculata*, or very simple, as with *Galium aparine*. Apart from stating the competitor in question only grows on dry land, in the most nutrient-rich habitats, etc., and therefore is excluded in the given habitat, few generalisations can be given.

Under water, in lake stands, the position is simpler. There are far fewer potential competitors (the shorter aquatics not being among them), and addenda like litter mats, root toxins, are barely there. It is up on the land, in the marshes and fens, that these may be crucial in determining competitive balance, or its timing. It is also up on the land that – apart from wet woodland – *Phragmites* was traditionally absent. Before much human impact, agriculture and managing marshes for wetland crops, reedbeds partly dry were succeeded by woodland. Case histories like various in this chapter would not have occurred before. This is useful to remember, when considering the different weapons now used: they were not necessarily evolved for the purpose!

The basis of the performance of *Phragmites* is the emerging bud. If these are many and wide, reeds will be dense and tall. If they are few and narrow, they will be sparse and short. *Phragmites* competes mainly by its thick canopy shading out all below. This means many large buds.

There is a minor control over bud width at ground level, as some species produce a root toxin which can narrow, and so shorten, the emerging shoots. Likewise, litter mats may be toxic. Even the litter mat of *Phragmites*, when thick, may be a hindrance. These are local (very thick *Phragmites* litter is but patchy). The ordinary *Phragmites* litter mat helps protect the bed from invasion, and does not poison it.

The major control, though, is through the horizontal rhizomes. They store organic food from the aerial shoots above and minerals from the soil and water around. Both can be in short supply, and future performance be affected. A few short reeds cannot supply the food to power a monodominant emergence a few months later.

In order to produce many large buds, therefore, the horizontal rhizome must have enough food to do so. It must also have signals from the rest of the plant about the environment. Thirdly, plant vigour may vary. An advancing margin will steamroller its way through most vegetation. Unless the hinterland has the vigour to maintain dominant *Phragmites*, when the primary vigour has gone, and all the competitors are trying to re-grow, the advancing margin is of little ultimate effect. Most stable stands are without advancing vigour, but unless the rhizomes are able to send up a potential canopy, there will not be a dominant stand. Where minerals (or food) are low, such vigour is often also low. It depends, though, on the competitors whether it is too low!

Spring burning (if not scorching the soil) increases reed vigour and buds (Chapters 3 and 4). This alone, apart from any unsatisfactory effect on other species, may alter balance in favour of the reed.

Conversely, unusual flooding or drought weakens reed, and lays the bed open to invasion. (Also see Burdick & Konisky, 2003; Windham, 1999, for saltmarsh patterns.)

# Chapter 8
# Malta: decline and endurance

## Introduction

The Maltese archipelago is very small (under 250 km$^2$) and geographically separate (lying between Sicily and North Africa) in the Mediterranean. As *Phragmites* occurs in the three larger islands, it forms a valuable case study for habitat and behaviour.

Malta is mostly limestone, and with an average of over 500 mm annual rainfall, used to be full of water, with many springs, marshes and rivers. Most of the land dries in summer though, and cultivation is in winter unless the fields are irrigated.

Water loss has been drastic, starting in the 1830s when the Government ordered the draining of the marshes for public health. (There was much 'fever'. At one point gastrointestinal diseases were a major cause of death. Cholera and typhoid are water-borne, and mosquito larvae live in water [malaria].) The inland, freshwater marshes were drained first, then the large estuarine ones, and finally, though not until well into the twentieth century, the small, more brackish marshes. These are of course a lower health risk. Though, as with all the marshes, if the soil was (or could be made) plentiful enough, the drained land could be cultivated.

Cultivated land is not reedbed!

Early, water for domestic use and irrigation was taken from springs, rivers and underground sources by gravity. This is reasonably sustainable. In the 1890s, however, the demand for clean water was such that water was pumped from aquifers, and the amount of extracted water has continued to increase. Despite about half the mains water now coming from Reverse Osmosis plants, more water is extracted than ever before, more for irrigation and industrial and leisure purposes (e.g., swimming pools) than for mains supplies. This dries springs and rivers, so likewise marshes; again, suitable land was cultivated. The consequence of all this is dry marshes. Streams carry little except storm water and rarely flow along their length. Most reed sites are now never flooded: an abnormal feature.

*Phragmites* now occurs in near-20 (known) sites, most in estuaries, with salt spray if not brackish water: reeds are encouraged by salt, as discussed in Chapter 6. These are mostly scattered around the coasts of all three islands. Some inland sites are on springs, seepage areas, or (as transplants) ornamental pond. Even in such a small area there is much of Malta and Gozo which cannot be seen from public roads so there may be more occurrences: but not many, since in the early twentieth century it was possible to walk almost everywhere, as did Professor John Borg. His *Flora* (1927) is the best record of reed distribution of that date.

In addition to water loss, there has been water pollution. Agrochemicals, run-off from roads, farm, industry, etc., and, on the lower sites, sea water seeping up through the Main (Sea Level) Water Table: all pollute the surface water. While, in the 1980s surface water was mostly clean, in the 2000s, even clear water is infrequent, and clean water is rare indeed. In the past, springs and spring-streams were calcium-dominant, which (see Chapter 6) is unfavourable for dominant *Phragmites*. Alluvium and salt are favourable. So it is not surprising most sites are coastal.

*Phragmites*, as the previous chapter also shows, is tough and adaptable. It is reasonable to suppose that what remains is a tiny fraction of the pre-1830s reed: but despite the dryness, despite the disturbance, despite the pollution, *Phragmites* is still there.

## Distribution

Unfortunately, the early Floras are incomplete, and *Phragmites* is one of the species not recorded before Borg's 1927 *Flora*. His data presumably reflect the 1905–1925 period, perhaps also earlier. Borg noted much drying during his lifetime.

The following are the Borg (1927) sites and their present *Phragmites* status:

### *A Coastal and estuarine marsh*

1. **Marsascala** Gone. Developed and disturbed (Figure 8.1).
2. **Marsaxlokk** Gone. Developed and disturbed (Figure 8.2).
3. **Bugibba** Gone. Developed and disturbed.
4. **Dweijra** (Gozo) Gone. Developed and disturbed (Figure 8.3).
5. **Marsalforn** (Gozo) Remnants at sides of lined brackish river. Look like medium-sized grass. Extraordinary it can survive like this, but for how long (Figure 8.4)?
6. **Ramla** (Gozo) Some 'field reed' (short, rhizomes cut when ploughed) in fields near sea, with bands of tall reed across the valley between the fields (Figure 8.5).
7. **Mgarr ix-Xini** (Gozo) A little field reed with fringing bands of tall reed (Figure 8.6).
8. **Ghadira** A saltmarsh was damaged, then bulldozed to make a bird reserve. *Phragmites* is in a discontinuous tall fringe outside the Reserve (Figure 8.7).
9. **Is-Simar** *Phragmites* grew beside the incoming stream for *c*. 100 m from the coast, until, in the 1990s, a bird reserve lake was constructed and was rapidly

filled by reeds. (In fact some needed clearing.) At least four clonal (genetic) types are present (Figure 8.8).

10. **Maghtab** A proper, though drying, marsh, in a hollow inland from the sea. Over the previous decade the habitat has dried greatly, the *Phragmites* is sparser, and *Arundo donax*, a dampland species, is increasing (Figure 8.9).

11. **Salini** In this site, an estuary silted up in mediaeval (?) times, *Phragmites* is scattered over the largest area (*c.* 1 km × *c.* 400 m). There is a *c.* 20 m wide band of dominant reed by the (sea water and polluted) river, narrow (slowly deteriorating) fringes to a ditch, field reeds in the fields near the sea, taller bands by the walls, etc., at their edges, and dwarf forms in trampled and indeed shady places (Figure 8.10).

12. **Many small marshes and bays** of which the following were found:

    **Marfa**, base of River Ramla. Small, reasonably good (1–1.5 m high) stand on 1 m high sand dune, presumably by an ex-marsh (Figure 8.11).

    **Paradise Bay**, in small patches both sides of (dry) tributary, above sea level. Surrounded by *Arundo donax*.

    **Mistra**. Reduced, since 1990, from a wide band across the estuary to scattered, very small tufts and a short, narrow, taller band (Figure 8.12).

**Figure 8.1**
Marsascala (Malta) *Phragmites* present in the 1920s, lost from disturbance and drying. 'Marsa' means marsh or harbour. Edge of a former fish pond, in an ex-marsh.

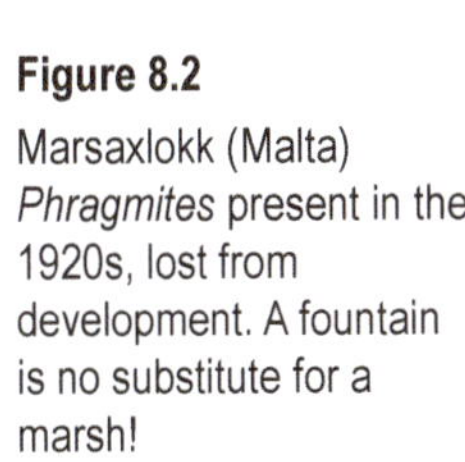

**Figure 8.2**
Marsaxlokk (Malta) *Phragmites* present in the 1920s, lost from development. A fountain is no substitute for a marsh!

**Figure 8.3** Dweijra (Gozo) (Perfect Advertising Ltd, Malta).

No longer suitable for *Phragmites*. Former marsh and river mouth middle front.

**Figure 8.4** Marsalforn (Gozo)

Very poor but persistent *Phragmites* at the edge of the river bed, which is dry except after heavy storms (polluted fresh water) or when sea water is driven inland.

**Figure 8.5** Ramla (Gozo) Narrow valley/river inland of 'sand dunes'.

**a)** Band of *Phragmites*, with *Arundo* behind (*Arundo* panicles are more tubular and longer) crossing valley between fields. Some field-reed.

**b)** Flourishing, though small stand. Panicles wider and shorter than in Malta.

**c)** Main emergence is in February. Their leaves are dying in November (date of photograph). Side shoots developed in mid-late summer – not spring, as is usual – so they are younger, and grow later than the main reeds.

**Figure 8.6** Mgarr ix-Xini (Gozo) Narrow valley river just inland of beach.

**a)** Vigorous, small stand, abundant flowering. Too dry for a litter mat.

**b)** Infructescence again more rounded than in Malta (tubular ones beyond are *Arundo donax*).

**c)** Dwarf reed, with hard and sharp leaves, on edge of beach.

**Figure 8.7** Ghadira bird reserve (Malta)

Bulldozed ex-saltmarsh. Tamarisk fringing the water, but *Phragmites* outside that (front).

**Figure 8.8** Is-Simar bird reserve (Malta) Bulldozed. Three genotypes seen in photograph, others further. They differ in habit, panicle shape, etc.

**Figure 8.9** Maghtab (Malta) *Phragmites* and herb layer are both abundant, since recent drying. *Arundo* on higher level, but poised to invade (two green patches, left). Panicles Maltese-type, left near reed have lost fruits, so are no longer 'fluffy'.

**Figure 8.10** Salini (Malta) Much dried estuary with dyke, farming, etc.

**a)** Ditch beside salt pans. *Legehalme* continue to cross the ditch in late summer, but are unable to establish on the further side. On *legehalme*, nodes may sprout, usually to aerial shoots, sometimes to *legehalme*. Where *legehalme* can establish the population can advance rapidly.

**b)** Well-grown *legehalme*, some of its reeds fruiting.

**c)** Field-reed and taller reeds by walls. Panicles on taller shoots with small ones on some field-reeds.

**Figure 8.11** Marfa (Malta) Small, reasonably good stand on 1 m high sand dune (right).

**Figure 8.12** Mistra (Malta) Disconnected small stands.

**a)** The largest stand. (Most 'reeds' shown are *Arundo donax*. Beware!)

**b)** More typical clump. Shoots short.

**c)** Retreated to crack in wall.

### B SPRING, SEEPAGE AND NON-COASTAL VALLEY SITES

1. **Gnien il-Kbir** Gone. Valley formerly with many springs, now with few, and more cultivated (Figure 8.13).
2. **Gnajn il-Kbir** Gone. Similar but wetter than the preceding. *Phragmites* present up to *c.* 1970.
3. **Santa Maria** (Comino) Healthy, several tens of metres stands, around a gully. This had less impact (drying, disturbance) than stands in Malta and Gozo.
4. **Ghajn Tuffieha** Patch on seepage-slope area in the 1960s, now lost: but may not be that recorded by Borg.
5. **Mgarr** Probably still there but smaller.

**Figure 8.13** Gnien il-Kbir (Malta)
A former (Borg, 1927) site. *Arundo donax* and *A. plinii* (the same height as *Phragmites*) are present, but unless the *Phragmites* is as constricted as that at Mistra, it is now absent.

### C SITES NOT RECORDED SPECIFICALLY IN BORG (1927), THOUGH (EXCEPT FOR THE EGYPTIAN STAND) THEY MAY BE INCLUDED UNDER HIS 'ETC.'

1. **Spring-pool near head of River Arkata** Pool neither dried nor stoned, but water brown and polluted by run-off, so nutrients raised from Calcium-dominated. Large and healthy reed fringe (Figure 8.14).
2. **Seepage/spring on slope, Fontana** (tributary of River Xlendi, Gozo) This wet site has shallow organic soil, a very small stand of reeds not seen a decade before, and four other wetland species. Under threat from *Arundo* and development. New site or enduring tiny reed now able to grow larger (?) (NB: shoots 10 cm high are easily unnoticed) (Figure 8.15).
3. **San Martin** This valley is like a drier Gnajn il-Kbir. The fairly steep slops are cultivated. *Phragmites* is still present in the stream bed (the stream now showing only after heavy storms). It is also present sparsely with the similar *Arundo plinii* near a spring further up the valley. This is a valuable site in that it may show a stage in the way *Phragmites* decreased and was lost in other valleys. Only a little further disturbance would kill both populations, and a little further competition with *A. plinii*, which grows in much drier habitats, would kill the valley one (Figure 8.16)

4. **Duck decoy pond** Towards south of Malta, facing migrant birds coming from Africa, planted with Egyptian *Phragmites* from the (wildfowl-shooting) El-Fayoum marshes. Potentially larger, already 2.5 m high, inflorescences sparkling golden balls, the leaves are wider and bend and fall differently, side shoots profuse in November (more, and three months earlier, than in Maltese genotypes) (Figure 8.17).

These are (almost) the entire populations of *Phragmites* in an archipelago, a nation state.

**Figure 8.14** Spring-pool near source of River Arkata (Malta)

**a)** General view.

**b)** Panicles to show Malta habit.

**c)** Inside dominant reed. Reeds dense. Litter mat present (it is wet enough). Several small late shoots.

**d)** Across the valley from the spring. The spring now drains to the former stream in the valley bottom.

**Figure 8.15** Fontana, Gozo

Drying seepage area. A small stand of reeds not seen a decade before (*Arundo donax* behind).

**Figure 8.16 (right)** San Martin (Malta)

**a)** *Phragmites* shoots in wall, among *Arundo plinii*. (*Phragmites* upper leaves straighter and more upright than those of the similar *A. plinii* shoots. And of course with ligules – not shown – unlike *A. plinii. A. plinii* panicles are also more tubular.)

**b)** Valley rather like Gnien il-Kbir. *Phragmites* panicles in dried stream bed at base of valley.

**Figure 8.17 (left)**
Duck decoy pond with Egyptian (El Fayoum marshes) *Phragmites*.

Not yet established but the habit differences already include potential height; wider, differently drooping leaves; near-spherical panicles with golden branches (e.g., Christmas tree bauble).

## Commentary

Only seven of the *Phragmites* sites named by Borg (1927) have been lost, four developed or much disturbed, one dried and disturbed, and the ones lost recently, disturbed.

Quantity is as important as frequency, and this is more difficult to assess. Borg (1927), the first *Flora* to mention reed, and Weber & Kendzior (2006), the latest before this book, both term reed 'frequent', but refer to different amounts. There has been much loss since *c.* 1920; e.g., in *c.* 1990 (Lanfranco, in Haslam & Borg, 1998), noted, near the Mistra coast '*Arundo donax*, great reed, beds by the river gave way to *Phragmites* near the sea'. Now there is no reed bed, merely scattered tufts and clumps. The Maghtab bed is deteriorating. The pattern at Salini clearly shows that, when wetter and less disturbed, reeds occurred over more widely and further inland of its present area. The speed at which, at is-Simar, a healthy, quality reedbed became established shows how good a *Phragmites* habitat can – and did – occur in Malta.

Some or many Maltese sites, on the other hand, would only have been small: those on or around small pools, seepage areas, etc..

So many *Phragmites* sites in such a small area, in such a summer-dry Mediterranean climate, needs explanation. Very favourable, slightly brackish water, is one reason! Another is that most herbs die back in summer, so are not available to compete (Chapter 7). At the salty end there are evergreens, but these are much shorter (e.g., *Salicornia* spp.) and there is no evidence of root competition. *Phragmites* elsewhere grows in drier places when competition is absent. In Malta it will grow where it appears to be 4–6 m above perennial water. (Rhizomes elsewhere grow to at least 2 m down: the relevant Malta sites could not be excavated.) Roots grow down at least 2 m further. Storm flows bring higher water intermittently, and partial flooding can be enough, Chapter 5. And, thirdly, genotypes may well be more tolerant to drought.

The only serious competitor is *Arundo donax*. Before much drying, this is also when *Arundo* was kept down by management, by being regularly harvested, and by being removed from places more needed for, e.g., grazing. In flowing rivers, *Arundo* cannot grow. On new-dried mud, it does. In flooded marshes it cannot grow. In dry ones, it can. Since *Arundo* is twice the size, it smothers out all in its path: including *Phragmites*. It has greatly increased in the last half-century. The two can, temporarily, co-exist.

In Ramla Bay (Figure 8.5), in the unploughed bands, *Arundo* is facing the sea, *Phragmites* the land: which is not a sustainable pattern. Formerly, no doubt there was a reedbed, far too wet for *Arundo*. In the ditch at Salini, *Phragmites* alone grows into the water, of course, but on the bank *Arundo* has much increased over *Phragmites* in the past forty years. *Arundo* is invading Maghtab. It is quite possible that *Arundo* competition helped to exterminate *Phragmites* from Gnien il-Kbir and Ghajn il-Kbir (like *A. plinii* at San Martin).

## Genetics and dispersal

Genetically, the Maltese material is very similar, potentially large, with large inflorescences. Yet there is a difference between Maltese and Gozitan material, and – as shown best at is-Simar – there are different genetic clones even within one marsh (see photographs). This suggests diversification over the millennia from a near-uniform invader, rather than multiple invasions.

The Egyptian material (see above) is strikingly different, both in appearance and in seasonal behaviour. The pond is on enclosed private land, for a probably illegal purpose (to shoot protected, as well as allowable birds, and out of, as well as in, the shooting season). How many such ponds are there? And how many of them are on springs which would keep them flooded when the owner loses interest? Will there be inter-breeding and establishment of either Egyptian, or half-Egyptian plants?

No (natural) propagation, to new areas either by seed or by rhizome, has been seen in recent decades. This is not to say it is impossible, but that recent habitats were unsuitable.

## Habits

Habits are partly shown in the photographs.

Field-reeds are recorded in Borg (1927) as often very common in ploughed fields in moist estuarine areas. This short reed, with fewer inflorescences, needs damp or wet to continue. In both the English Fenland (in the past 40 years) and in Malta, with drying, this cut-up form decreased. It now occurs up to *c.* 600 m from the coast at Salini, and *c.* 100 m at Mgarr i-Xini, and less at Ramla Bay. Not 'very common'! At the unploughed edges reeds grow 1–2 m high, and flower freely.

Dwarf reed where even the 'King' reed (on a first-year rhizome, Chapter 3) is barely 15 cm, and other reeds may be 5 cm, is frequent – and easy to overlook. Where salty and trampled, the leaves are tough and sharp, able to scratch skin. This is the extreme form. With less salt, the leaves are softer, and are softer again in shade. Trampling leads to horizontal shoots.

*Legehalme* are common in the flooded ditch at Salini, but otherwise habitats are not suitable, so the potentiality (Chapter 3) is not known. At Salini, they are fewer and less healthy than in the 1960s.

At Marsalforn the remnant plants have the grass form of, say, water-swept *Dactylis glomerata*. The smaller ones at Mistra are similar, though vertical (not water-swept). Not at all typical. 'Typical' shoots are up to *c.* 2.5 m, and flower profusely. All drop their leaves in winter, but in early spring larger shoots may grow side shoots in their second and even their third year. Such 'bushiness' is more common going south. In the north (whatever the potentiality) shoots are killed by cold, so second-year growth cannot occur.

Salini reed can grow through tarmac over 10 cm thick, and do well! The emerging buds were narrowed, as when growing through plant tussocks (Chapter 7).

## Discussion

What a plant! What toughness, what variation, what range of habits and habitats! Looking at the departing stands of Marsalforn and Mistra evokes admiration that any reedswamp plant can grow like that!

In number of sites and, even more, area covered, *Phragmites* is much reduced from the 1920s. So, probably, much again from the 1830s.

Its uses included thatching, wind-breaks, screening, pipes (for water, music, and tobacco) and general farm and domestic purposes. For most of these *Arundo* is preferable: which shows the comparative abundance of the two in the past. There are now, a few wind-breaks, tourist parasols, and suchlike. A 1680s picture shows a man sitting smoking, the tobacco in a little clay pot on the ground (hardly bigger than the cup of a later pipe) with a *Phragmites* reed a good one metre long reaching up to his mouth.

Once a population becomes small, as in so many of the Maltese stands, it is easier to eliminate, either just by the current conditions continuing, or by a new factor such as encroaching cultivation, *Arundo*, roads, picnic places, or major development or disturbance. The plant is indeed tough, but a small population is fragile. Quite small projects, like the is-Simar bird reserve, can have a major conservation impact on *Phragmites* – which was quite unintentional.

However, Malta's usual 'rehabilitation' projects are generally intended to make places look like urban South France, so *Phragmites* is more likely to prosper, or at least endure, with neglect and drying than with rehabilitation.

How long can many of these populations endure? Another half-century? Hope could come from the worst-case scenario for mains water supply (from pollution and lack of water) which in 2010 was that before 2020, extraction would have to stop. If rain was allowed to refill aquifers, reed could again flourish.

# Chapter 9
## Uses of reed (Thatching is described in the next chapter)

*IF YOU THINK ABOUT WETLANDS...*
*...think about peat*
*Peat is everywhere*
*Peat is what we eat*
*Peat is what we breathe*
*Peat is in our heat*
*Indeed...peat*
*Peatlands are everywhere*
*From the tundra*
*To the tropics and*
*To the uttermost part of the Earth*
*From the mountains*
*To the sea*
*But peatlands are hidden*
*Peatlands hide under meadows*
*Peatlands hide under forests*
*Peatlands hide under frost*
*Peatlands hide under mangroves*
*Peatlands hide under swamps*
*Peatlands are water*
*Peatlands are climate*
*Peatlands have biodiversity*
*Peatlands are biodiversity*
*Peatlands provide*
*Peatlands protect*
*Peatlands preserve*
*Peatlands prevent*
*Peatlands erode*
*Peatlands decrease*
*Peatlands decease*
*Peatlands revert*
*Peatlands desert*
*Peatlands degrade*
*Peatlands deflate*
*And therefore...*
*Peatlands depend*
*Peatlands depend on...*
*– attention*
*– protection and wise use (urgently in tropics)*
*– integration into river and water management...*
*Peatlands depend on YOU...*
*So if you think about wetlands*
*Think about peat.*

IMCG *Newsletter*, 2007

### Introduction

Since *Phragmites* occurs so widely, and often reedbeds are large, in all ages and places people have made what use they can of a resource that is just there, and – until recently – not easy to be rid of. So uses are old and new, sophisticated and simple, for the reedbed itself and for the plant parts removed from the reedbed. Europe and the Near East, the centre of distribution, have naturally enough, widespread uses. Asia and Africa follow, with locally an extreme number of uses in the Americas, where reedbeds are fewer. Least are recorded in Australasia, where there is little *Phragmites*. That is, when it is there, uses must be found, but *Phragmites* is not always so obviously a valuable plant as to be sought for. It is not, and never has been, crucial for food, clothes or shelter for the main population of a country.

## Definition

European wetlands can, and have been mapped and recognised, both those now present, and a good deal of those extant from the post-Glacial period. It may be less easy to determine how much was reedbed. If peat is present, and in good condition, *Phragmites* peat is easily recognisable, but if peat has been destroyed, identification may be dubious. *Phragmites* peat is not the only fen peat (see Chapter 1) and fen peat is less than bog peat.

In historic times also, identification may be difficult. The English word 'reed', as in 'reedswamp', meant any tall monocotyledon in a sward in a wet place. Even Dr Johnson (1755), when plant names were becoming standardised, defined reed as a hollow knotted (noded) stalk, growing in wet grounds. This excludes the long-leaved *Typha* spp., *Carex* spp., *Iris* spp., *Cladium mariscus*, etc., and the tubular stems of *Juncus* spp. and *Scirpus* spp., but includes other tall grasses like *Glyceria maxima* and *Phalaris arundinacea*, which may dominate in reedswamps, but are small in both space and time, and whose peat is rare.

A long-standing reedbed may have a mix of other important species, used with the *Phragmites* when feasible, or be monodominant, with other species very subordinate.

## Within the traditional reedbed

People can go into reedbeds to collect materials from fish to seeds, and they can live entirely within a reedswamp with little or no materials from outside, or be anything in between.

The Marsh Arabs of Iraq were described in Chapter 1. They are one of the peoples with reedbeds as their historic home. Thesiger (1964) records use for:

- homes and community buildings (houses may be reed and earth layers);
- raising level for houses ('swans nests');
- canoes (bitumen-coated);
- fences;
- water buffalo fodder, *in situ* and harvested;
- furniture, mats, fans;
- sugar (sucking young reeds), and presumably medicine, drinks, salads;
- rice cultivation where silted (nutrient rich);
- fish, fowl.

The peoples of wetlands tend to form tight, fiercely independent communities, proud of their isolation and self-sufficiency. If left alone, all is well. But when greedy eyes, or possessive power look from the mainland, the marsh peoples object in vain to the loss of their way of life.

*They hang the man and flog the woman*
*Who takes the goose from off the common*
*But let the greater villain loose*
*Who takes the common from the goose.* Anon.

Developers take the wetland from the people. In England, the peoples of the Fenland and the Somerset wetlands became, by the nineteenth century, squalid and drunken with no gentry. There was therefore a moral case for drainage, destruction of an independent way of life, and making the people healthy and industrious, depending on wages. But at that time, what, other than alcohol, could remote country folk spend their disposable income on? Highlanders in remote (but healthier) bog-lands with a clan structure and cultivated families at the top became a cultivated people: surely not just the difference between fen and bog.

Others went to the reedbeds to bring sanctity to the demon-ridden wastes, as in the English Fenland. The earliest (eighth century) English poem, *Beowulf*, describes the monsters of the fearsome wetlands. St Guthlac (674–716), a hermit living in the Fenland, probably lived among reedbeds (Chapter 11; Haslam, 2003). A culture of the holy developed in the fastnesses of Eastern England wetlands, hermits, men and women, monasteries on higher ground (e.g., Crowland) – building stone, people, food and news were carried by boat. Greedy eyes came. The Danish Vikings arrived, and monks and nuns were no match for pagan swords.

The Vikings were the only invaders of England who travelled inland by boat, so the wetlands suddenly became easily accessible.

The third reason for living in the wetland is flight. Those fleeing could hide, and live in unknown exile, or gain respite, and emerge again. Alfred-the-Great, in 878, suffered a major defeat by the Danes, and retreated into the fen fastness of Athelney, Somerset, from where he emerged, gathered fighters, and defeated the Danes. The Danes sued for peace from the king who had been throneless a month before. When William the Conqueror was 'pacifying' England after his conquest of 1066, one of the Saxon leaders resisting him was Hereward, known (for his watchfulness or for a later boast) as 'The Wake'. Hereward retreated into the Fenland, but was eventually successfully pursued by William.

However, not all wetlands are reedbeds! Figures 1.4, 1.5 and Table 1.4 show a variety of peat and alluvial substrates in fen. *Phragmites* is only one, and one that needs a supply of nutrients without excess calcium (Chapter 5).

Smaller reedbeds and other wetlands may be uniform, but large ones are not. There are open pools too deep for *Phragmites*, there are islets or even islands rising up from the water. There is, in short, habitat diversity. When islands are big enough for ordinary villages, there may be farming: 'winter land', that flooded rarely, around the village, and 'summer land' further downslope, exposed only in summer, but good grazing. Marsh species grow in the real wetland below.

## REEDS AND RHIZOMES REMOVED FROM THE BED

### USES OF HARVESTED PHRAGMITES

Harvested *Phragmites*, like the reedbeds themselves, have existed from earliest times. Täckholm & Täckholm (1941) record ancient Egyptian uses. Reeds filled walls in the first and second dynasties – as also in mediaeval England. They were in mortuary bouquets in the eighteenth dynasty, and as a motif for columns in the, e.g., eighteenth and nineteenth–twentieth dynasties. Panicles are a common hieroglyphic sign. Reeds are in pictures on old monuments, e.g., Rameses III with a lion fleeing through reeds. Coptic culture used reeds for part of weaving looms.

Arrows and crates, leaves for sleeping mats, rhizomes to make drinks diaphoretic and diuretic, all are recorded. Fuel and fodder are valuable. Reeds are tied in bundles and sold for roofs and fences (and see Chapter 10). As described in Chapter 1, writing pens were made, as for millennia and centuries later. Egypt could use a giant reed, but where only smaller were available, smaller were used. Given the sparsity of *Phragmites* in North America, a surprisingly long list of uses is given by Kiviat & Hamilton (2001) (see also Lefeuvre, 1998).

Recorded uses include: alcoholic drink, apron, arrow shaft, art, basketry, baskets, beads, beds, bird traps, bird nets, canoe (other boats), cigarette, 'coffee', construction, containers (various), cordage, crafts for tourists, digestive (root), door covering, drinking straw, emetic (whole plant), farm litter, firelighter, firing (of pottery, etc.), fish traps, flageolet, flute, fodder (cut and grazed), food (rhizomes), frames, games (dice, counting sticks, etc.), green thatching (temporary), hat, hides, knife, letters, looms, malting, matting, mattresses, medicine, ornament, paintbrush handle, pens, pillows (stuffing), pipe stems, plaster (mixed with reeds, to harden), rabbit traps, rattle, seed sack, shelter, sieves, spears, stuffing (seeds), sugar (sap, young shoots), thatch (roof), twirling wand, under floors, walking stick, wall-filling, whistle.

Larger-scale, usually more modern, uses include: erosion (planted live reeds, bundles of dead reeds), insulation mats, paper, reedboard.

### PAPER, CELLULOSE

The early trial of paper-producing reed was in the Danube Delta (see Rudescu *et al.*, 1965). This did not prove viable long-term.

The cellulose content of the giant reed *Phragmites* is 27–67%, and the mean length of fibres is 0.9–2.0 mm, being longer in taller populations. Most commercial fibre species have longer fibres. Used alone, Danube reed paper is of poor quality. Used at 80% (coarse) brown paper is adequate for wrapping, but good-quality writing paper cannot have more than *c.* 20% *Phragmites*.

However, in the early twenty-first century, with technology presumably improved, China is producing quality paper efficiently, and the harvesting and tendance gives a good living to farmers. The best reed is that found near parts of the coast, *c.* 1 m high, but inland 2–2.5 m reed has adequate fibre. Is there perhaps a difference with size? It is, anyway, excellent that such a useful harvestable crop exists (C. Zhou, pers. comm.).

#### *Reed board*

This lightweight, pressed-reed board was widely in use in Central and Eastern Europe, but, as with so many commercial reed projects, *Phragmites* is not the best possible material.

### ***Phragmites* peat, *in situ* or removed**

Commercial:

- dried and cultivated (is fertile);
- fuel (most is not *Phragmites* see below); horticulture (most is not *Phragmites*).

Scientific (peat is buried history):

- artefacts (objects for human use, archaeology);
- pollen analysis (vegetation history);
- preserved plant and animal remains; (and see below).

#### *Fuel*

Ever since people used fire, dried peat turves were presumably used for fuel. *Phragmites* peat is slow growing. First, it has to be under water perennially. Summer-drying oxidises (blackens) and undoes the winter accumulation. Growth has not been measured, but 1 cm in a decade is probably a conservative estimate. *Phragmites* peat is unduly fibrous, making digging more difficult (see Figure 9.1) but it burns properly. Burning for domestic use in isolated hamlets is reasonably sustainable.

**Figure 9.1**
Peat cutting in Fen
(Y. Bower in Haslam, 2003).

The Danes brought commercial extraction to eastern England in the ninth century, they having done this at home since at least 500 BC. Curiously, in the east of East Anglia, they dug huge pits, later flooded and known as the Norfolk Broads (e.g., Lambert, 1951; Jennings & Lambert, 1951), but in its west, the Fenland, they took the top off large areas, lowering land level slightly, all over. Some was *Phragmites*. The same was done elsewhere in Britain, and in The Netherlands, where again pits were dug, and in other countries. The peat dates back up to 10,000 BC, the late Glacial period (Chapter 1).

The value of, e.g., the Broads, now outweighs the destruction done a thousand years ago to make them. But present-day commercial extraction, with huge machines, and taking most of the peat is quite different, and highly deplorable: particularly when used just for luxury, as garden soil. Finland and Canada think their extraction (for power stations and horticulture) is sustainable, but whether it is in the long term remains to be seen.

## Reedbeds

### *Drainage*

Usually developed for drainage of flooded land. Transient. Possible as permanent use.

### *Fodder and food*

Young shoots are palatable, and are eagerly grazed by domestic livestock and, of course, wild herbivores. Larger animals like cows can eat shoots up to 50–75 cm high, after which the lower stems are too tough, and the tops, too high.

Grazing, however, is not sustainable. Unlike the fodder grasses, the growing point of reed is not at ground level, where new leaves can continually grow, but at the tip. So if the tip is removed, the reed cannot grow further. If a new reed (new bud) is produced from the rhizome, it has less food than its predecessor (see Chapter 3) and the population declines. Hence there is often a sharp line between dominant and sparse, short *Phragmites*, being where livestock are kept out by deep water, wall, or fence. When a wet patch of reed is drained, livestock can return, and within two years there are dead stumps of reed invaded by grazing-tolerant species. Rough grazing, with a very low stock density, may be sustainable.

In the Eastern England Breck fens, there used to be many horse (marsh hay) meadows, to supply London's horses. These were mostly the (sustainable) *Glyceria maxima* and *Phalaris arundinacea*, *Phragmites* being relegated to the surrounding ditches. A century after they were abandoned, *Phragmites* is, in some, now dominant.

Considering most of the peat is *Phragmites*, this is a reversion, or restoration of the pre-impact vegetation, though now in dry, not flooded peat. Such fens may be in conservation zones, or be kept because the owner likes them, but are at risk from the internal cause of invasion by trees, or the external one, of conversion to another use.

Reedbeds can support an abundance of wildlife, animals that find food, shelter or nesting/breeding habitat in reedbeds. These are both herbivores, and carnivores living on the fish, fowl and invertebrates. Wildfowl, waterfowl and other large birds (e.g., bittern) and small ones (e.g., reed bunting, bearded tit) all live in reedbeds, and formerly (before over-exploitation and habitat loss) in almost inconceivable numbers (see Vajda, 1998, for herons, in Hungary).

In flooded beds, mostly in lakes or large (undredged) rivers, fish can find cover, shelter and food within the beds, and are themselves food for other fish, birds, mammals and people (e.g., Micovski, 1998; Haslam, 1991). In the (partly reed) Fenland, eels were sufficiently abundant to be used as currency for centuries (Darby, 1968 and 1983). Reedbeds in lakes form good nursery grounds for fish, and can be mimicked in constructed ponds (e.g., Micovski, 1998). The large Czech fishponds (see Haslam, 1991) have wide fringing reedbeds.

Since beavers were lost from Britain in mediaeval times, no water mammal has been hunted for fur. American mink arrived in Britain (commercial fur farms) in 1929. Since the second half of the twentieth century (*c.* 1956), escaped mink from these fur farms have colonised many rivers to the great detriment of native water voles, duckling, etc.. Unfortunately they are not trapped for fur. (Nor were their similarly-escaped predecessors, coypu, which were exterminated in the 1970s because – unlike mink – they damaged property, as well as reedbeds and river banks.) Beaver, mink and muskrats of various kinds may still yield fur. However, fur (particularly water-fur) in any event, is much less popular now, because of the larger variety of clothing materials, and the way too many fur animals are kept in bad conditions.

### Cleaning the water

This occurs in any reedbed, but since about 1980 the use of specially made, constructed wetlands has spread. These usually use *Phragmites* in Europe, where *Phragmites* is at the centre of its range, but *Typha* spp. in North America, and, often, floating species in the tropics, for the same reason.

*Phragmites* beds, like other fens, marshes and reedswamps filter, transform, and vary in absorption and storage (Reddy & Smith, 1987).

For rapid breakdown, appropriate microbes, good substrates for them, and good other habitat conditions are needed. Conditions for breakdown are different for different substances. A variety of plant species and of organic materials give 'hot spots' for the speedy decomposition of a multitude of substances. These include, e.g., aerobic/anaerobic and acid/calcareous conditions. Transformations are usually most when there are mosaics as in an anaerobic, waterlogged soil criss-crossed by roots of different species, all with oxygenated rhizospheres, each with different substances exuded from their roots. Alternatively, rising and falling water brings anaerobic/aerobic conditions. In this demanding habitat, many interesting, and often unique, defensive and aggressive chemical and biological-chemical processes have developed in and around the rhizosphere. These influence plant–pest and plant–plant competition (Neori *et al.*, 2000).

Oxygen in anaerobic soils comes in mainly through the root tips, before they are thickened further back. Oxygen diffuses from the root and reaches the root and shoot by a variety of physical mechanisms (Westlake *et al.*, 1998). Roots with through-flow have longer lengths giving off oxygen, so are the more effective (e.g., *Phragmites*, *Typha* spp.). Wind powers this convection, and increases gas flow in the growing season (Armstrong *et al.*, 1992; Moshiri, 1993). This oxygen controls many biochemical and biological activities, e.g., methane release from peat, perhaps (Neori *et al.*, 2000).

While roots are commonly thought of as absorbing water and minerals, they also absorb a wide range of other substances, including toxins (Table 9.1).

Many chemicals are bio-active, altering the behaviour and activity of organisms. In wetlands these occur in the soil, the plant and the plant exudate (rhizosphere) (Neori *et al.*, 2000). The chemical structure of humic material influences breakdown. Herbicides, for instance, may be degraded via microorganisms, via the non-organism route or be transported down the soil layers, and which of these happens depends on the humic material and how it is linked to mineral soil present. Glyphosate, when absorbed in dissolved humic substances, may be easily transported, so is likely to reach fresh waters (Piccolo, 1994).

Different root exudates, so different microbial populations, are found in different species. The rhizosphere organisms influence nutrient availability and plant growth regulators. They stimulate growth. The root environment increases growth and produces allelochemicals (Gunnison & Barko 1989; Reddy & Smith, 1987).

Exudates include, among many more (Shimp *et al.*, 1993; Neori *et al.*, 2000): amino acids, derivatives of fatty acids, elemental sulphur, enzymes (e.g., ferric iron chelators), ethylene, flavonones, glutanases, growth factors/other hormones (gibberellin, etc.), hydrogen cyanide, invertases, long-chain fatty acids, nucleotides, organic acids (e.g., binders of heavy metals), peptides (metal-binding), peroxidases, phosphatases, phytoalexines, proteases, rotenones, steroids, sugars. They vary with: age of root, competition, light, microbes, nutrients, plant species, temperature, water.

Microbes are much denser in rhizospheres, because of the good substrate and habitat. Mycorrhiza may help in degradation, e.g,. *Salix* spp., *Populus* spp. There are nitrogen-fixing microbes in root nodules of *Alnus glutinosa* and *Myrica gale*.

Soil microbes are very widely distributed. Generally, given proper conditions, the species explode. Very toxic substances (such as effluent from chemical factories) can be degraded, as can glyphosate (herbicide) and many aromatic compounds, by *Rhizobium* (Shimp *et al.*, 1993).

Alternative microbes may be available to do the same process in different conditions, for example *Azotobacter* fixes nitrogen well at over pH 6 and with abundant organic matter; *Clostridium* fixes it at low pH.

In wetlands flooded enough for a good algal flora to develop, periphyton – thick on plant parts in moderate pollution – add an extra purifying component. The algae remove and store many nutrients, heavy metals, etc., and these return to the system on death, often in more acceptable forms (e.g., Lakatós & Biro, 1991; Hammer, 1989).

Roots take up nutrients, even surplus ones. They take up metals, organic compounds, etc.. Many metal and organic pollutants stay mainly in the roots, although nutrients, necessarily, move up and are used in the growth of the shoots. The least mobile molecules hardly enter roots (those with low water solubility and low fat solubility, like polycyclic aromatic hydrocarbons).

A very wide range of pollutants has been tested at the in-and-out level of Table 9.2, and the best results from wetlands are better than those from ordinary sewage treatment works. Obviously effectiveness varies with wetland type, plant species and ecotype, design and construction of the wetland and its maintenance, and the climate. (See volumes edited by Athie & Cerri, 1987; Cooper & Findlater, 1990; Mitsch, 1994; Moshiri, 1993; Reddy & Smith, 1987; Rubec & Overend, 1987; Vymazal *et al.*, 1998; Vymazal, 2001.)

There are many non-quantified data, for instance that atrazine is relatively stable, and that cyanazine degrades faster in nitrate-reducing conditions (Gee *et al.*, 1992).

When an effluent changes, as when a chemical factory switches to a different product, it takes about three weeks for the microbes to change and be as efficient purifiers of the new as of the old effluent.

The time the water stays in the system to be worked on, its residence time, is important. So, of course, is whether the system is intended to deal with solids as well as liquids (e.g., raw sewage), with just liquid effluents or as the third stage (tertiary) treatment, to improve inadequate effluent to a polished, good one.

Design of constructed wetlands varies from the simple pond, to those with flow control, to the complex, with specified and varied substrates, and flow directed downwards or sideways in the soil. Unfortunately, there is very little information on the associated flora and fauna, or even whether the habitat is or should be mown annually, so removing a small but significant repository of nutrients. (Mowing of course removes less of those substances staying mainly in the roots, such as heavy metals and various organic pollutants.) Although processes are slower in the cold, biological degradation occurs at less than 5°C (Mitsch, 1994), and in really cold climes as in Canada, effluents are stored in winter (Mitsch, 1994).

Such constructed wetlands are used for: domestic sewage, urban run-off, livestock unit effluent, agricultural run-off, factory effluent (milk, chemical, vegetable, potato, meat processing, sugar, jam, paper, etc.), mine drainage (acid, neutral, sediment-rich), farmyard run-off, dairy run-off, manure, silage run-off, landfill run-off and leachate and similar.

As Worrall (1998) writes: Water treatment or conservation? The implications of reconciling the two and planning for both or either, need to be considered.

Good removal occurs for nitrogen because it leaves for the atmosphere, and for toxic complex products that are broken down to simple ones (e.g., into water, nitrogen, carbon dioxide, methane), substances either going to the air or becoming normal constituents of natural wetlands and not altering properties. Particles of all kinds are filtered out well, but their accumulation may alter the system and its functioning, especially the through-flow. Then there are the substances immobilised in some way, such as heavy metals, which may require soil treatment or removal at intervals to prevent the soil becoming over-saturated and so no longer removing the substances from the incoming flow. Phosphorus has been much trouble, but recently designs do seem to be satisfactory. Dosing with alum (or alum plus lime) increases phosphorus removal, with the bed expected to last for 40–80 years (Moshiri, 1993).

It is suggested that clay is the best substrate: good for adsorption, ion exchange and chemical reaction to inert insoluble forms. There is little removal in a gravel or sandy bed.

Another criterion determining design is whether the incoming flow is steady, as with sewage, or very variable, as with storm water run-off, which may indeed be grossly toxic, but comes only after rain, and the system must accommodate such fluctuations. Such need careful design (e.g., Mitsch, 1994).

Mine effluents are, of course, metal-rich. Long-term metal retention depends on transforming oxidised metals to reduced mineral phases. Chemical treatment of the effluent may be needed before it passes to the wetland (Cooper & Findlater, 1990, and see Peverly *et al.*, 1992). Complex patterns may be needed to remove the trace metals from acid mine drainage, e.g., wet meadow, sedges and grasses, and *Typha latifolia* with straw at the bottom for increased sulphate reduction (Moshiri, 1993).

The plants are a necessary part of the constructed wetland proper. *Phragmites* is particularly useful in collecting and converting sediment and providing pathways for filtering down from the top. The root system grows and extends each year, and dead as well as living roots keep the filter permeable. The plant litter brings in cellulose, etc., so brings more humus. The oxygen supply to the soil comes from root oxidation, and pores formed by the combination of root presence and growth, and from the water moving up to the shoots for transpiration. There is a large surface for microorganisms. *Phragmites* mediates aerobic, anoxic and anaerobic zones, giving best conditions for mineralisation, nitrification and denitrification. The plants are indeed valuable (Hofmann, 1991)!

If wetland-treated water is not fully clean, it may yet be good for hydroponic crops. (Section from Haslam, 2003.)

Nuttall (1997) reviews designs and management of constructed wetlands. See also Somerset County Council (1998).

Oddly, water from reedbeds – or any other place – has never been analysed. Only a dozen or two substances, or even less, are recorded: and even then from very few reedbeds. These analysed substances generally add up to about 10% of the total dissolved solids. There are certainly tens of thousands, probably hundreds of thousands of substances in that. And the functions of the substances are also unknown.

Over the past two centuries, in particular, fresh water has become far more polluted (pollution used to be localised, see Haslam, 1990), and also far less in quantity. Main river drainage and field drainage and under drainage have greatly dried the land. Both contribute to the poor state of fresh water. In the past, in most types of riverscape, there were small wetlands where early run-off gathered, where springs rose, where run-off water was halted by a line of bushes, a wall, a change in level, beside the small brooks which have gone (Haslam, 2008). These small wetlands received water, and largely cleaned it.

This could be somewhat remedied. In parts of North America little wetlands are being inserted into (now dirty) riverscapes, to clean the water. These may, or may not, be large and wet enough for reeds. Reedbeds can be returned to floodplains, if water is partly returned. Cleaning should be considered on a riverscape scale.

Constructed reedbeds are being increasingly developed for small tertiary house filtration systems in Britain. Large beds are also being created. The north-west of England, and Scotland, are the main growth areas, and reed transplant 'plugs' are ordered by the hundred thousand (British Reed Growers Association). Germany has had factory reedbeds for over two decades (e.g., Winter & Kickuth, 1985 and 1989) and it was in Germany that Seidel (e.g., 1956 and 1966) noticed, and pioneered, the use of reedswamp plants for cleaning.

Best management practices to reduce pollution are given in Table 9.3.

### *Water storage*

In managed lands, flood water can be diverted into reedbeds and other marshes, and released slowly so that downstream damage is averted. Also, where most rain falls in winter, reedbeds can store water for use in summer as reservoirs. However (Chapter 6), reedbeds require a consistent water regime: rising 1–2 m in winter and falling in summer is satisfactory, but rising 1–2 m in different months in different years, is not.

### *Inspiration for the arts*

Down the ages, reedbeds have been an inspiration. The Egyptian sculpture and ornament are described above. There, people went into the Marshes and came out to make the graven pictures. Those living entirely within the reedbed are usually too near subsistence level to indulge in the Arts. It is those living comfortably on the land who do so (but see Iraqi mudhifs, Chapter 1).

Poems and novels are strewed through English literature. *Beowulf* (Anglo-Saxon) is set in gloomy and fearsome Fenland. *The Reed* (Chapter 1) describes the reedbed and uses of the cut reed, again in Anglo-Saxon times.

In the Fenland (part-reed) an eighteenth century parson was sufficiently impressed by the unhealthiness to write:

*The moory soil, the wat'ry atmosphere.*
*Thick stinking fogs, and noxious vapours fall*
*Agues and coughs are epidemical;*
*Hence every face presented to one's view*
*Looks of a pallid or a sallow hue.*

With drainage, both fogs and the pools with mosquito larvae and other waterborne disease drastically declined, and health improved. However, Defoe (1724–1727) was, in contrast, pleasantly surprised by the *health* of the fen peoples. No doubt health varied from place to place (boggy areas, being slightly disinfectant, had less disease then fenny ones).

Reeds inspired Elizabeth Barrett Browning, in *A Musical Instrument* in the mid-nineteenth century:

*...What was he doing, the great god Pan*
*Down in the reeds by the river?*
*He tore out a reed, the great god Pan*
*From the deep cool bed of the river,*
*The limpid water, turbidly ran,*
*And the broken lilies a dying lay,*
*And the dragon-fly had fled away,*
*Ere he brought it out of the river...*
*Yet half a beast is the great god Pan,*
*To laugh as he sits by the river,*
*Making a poet out of a man:*
*The true gods sigh for the cost and the pain –*
*For the reed which grows nevermore again*
*As a reed with the reeds in the river.*

Pipes were obviously still made from *Phragmites* at this time! It is the great god Pan, however, who dominates the river, not the reed fringes.

In the early twentieth century, Rudyard Kipling, discussing the stability and slow-moving character of the English, wrote on Magna Carta, the basic charter on freedoms and rights (and numerous other regulations), signed by King John at Runnymede in 1215, by the reedbeds of the River Thames:

*And still, when mob or monarch lays*
*Too rude a hand on English ways;*

*The whisper wakes, the shudder plays,*
*Across the reeds at Runnymede.*
*And Thames, that knows the moods of kings,*
*And crowds, and priests and such-like things,*
*Rolls deep and dreadful as he brings*
*Their warning down from Runnymede!*

Novels may have fenny reedbeds as a protagonist, especially in the nineteenth century, when drainage and good roads first made them socially interesting, e.g., *Lorna Doone* by R.D. Blackmore (Somerset and Devon), *Hereward the Wake* by Charles Kingsley (Fenland). Minor works are frequent, e.g., *Death of an Effendi* by M. Pearce (El Fayoum Marshes, Egypt: tall reed, soft oozy substrate, even *Convolvulus* sp.).

With the late coming of the reedbed as pleasant, rather than the home of disease, mud, drowning and perhaps 'fearsome savages', it is not surprising art has been more in photography than in painting (though the Norfolk and Norwich museum, in the capital of Broadland, shows many paintings. Large reedbeds, and seeing the sails of a Norfolk wherry (barge) apparently on land in a reedbed (on a reed-fringed river) make striking paintings). The reed fringes, unfortunately, have gone, a victim of increased leisure boats in the 1970s. In, e.g., the Lake District and Scottish lochs, reeds may dominate shallow bays and marshes behind, but make only a small, though valuable, part of the whole lake picture.

A smaller number are inspired to work within the reedbed. An unusual people, but those who find an extraordinarily special and unique 'Sense of Place'. *Phragmites* hides people, unlike bogs, poor fen and wet pasture, but is on a human scale, unlike *Arundo* or sugar cane. Reeds surround you, their soughing and bird song the only sounds, and the sky and clouds are clear above you. Reedbeds are for those working alone. The atmosphere is lost with company, and is lost to those beside, rather than within, the reedbed.

### *Prevention of river bank erosion*

A good fringe of *Phragmites*, binding loose soil with its dense underground parts (Chapter 3) is very effective for preserving banks (e.g., Caffrey & Beglin, 1996, describing planting reeds for stabilisation, and its environmental benefits, though, in this instance, poor survival). In the traditional floodplain, reed fringes are wide, and very difficult to erode, whether by boats or storms (e.g., River Tisza lakes in Hungary). When rivers are dredged the fringes are necessarily narrow, and may find anchorage difficult in steep banks (e.g., Norfolk rivers), or impossible when scour, or water fluctuations are too great. When boats are added to an already fragile habitat, reeds are lost.

Bundles of (dead) reed can, however, be used to line banks, and if placed correctly, are generally useful, for quite a while. Osiers, in bundles and mats, are also used successfully.

### Managing reedbeds for thatching reed

Reedbeds for thatching reed are now mainly in England (see Chapter 10), and the oldest have probably been harvested for at least 1,500 years, quite possibly much longer. Water regime has varied (Table 1.3), so the beds have been wetter and drier over the centuries, but have become much drier in the past 150+ years. If left alone, many would be dry most of the year, and all these would be liable to tree invasion, or Tall Herb spread. If wet, they could develop into bog (Chapter 1). It is the regular, seasonal flooding (season varies!) that keeps the reed potentially healthy, and the regular clearance, not just of reed, but also of any invading woody growth, that keeps reed actually dominant and harvestable. Water regime may be controlled by dykes and sluices, or it may be left to irregular winter floods. A reedbed in poor condition may be curable by one or two years harvesting or, if not, by a late winter or early spring burn (before the reeds are *c.* 30 cm high, Chapter 10). Both treatments encourage increased density, by bud stimulation by temperature fluctuations and the formation of even-sized buds (Chapter 5). Oddly, straightness is also encouraged. As the beds are old and drying, the substrate has been firmed, and access is easy, unlike many wetter lake reedbeds, where the soft oozy peat or mud means both harvesting and reed removal is difficult.

The traditional marshmen, now unfortunately extinct, could produce any water and burn regime that was needed, whether for good thatching reed or for research. Reed could be burned over a large or small (e.g., 20 m x 10 m) area, with a 1–2 m border, and, at request, at *c.* 20 cm above ground, *c.* 5 cm above ground, or scorching the peat itself.

### Drying flooded land

This use of reedbeds is sporadic, the main example being the new polders in the erstwhile Zuider Zee, cut off from the sea as the IJsselmeer. After the Second World War, more land was wanted, and these large polders were formed, and pumped. This left land still too wet for cultivation. Reed seed was scattered by aeroplane, the seeds – in good conditions of stable wet mineral soil (Chapter 2) – germinated and did well. *Phragmites* has a very high transpiration rate (Chapter 5) and, with incoming run-off cut off, the polders dried. The *Phragmites* was eliminated, and crops and farms could follow (Bakker *et al.*, 1960). This was all done well and safely.

The differences to a natural succession should be noted. *Phragmites* reedswamp invades shallowing water, as water leaves with changes in sea level or build-up of peat. The polders water was pumped out. *Phragmites* was placed by people. It only dried wet soil by evapotranspirations. When in a natural bed, succession is to woodland or bog, in the polder, manmade disturbance allows crops to grow on the former reedbed.

### Conservation

(Also see Chapters 11 and 13, and incidental commentary elsewhere.)

Conservation can perhaps be said to be a human use. This is both the conservation of existing and traditional reedbeds, whether or not harvested or used for any other purpose (e.g., recreation, fuel), and the creation of new ones.

Reedbeds can be used for boating (manual, sail, power), fishing, nature study, photography (art, etc.), picnicking (usually beside!), rambling beside and walking within, recreation, shooting (hunting), swimming and paddling (open pools), viewing the scene (and see Chapter 10) (Bahrman & Gorenflot, 1983; Bakker *et al.*, 1960; Brookhouse, 1998; Haslam *et al.*, 1998; Haslam, 1969; Kiviat & Hamilton, 2001; Ksenofontova, 1989; Květ, 1973; Letts, 1999; Thesiger, 1964).

There is no way *Phragmites* peat can be recreated, with its store of history and archaeology. However, a basic reedbed, with its fish, fowl and (non-rare) associated plants is one of the easiest communities to recreate. (Recreating a specific reedbed, with its inbred uniqueness of history, substrate, flora, fauna and use, is another, and near-impossible process.) Suitable water, enough nutrients, a nearby source of *Phragmites* (or transplants or seed sowing) and, hey presto, the reed bed is there within 2–6 years. Recreating bog, *Schoenus nigricans* or *Carex davalliana* communities is, at best, tedious.

The most important reedbeds to conserve are those:

- on *Phragmites* peat, with recent or future potential for more peat growth – reason: vegetation and history;
- where a sustainable use has been present for centuries, e.g., old commercial reedbeds for thatching – reason: use, history, vegetation;
- on lower nutrients (not newly-made!), since these have more rare associated species, and are more fragile – reason: vegetation, habitat;
- small areas within other habitats, such as sheltered bays in lakes, flushes in nutrient-poor vegetation, patches in wet woodland – reason: vegetation;
- places giving aesthetic experiences to visitors – reason: vegetation;
- areas manageable for the conservation of rare or valuable other species, e.g., the recent EU Bittern Project, which led to many new reedbeds in order to attract and keep bitterns (Chapter 11) – reason: wildlife
- areas with particularly valuable other native species, communities or genotypes (this does *not* mean planting alien, or even non-local, 'wildlife seed' to replace and destroy native biodiversity!) – reason: native species
- patches, constructed wetlands, etc., for purifying water (dirty water should not be led into beds significant for native features) – reason: cleaning water.

Obviously as much reedbed as possible should be preserved, conserved, restored and created!

(A note of 'vegetation' above includes the animals typically associated with that vegetation.)

Conservation of birds in England has a cyclical bad effect on *Phragmites*. As new waves of conservation officers come along, they are too sure that traditional management methods are wrong, as must be the research of their predecessors, despite the fact it supported innumerable birds. They believe the only way to save birds is to stop reedbed management. Since the long-standing harvestable beds can now only exist with management (see above, they need water, cutting and perhaps burning to survive), they deteriorate, until these conservation officers notice and do their research. Then all is well until the next wave are appointed.

### Recreation

Recreation usually comprises:

- **In the wetland**: viewing the scene, rambling, picnicking (camping, etc.), photography (recording, etc.), nature study (including bird watching), shooting (hunting).
- **Beside the wetland**: the activities done in the wetland and boating (manually, sail and power propelled), swimming, fishing.

The demand for recreation rises with the prosperity of the visiting population, and the amenity value and ease of access of the wetland.

Mutually compatible activities usually involve few people. They include nature study, fishing, viewing, rambling, picnicking, photography and painting. There are three main types of wetland likely to be used:

1. Neglected or untouched wetlands in accessible areas, or small ones with a potentially somewhat higher visitor density. Small wetlands are aesthetically pleasing, but most visitors remain on the dry land alongside. Invasion by trees should be prevented, because of the loss of habitat diversity, the fundamental change in physiognomy, and the development of acid peat under conifers, but no other management is required unless changes are impending in habitat or visitors.
2. Wetlands with traditional management for economic use. Limited recreation is satisfactory, though extra facilities may be desirable (see below).
3. Nature reserves, which may also be (1) or (2) above. Their primary purpose is the conservation of all or part of the ecosystem, ideally all.

However, a reedbed managed for plants may permit destructive shooting. Eutrophication and flooding may be allowed to alter the natural vegetation of one managed for waterfowl. Visitor density varies with the aim of a reserve, e.g., a reedbed managed for *Phragmites* and moths tolerates more visitors than the same bed regarded as a bird sanctuary. Wardens and car parks are needed if visitors are frequent.

Management should provide the maximum educational and aesthetic benefit with the minimum damage. Nature trails suit both purposes and can provide essential instruction on the importance of wetlands. Specialist leaflets are recommended (e.g., elementary and advanced, botanical and entomological, activities for schools). Escorted tours provide more information and protection, and may be organised for general viewing (more glamorous by boat) or study. Unescorted visitors should be directed by ditches, wet soil, *Urtica dioica,* etc., rather than by 'No Entry' signs. A summer-dry path with little use needs mowing perhaps twice a year. With increased use, it must be widened or protected by boards, gravel, etc.. In wetter places paths may be raised on e.g., rubble, or constructed on stilts. Unescorted boats should be prohibited in reedswamps, because of mechanical damage (e.g., Sukopp, 1971). Escorted boat tours are advantageous in areas where substantial damage is possible.

Bird watching is often the main form of recreation. Birds and larger mammals tolerate less disturbance, and so need more protection, than other organisms. If visitors are numerous, it is advisable to have, e.g., hides (blinds, which also aid viewing), screened paths, and visitors limited to certain seasons or days. Occasional major disturbances are preferable to continual minor ones. High densities of visitors restrict birds to tame species such as ducks.

Angling should be restricted to specified sites and seasons, e.g., through club membership. Wetland fishing is mainly for recreation, as reedswamp and shallow water are unsuitable for large nets. It may, however, have the best fishing, as the fish, their eggs and their food supply are protected (see below). If the reedswamp surrounds a much-fished lake, it may be stocked. Access, and prevention of boat damage to reedswamp, must also be arranged. In more remote areas, wetland fishing may provide animal protein for the people living near.

Shooting is incompatible with other activities at any one time. Larger mammals (e.g., deer, fur-bearers) may be shot for sport or limitation, but large-scale harvesting is mainly trapping of fur-bearers. Management of populations, and of the proportion of open water, etc., improves both conservation and crop. Hides, screens and access are needed in more populous areas.

The shooting of duck, geese, etc., is now mainly for recreation (although it sometimes forms an important source of animal protein). In some countries any birds may be shot (e.g., Malta). Where conservation laws are enforced, in remote parts and sanctuaries, shooting is negligible. The third extreme is when a small specified number of named species may be shot at specified times by licensed, brightly coloured hunters (e.g., parts of North America). Intermediates, of course, occur.

Pheasants, whether or not specially reared for sport, may depend on wetlands for shelter in winter and cover for nesting. Grazing or other removal of standing dead vegetation decreases nesting. Wetland drainage may decrease the crop of pheasants.

Waterfowl live not only on reedbeds and other wetlands but also on adjacent open water and dry land. Many are migrants, and are thus targets for wildfowlers in many different places. Hence conservation is needed for each flyway as a whole. Coordination is easier if only one or two countries are concerned, e.g., North America. Elsewhere, a network of sanctuaries is recommended along each flyway.

In spite of much drainage in North America, particularly in the breeding areas, waterfowl populations remain adequate for preservation, photography, etc.. Management is required to produce a surplus for shooting. This includes preserving small nesting wetlands in the Prairies, etc., keeping them from drainage, infilling and agriculture. The Canadian Wildlife Service and Ducks Unlimited are mainly responsible. Diversity is, however, decreased as their management aims at shallow-flooded, sparse, eutrophic reedswamp around open water.

Activities involving many people and much disturbance include boating, camping, and any other activity which happens to have many adherents. Sites are usually in populous societies. They usually have open water; the reedbed is inessential, but adds to the scenic value (Haslam *et al*., 1998).

### CARBON STORAGE AND SUBSEQUENT LOSS

Peat is made up of plants (almost entirely). Plants make and store carbohydrates, so peat stores huge amounts of carbon. When the peat is destroyed by burning, drainage and conversion to arable and oxidation or degradation in gardens, the carbon is released. Together these give 3,000 megatonnes of carbon dioxide emissions each year (Joosten, 2007). Only part of this, of course, is from *Phragmites*, but it highlights that cultivating 'biofuel' species on reed or any other peat is putting more, not less carbon dioxide into the atmosphere. On mineral soils, it is all right. Greenhouse gas emissions are greater from nutrient-rich peats, of which *Phragmites* is one (Couwenberg, 2007). Using UN, etc., data for traded carbon, preventing the present drainage of peats world-wide, would be worth 29 billion euros annually.

However, it is possible to grow plant species that:

1. thrive under wet conditions;
2. produce biomass of commercial use, in purpose, quantity and quality;
3. build more peat.

*Phragmites* is one of the most important for (1) and (3) above. A reedbed may be a source or a sink of carbon (Brix *et al*., 2001). With a productivity of 3.6–43.5 tonnes dry weight per hectare per year, it certainly produces the biomass. It equally certainly gives rise to expectations of large-scale commercial use for biofuel, and small-scale for thatching and reedboard (Wichtmann & Joosten, 2007). However, such expectations in the past have not come to pass because of technical or other economic costs (e.g., haulage costs, product less saleable than expected). We may hope that this time, for the sake of the planet as well as conservation, that it does so. For conservation, of course, reedbeds must not be planted where other, and rare and valuable, wetland communities should be growing.

## COMMENTS

For a relatively minor, though especially widespread plant, *Phragmites* has a quite exceptional number of uses. Interestingly and unusually, as the old uses fade out, e.g., pens, baskets, arrows, which have been replaced by new technology, other uses come along for which, as the others in earlier times, *Phragmites* is the best available plant. Thatching (Britain) is if anything increasing, and as reed is now imported from all over Europe and beyond, some reedbeds in these countries are conserved because this new use for them has been found, and profits made. Purification in constructed reedbeds, and in reedbeds planted for this particular reason are increasing at a great rate. This is a use not really considered in the past. (How far did communities putting waste in the nearest marsh realise it improved dirty water, and how far was it just a convenient dump?)

Regrettably, in the view of governments, *Phragmites* is much too unimportant for its research to deserve funding. The International Biological Programme of the late 1960s to early 1970s was the only major international scheme to encourage *Phragmites* research. Neither national nor international money is readily available. (The EU Bittern Project was to create habitat, not to research reedbeds.)

**Table 9.1** Some organic compounds taken up by the roots of land plants (from Shimp *et al.*, 1993)

| Substance | Species | Substance | Species |
|---|---|---|---|
| Sulphonamides | Broad bean | Polycyclic aromatics | Many species |
| Anthracene<br>Dichlobenil | Bush bean | 2, 3, 7, 8-Tetrachloride benzo(p)dioxin | Many species |
| Organochlorine insecticides | | Bromacene<br>Dichlorobenzonitrite<br>Phenol | Soybean, barley |
| Benzenes<br>Substituted benzenes | Soybean | Nitroquanicline | Soybean, fescue, brome grass |
| PCBs | Many species, e.g. Loosestrife | Nitrobenzene | Soybean, barley, poplar, Russian olive, green ash |
| Atrazine | Barley | | |
| DDT<br>Chlorobenzene<br>(*o*)-Methyl carbamal<br>Phenylurea | | | |
| Asulam<br>Bromacil | Maize, bush bean | Dieldrin<br>Chlordane<br>Heptachlor | Eight field crops |

**Table 9.2** Examples of chemical removal in constructed wetlands
(See e.g. Athie & Cerri, 1987; Hammer, 1989; Reddy & Smith, 1987; Rubec & Overend, 1987.)

| | % removed |
|---|---|
| Nitrogen | 25–97 |
| Biological Oxygen Demand (B.O.D.) | 55–90 |
| Suspended solids | 54–98 |
| Iron | 82–99 |
| Toluene | 99 |
| Chloroform | 32 |
| Benzene | 99 |
| Tetrachlorethylene | 25 |
| Phosphorus | 20–99 |
| Chemical Oxygen Demand (C.O.D.) | 85–95 |
| Manganese | 9–98 |
| Pathogens[1] | 86–99 |
| *p*-Xylene | 99 |

[1] Faecal streptococci remain the longest, so they, not *Escherichia coli* should be measured.

**Table 9.3** Selecting best management practices by pollutant: rules of thumb (Haslam, 2003, after Novotny & Olem, 1994)

| Pollutant | Methods of control | Vegetative |
|---|---|---|
| Sediment | Control erosion on land and stream bank<br>Use best management practices that capture sediment<br>Dispose of sediment properly | Cover crops and rotations<br>Buffer strips |
| Nutrient and miscellaneous effluents and run-offs<br>Animal waste | Minimise sources<br>Take into crop all that is applied to the land or contain and recycle/re-use | Crop rotations and management<br>Cover crop, buffer strip, change crop or grass species to one that is more nutrient demanding |
| Pathogens (bacteria, viruses, etc.) | Minimise source<br>Minimise movement so bacteria die<br>Treat water | Buffer strips<br>Constructed wetland/microbial filter |
| Metals | Control soil sources<br>Control added sources<br>Treat water | Crop/plant selection<br>Crop selection<br>Constructed wetland/microbial filter |
| Salts/salinity | Limit availability<br>Control loss | Crop selection, saline wetland buffer, land-use conversion |
| Pesticides and other toxins | Minimise sources<br>Minimise movement and discharge<br>Treat discharge water | Plant variety/crop selection<br>Buffer strip, wetland enhancement<br>Constructed wetland |
| Physical habitat alteration | Minimise disturbance within 30 m of water<br>Control erosion on land<br>Maintain or restore natural riparian area vegetation and hydrology | Buffer strips<br>Wetland enhancement |

# Chapter 10
# On the roof: thatch

## Introduction

Putting plant material on the top of a stone, wooden, wattle-and-daub, etc., house was a normal and usual thing to do, easier than putting heavy stone as a roof. The plant material used, though, was whatever was easily available, from heather and furze on the hill, to grain stems or flax on arable land, to reed and sedge for a wetland. Thatch lasts anything from one season to maybe 80 years. It could be used on the manor, church, cottage and all outbuildings. The quality of the thatch on the manor would differ to that on the peasant's pigsty!

Thatch can be set alight too easily by householders, careless with chimneys on fire, sparks from fires or flames from candles and lamps. When less flammable materials, tile and slate, became available (and easily available once railways came in the mid-nineteenth century) it was easier to ban thatch in towns, and, later, in other places considered a risk. This applies today.

Because the poor were stuck with thatch when the rich could afford tile, on most of the continent, thatch came to be despised, and was replaced. The culture of England, however, came to differ. There, thatch became the symbol of idyllic rural life (e.g., Moir & Letts, 1999) and thatched cottages became viable, many still remaining, and indeed new ones being built. Those familiar with descriptions in the literature of appalling conditions in such cottages should see Figure 10.1, showing it was not the thatch, but the maintenance which was at fault. Since 1790 the number of thatched houses has indeed dropped, but is still a feature of the countryside. Meanwhile the number of other thatched buildings, sheds, small manufactories, barns, stables and suchlike (Figure 10.2) has diminished to vanishing point. In 1790 these greatly outnumbered thatched houses but, not forming part of the country idyll, they have gone (Moir & Letts, 1999).

While the continent of Europe has effectively dropped thatch, the large area of China has not. There are many wetlands, so many sources of reed for thatch (C. Zhou, pers. comm.).

Reed, from Neolithic times until now, in England has remained at about 20% of thatching material (less 1700–1900; more since 1950; Letts, 1999). Until twentieth-century lorries, local material was used: heather in the Scottish Highlands; reed near the Norfolk Broads and other large reedbeds; and wheat (or other grain) in agricultural lands: the commonest of all. (In 1796, 20 miles (32 km) was the distance haulage costs equalled thatching costs, though in fact reed was taken 40–50 miles (64–80 km), for use within Norfolk; Moir & Letts, 1999.)

Figure 10.3 shows a present-day cottage with thatch and various reed features in thatch. (The thatch of alternate layers of reed and mud has never been common in recorded times, or in historic remains, but did exist.) (This book is not a thatching manual. For that, see, e.g., Cooper, 1972, and Moir & Letts, 1999, and information from the various Master Thatchers Associations.) All nutrient and strength tables are gathered in Tables 6.6–6.9, to keep this difficult material together.

**Figure 10.1 (above)** The Victorian hovel and the model cottage (straw thatch) (Gray in Moir & Letts, 1997)

**Figure 10.2 (Left)** Thatched industrial building, late nineteenth century nail shop, Birmingham (Pope in Moir & Letts, 1997). Hardly first class thatching!

**Figure 10.3**

Thatch, England

a) Thatch in good conditions, netted against animal and wind damage. The pattern on the roof ridge is made with *Cladium mariscus*.

b) North-facing thatch with lichens and mosses. Later, slimy (blue-green) algae and fungi are likely to spread.

c) Ready for re-thatching.

d) Re-thatching.

## ANCIENT THATCHES AND THOSE MADE BY REEDBED-DWELLERS

The sophisticated culture and reed-use of the Marsh Arabs up to *c.* 1965 was shown in Figure 1.8. Thesiger (1964) considers this craft had been much the same for at least 5,000 years, which would be early Neolithic times. At the other extreme, reedbed-dwellers in the Danube delta in the 1970s lived a squalid life in squalid reed shacks. The quality of both should be remembered when considering past skills and lifestyles. Neither had homes which were intended to last – 20 years would be exceptionally long. This was not picturesque countryside.

Mediaeval thatches are the oldest commonly remaining for the study of methods and longevity. At that date (Figure 10.4) smoke from the fire stayed indoors, and blackened the inside of the thatch (described in, e.g., *The Canterbury Tales*). When re-thatching with reed, until very recently, the remaining reed – the under part – was left on. This means a time-capsule was left, from (perhaps) smoke-blackened reed to nineteenth-century thatch. These provide the thatching methods, type of reed (or other material) used, the pollen and other plant record, artefacts: of extreme value to archaeobotany and agricultural history (Letts, 1999).

**Figure 10.4**

A mediaeval hall with an open fire and smoke-blackened thatch on wattle, battens and water reed 'flecking' (Harris in Letts, 1999)

Regrettably, although in the 1920s expert opinion said reed should be re-thatched *over* existing thatch, in recent decades less experienced thatchers, who are not concerned with conservation value, may remove the whole historic thatch: and discard it without expert analysis (Letts, 1999).

So the evidence for thatching tradition, and past history – plant remains of unparalleled diversity – will be and is being destroyed. Useful and interesting ethnographic data is being lost. Specialists should be consulted whenever ancient thatch is at risk of being damaged or lost. While the English are busy destroying, the Japanese (also with a history of thatch) are busy artificially smoke-blackening to preserve historic or other valuable thatches (Letts, 1999). Smoke blackening preserves. It decreases rats, insects and microorganisms and maybe flammability. If still present, it is now usually under a metre or more of later thatching (Letts, 1999).

## In the bed: thatching reed

Any reed can be used for thatching, but not all reed will last 70 years, let alone 80.

Durable thatch is prepared with dry reed that has matured during autumn and early winter, the woody tissues (sclerenchyma and xylem) have hardened, the soft ones have died and emptied, and the food-rich sap has been transferred to the rhizomes. Soft, food-rich tissue does not last well!

In Europe, reeds become dormant in winter, losing their leaves, and becoming hard. In the far south, larger ones do not lose all their sap, but the hardening is at least as much as further north. South again, and in the tropics there is little or no dormancy, so thatching reed must be used with food in the reeds and even with soft green leaves on. This is therefore a short-lived thatch. Such is adequate where houses are small, and reeds nearby and free, so it is easy to replace the roof frequently. The same was done in Europe on some outbuildings – free, quick to replace. But that will not do for more sophisticated thatching.

European hardened dead reed has always been used (since thatching began). It is cut from large reedswamps and small ditches. With England the main importer, reed is cut from accessible large beds anywhere from Sweden to France, The Netherlands to Turkey, and exported. These are large beds and, because they are commercial, mechanical in operation. The reed is cut; if it is too large, it is cut high so the butt (base) is not used (see below), if it is a nuisance to collect, it is left for months on the bed in contact with all the decaying organisms of the soil and litter (Figure 10.5) and, with or without unpleasant pests or diseases, on to English roofs it goes.

Although occasional imports occurred in the early twentieth century, it was not until (in the 1970s) short-stemmed wheat varieties were used that it became a major industry. The stalks of these wheats are too short for thatching, so the supply of thatching straw nearly dried up, and if thatching was to continue, roofs had to be converted to a steeper pitch, and be thatched with reed. The English reedbeds are now too small and too few to meet this extreme increase in demand. It is also doubtful whether, when there were more, and larger, beds they could have met it: and the profit is less than that from arable, on land suitable (with drainage) for either.

**Figure 10.5** Winter-harvested bundled reed left on wet soil attracts decay, Camargue, France

This reed has been properly harvested, cleaned and bundled, and looks eminently suitable for sale. However, it has been left on the bed for several months, and may decay over-rapidly. It should be removed as soon as the bed is harvested, and the bundles be stored above ground or indoors.

England is unusual in that the Broads thatching reedbeds, or some of them, have been in existence for centuries. During that time, water level has dropped (sea level, build-up of peat, and, latterly, drainage). This means the peat or alluvium is consolidated, and most are firm, giving good access. (Elsewhere in the Broads there may be, or have recently been, soft substrate quaking fen – with a layer of water below, and hover [loose reedbed] – on the fringe, over water.) It also means the beds are fully dry enough to have become wood, bog or Tall Herb. The management – of harvesting (preventing invasion), and annual flooding (preventing bog and invasion) – has maintained the beds as reed.

Water regime is controlled by the beds being dyked, being criss-crossed every 100 m or so with ditches large enough (1.5+ m wide, and correspondingly deep) to take small barges. These, with sluices and, if necessary, pumps, controlled the water level, they allowed access and egress for the cutters and the reed bundles, and they kept some small water movement within the bed. This last improves reed growth, it is traditionally said. Of course, until the mid-twentieth century, labour was cheap compared with commodities, the opposite of the present position. Therefore marshmen were plentiful (maybe doubling as fishermen in summer), and sluices, dykes, boats, etc., had frequent attention and were kept in good repair. The floods and droughts now afflicting various beds were much mitigated when sluices had daily checks (and when agriculture was less intensive, soil preservation more important and development less, so that flash-flooding was less). (Also see McDougall, 1972.)

Commercial reed also comes from coastal beds in the Tay estuary (planted in Victorian times against bank erosion) (Ingram *et al.*, 1980) and around the Essex and south coasts, but East Anglia, Broadland and coastal (e.g., Cley, Dunwich, Walberswick), is the main source. Past managed coastal beds were along the coasts of Sussex, Romney Marsh, North Kent, East Anglia, Lincolnshire, and Lancashire; and inland ones in the Broads, Fenland, Somerset, and Trent areas. Natural reedy estuaries were mostly in the south. Locally, anywhere, there could be reed in ponds, wetlands and ditches. Quotes show the crop value was as good as arable. The Fenland reed was rapidly decreasing in the eighteenth century, and everywhere in the nineteenth and twentieth centuries (Moir & Letts, 1999).

Wicken Fen, inland near the edge of the south Fenland, has a commercial bed. The peat is not *Phragmites*, but sedge and organic mud: sited with calcium-rich water (once) running off the chalk, this was too calcium-dominated for anything but sparse (subordinate) reed. However, there is the reedbed! During the 1939–1945 World War this was turned over to arable, ploughed (disturbed, releasing stored nutrients) and fertilised (raising nutrients more). So, when in the 1960s the area became National Trust land, reed invaded at great speed, advancing by *c.* 10 m a year. Temporarily, Old Buckenham Fen – low-medium nutrients (Tallantire, 1954) – was fertilised for reed.

Usually, now, reed bundles are just that, reed (Figure 10.5). In the past the reedbeds bore more of other species, particularly *Typha* spp., bulrushes, and some sedges (*Carex* spp. or *Cladium mariscus*). Up to a quarter of non-*Phragmites* was acceptable. Indeed, some *Typha* was considered an advantage, since the leaves are 'spongy' so air can pass freely, and aeration preserves thatch (see below). In the interests of neatness this is now unwanted, and other species are either eliminated on the bed, or 'cleaned' out of the bundle before sale. Short 'messy' material, reed or other, was no doubt usually removed in the past as well. However, sheds and suchlike might well have been thatched with mixed reed, if that was what was available locally, and replacement was cheap. Efficiency before appearance.

When considering whether a reedbed is suitable for commercial harvesting now, there are several questions to be asked.

*Is there a large enough area to justify the expense of the cutter, the harvester, and haulage?* The answer varies not just with actual size, but with the nearness of other beds, the trader or the thatcher buying direct (and whether the cutter is doing so free for his own sheds, or it is a commercial enterprise).

*Is the reed dense enough to give a good crop? And, if not, can it be made so?* If it is sparse because nutrients are too low (Chapter 6) then in the interest of conservation it should not be used, such habitats being valuable and increasingly sparse (with disturbance and pollution). The same applies if the other vegetation is particularly important, e.g., tussock sedge; or of historic significance, e.g., with millennium-old records. Otherwise, good management will produce good reed. A stable water level, with flooding for a few months in the year at the same season (and not when harvesters need access), minimum aeration (from ditches or dykes), removal of woody growth, harvesting for 1–3 consecutive years (variable); and burning in late winter, above ground level, if these others will not do it (see Chapter 5). The aim is to produce many below-ground buds at emergence, and to have early emergence. This comes from good water regime, and temperature fluctuations, and leads to dense early growth which both gives a uniform and dense reed (many buds are more uniform, Chapters 3 and 4) and an early canopy which can shade out most competitors before they themselves can grow well. Harvesting clears the bed, so the upper rhizomes have no insulation from temperature variations and spring frosts. This (Chapter 4) leads to more, and more uniform buds. A shock of some kind, usually a (permanent) change of water regime or burning, is an effective treatment for many ills of an established bed, also.

*Is there access or can access be provided simply?* Access is now wanted by land, therefore not on deeply flooded land, and not by boat. Also the substrate must be firm enough to take machinery, preferably not just a modified alanscythe but a heavy harvester, both cutting and bundling. Of course a heavy machine squashes rhizomes below, and so next year's reed is shorter, sparser and maybe bent in the affected parts: the bed is uneven and patchy.

*Does the reed appear to be such as to make good bundles? Is it straight?* Reed can be very straight, or can bend at its nodes, technically termed dog-legged. Dog-legged reed is useful only for some filling, not for the main coat on the roof. Dog-leggedness may be: (a) strong winds during growth. This is usually sporadic, and creating a shelter belt is rarely justified; (b) is 'wild'. This is the term used for unmanaged reed, with an untidy 'look' to the bed, not the neat uniformity of the commercial reedbed. This may be remedied by harvesting, and if not, it will be by burning; (c) it may be a strongly clonal, genetic character that will persist and is incurable. Fortunately this is rare and usually found in small patches of single clones which can be ignored.

*Does it make up into usable bundles?* (As a traditional craft the measurements are in imperial, not metric, so look odd in the latter.) Cut reed is tied in bunches or bundles, between 23 cm and 38 cm (9–15 in) from the butt and at least 60 cm (24 in) round the point of the tie.

The bundle should be fairly even, not with reeds of all heights all over the place. Looking at the main length, 91–120 cm (3–4 ft) is Short; 120–170 cm (4–5 ft 6 in), Medium; and 170–200 cm (5 ft 6 in–6 ft 6 in), Long. That should be around 180 cm. 200 cm+ reed should be notified to the buyer. The butt is the hardest and toughest part of the reed, and therefore as much as possible of it should be in the bundle, on the roof. When very tall (2.5–6 m) reed is being cut, mainly on the continent, it is often cut well up the reed, so none of the butt is present. This should be checked. Harvesting at 10–30 cm from the ground is acceptable.

For a more detailed assessment, with possible management for various 'faults' in thatching reed, see Table 10.1. Much of this is 'gilding the lily'. All reasonable quality British reed can be sold for thatching – whether for the main roof or, for the badly-sized or shaped, for filling. So changing it is rarely necessary. If reed should again be in surplus, and experienced thatchers be able to pick and choose, then minor qualities may again become important.

*Single and double wale?* Commercial reedbeds are usually cut every year, in East Anglian terms, single wale or weyl. Given that the Danes dug out the Broads, it is very possible 'wale' derives from Old Norse, just possible from seventeenth-century Dutch. Double wale, cut alternate years, is spoken about more for East Anglia than from other countries, and indeed, other parts of Britain.

Cutting alternative years is less of a drain on the nutrient reserves of the rhizomes. By the second winter all movable nutrients are in the rhizome, and, more importantly, the protection of the first year's dead leaves lessens temperature fluctuations, thus buds, thus nutrients. Reedbeds on good alluvial mud are strong with ample minerals, and single wale cutting for indefinite periods works well. On peat with lower nutrients, however, double wale has been traditional, elderly marshmen of a half-century ago explaining the bed would not take single wale.

This traditional wisdom has been vindicated in one such reedbed. It was said, with the authority of centuries, the bed would not tolerate single wale. The farmer checked: and after three years the reed yield was down. After a change in ownership, single wale was decreed. At first, the reed crops were fine. But the reed became thin, and double wale was recommended. Tradition was correct!

This could also apply in the smaller nutrient-low reedbeds of the Highlands and Welsh mountains, but records are missing.

The bundle may be tapered. Too much taper can contribute to slippage on some roofs. It may be cylindrical, straight, which is often wanted for large areas of roof and slacker pitches. Thirdly it may be big-topped (apparently widening at the top, actually with some bending above), which is – being loose – often wanted in difficult drying conditions, especially when it is long and coarse, which also helps aeration and drying.

The bundle should have no more than 10% of non-reed, and that should not alter performance. Short reed and non-reed should be dressed, cleaned before sale (P.W. Brockett, British Reed Growers Association).

*Can it be cut when mature?* The leaves should have dropped, and the reed itself hardened ('rattles when shaken on the bed'), which happens in January and, usually in hard winters, late December. Reed cut in November is unsatisfactory, so if some practicality (of flood or expense) means winter cutting is impossible, the bed should not be harvested. Cutting can continue until the young shoots, in spring, are tall enough to be removed by the harvesting. This varies with weather, climate and reedbed, but is probably April.

*Is the diameter appropriate for the use?* Generally, 5.6 mm reeds and over are coarse, and while drying well are not as showy. Medium reeds, 4.8 mm, are all-purpose, and narrow ones, 3.2 mm, look good but do not dry as well.

*Is it bright and slightly golden? Or dull or pale and straw-like?* Colour itself does not, it is thought, affect performance, but a sheen or shiny wax surface helps repel breakdown organisms.

*Is it strong, is it difficult to break?* (Also see below.) Brackish water reed is often hard but brittle. Inland reed varies from being tough and fibrous to crumbling in the palm of the hand (P.W. Brockett, British Reed Growers Association).

*Is there dry storage space with lorry access where the bundles can be stacked until collection?* These stacks should be loose (for drying) and, if to be left for over a few weeks, be themselves thatched or covered (to repel rain but to allow air movement).

The quantity of reed, the yield, is also important! Dominance of reed ensures an adequate harvested crop, but not the maximum possible. Improving water management (water level, some water movement) can double yield – in a non-managed bed (e.g., Rudescu *et al.*, 1965, though burning was used also). Nutrients (as discussed elsewhere) are either adequate or the bed should not be used for reed, so are not a problem. Weather alters yield, but is not itself alterable. Good amounts of water on the ground and plenty of hot, sunny weather particularly up to July, give the highest yields. Flooding of stubble before spring when next year's shoots are above water level does lower yield. With stubble under water, oxygen cannot go down the cut stems into the rhizome and root system. This lack of aeration is harmful, and can usually be avoided. If water level is controlled, it can be kept below the stubble. If uncontrolled, late harvest and taller stubble (30 cm or even more) decrease flood effects.

However, management should aim at increasing density, not (over *c.* 1.8 m) height. Over-long reeds are not good for thatching, and if cut high on the bed mean the base of the reed is softer, so wearing worse, than the butt.

## Off the bed: storage

Whatever the state of the reed in the reedbed, it can be ruined by bad storage in a dealer's premises! Something often forgotten when the reedbed or the thatcher is blamed for early breakdown. Leaving reed exposed to bright sunshine all summer starts the brittle-type decay that will continue on the roof. Similarly, leaving it wet will start microbial breakdown. Stacks should not touch the ground, but have at least 7–10 cm of space, both to allow air to circulate, and to prevent infection from soil below. Reeds should be stored for at least six months, as once off the bed the reed dries (by some 5%) and shrinks, and changes in strength. It is better these alter before being fixed to the roof.

With proper treatment in stack, there is a loss, in eight months, of hard components (Rudescu *et al.*, 1965, found 4–7% of lignin, 2–3% of ash in cellulose) and an increase of softer ones (e.g., 1–3% of cellulose). Poor conditions lead to more decay (e.g., 70% in weight in 6 months; Wazny & Wytwer, 1963).

## On the roof: thatching

Thatching methods in general, and for reed, have altered down the centuries: a) with purpose (church or shed?); b) with material (10% non-reed now, 25% earlier); and reed now harvested only for thatching, in Britain, instead of for many uses (Chapter 9). (This means putting material on the roof now which is very-weak, -thin, -tall, etc.. Formerly this could be used elsewhere); and c) with fashion (having ridges and gables nicely patterned with fen sedge, *Cladium mariscus*, is a fashion which started in Norfolk, for instance) (see Figure 10.6, flecking).

Reed roofs should have a high pitch (angle) of at least 50°. This is steeper than is needed for straw, and when those roofs have to be re-thatched with reed, they need to be fitted with tilting fillets to create the necessary tension in the newly-laid course. Otherwise the reed is liable to damp and decay. Reed is not just a variety of straw (Figure 10.7)!

Rain only wets the upper few centimetres of the thatch, which is also the part most easily dried. Damp penetrates further in, and on north-facing slopes with no, e.g., drying winds, may persist for months, or even all year.

Generally, thatch breaks down in patches, and in the past was patched, with reasonable luck not until after 50 years. Local breakdown can be from accidents; from damage by birds or rats (particularly if the roof was not netted); by the chimney if there is a fire below and the reed crumbles from the heat; in other very sunny places and, conversely, in particularly shady damp parts where microorganisms flourish inordinately. Moss should be removed from the damper slopes, as this absorbs and stores rainwater, adding to the damp (see Anthony, 2000; Kirby & Rayner, 1988 and 1989).

**Figure 10.6 (left)**
Reed flecking (Moir & Letts, 1997)
Top – Post mediaeval.
Middle – Late eighteenth century.
Bottom – Early nineteenth century.

**Figure 10.7 (right)**
The difference on the roof between reed and wheat straw (Moir & Letts, 1997)
Top – Combed wheat.
Middle – Reed.
Bottom – Long straw.

Now, in the interests of neatness and showing off wealth, patching is less likely than stripping off all, including the still-good base layers, and so destroying valuable heritage (see above). These base layers, a century and more old, often have reed stronger and better quality than that about to be applied. It is, though, accepted that longevity is increased by regular cleaning, re-dressing and repair every 7–10 years.

Thatching is done by laying bundles with their butt downwards and outwards, exposed to weathering and animal and microbial attack. The reeds are laid at least 30 cm thick. The butt is, as noted above, the hardest part. So wearing it away is slow. Once the butt has gone, the middle part of the reed is exposed. It is less strong, so wearing away becomes faster. The hardness of the butt may extend only *c.* 10 cm up, or this may be nearer 50 cm. The longer the hardness, the better the roof. Reeds may lose up to 60 cm during the life of the thatch: showing the importance of a hard base to start with. (See Cooper, 1972; Letts, 1999; Moir & Letts, 1999; and Master Thatchers Associations.)

## Thatch decay

All thatch decays and eventually needs replacing. However, some thatches last better than others. However skilled the thatcher, poor reed already decaying in storage will decay earlier on the roof; and *vice versa.*

The reed must be matched to the roof. The most easily-drying reed is coarse, long and with a good proportion of *Typha* sp. This may not suit current fashion, but it is the best for wet places or high-rainfall areas. The reed retaining most moisture is narrow and fine, settling in close together, so air does not move freely.

The east of England has less rainfall than the west. Flat open areas, given the same wind and rain, dry quicker than narrow lush valleys. North-facing slopes are damper than south-facing ones. South-facing slopes receive more sunlight including ultra-violet. This must be taken into account when choosing the type of reed for the thatch.

Driving rain and gales obviously can cause damage, and where these are common, thatch will decay faster, other factors being equal. Reed already weakened by any of these factors, whether weak to start with, subjected to heavy storms, or a plague of rats, is obviously more likely to decay faster when other things also are wrong. Decay is from a *combination* of adverse factors.

Standard management of water regime, harvesting and perhaps burning will produce reeds of suitable height, density and uniformity. Standard practices of storage and thatching produce a well-thatched roof. The variable which is not known and standardised is the strength of the reed in the first instance. And all the rest depends on that! A reed already crumbly will suffer more in storage, with an incompetent thatcher and on a very damp or ultra-violet lit roof. Reed decay *is* it becoming crumbly, actually crumbling, and vanishing.

Reeds can be influenced during growth by the habitat both above and below ground. Their predecessor buds are controlled by the horizontal rhizome, in turn fed by the shoots of the previous year. The opportunities for influencing factors are thus many! In midsummer, young reeds, as would be expected, easily bend in a greenstick fracture (brittleness), are easily squashed (hardness) and have much twisting strength (fibrousness). By August hardness and non-greenstick breaking have developed and fibres are full-grown (Rudescu *et al.*, 1965). However, in September, future strong shoots are stronger than in August, but those that will develop weak, are weaker than in August.

Young shoots, therefore, seem strong, then some weaken and others strengthen. Odd! Fibres?

Very poor reed may be weak or waste. The two differ. Waste reed is that discarded from the bunch by the cutter, cut, or combed ('cleaned') from the bundle. Waste reeds are usually short, making the bundle untidy, and are usually weak. It is never blamed by thatchers, as it never reaches them. The proportion ranges from almost none, to about a third. It varies with the reedbed. A crop with many short shoots loses more as waste than one with few. A crop with much crumbly reed, crumbling even on being combed, loses more than a strong one. Secondly, it varies with the view of the cutter, what he thinks suitable for sale. No one knows its effect on roof longevity, but excess short reed in a bundle is awkward for thatching.

When strong and weak clones grow in a mosaic pattern in a bed, having strong reed is irrelevant to competition. Vigour, bud density, etc., are important, and determine clone advance or retreat or a stable pattern. Strength is not. This may be the crucial argument for the irrelevance of reed strength to the plant: that it does not help in competition with other reeds.

The importance of the combined effect of microbial decay and ultraviolet light was also stressed by Bosman (1985). He considers that silica and lignin do protect, particularly when the outer stem layer has been peeled off (by ultra-violet light) and the support of the ground tissue has been lost.

There is therefore much that can be done to slow decay – choosing the appropriate reed, dealer and thatcher, and using simple measures like netting the thatch. This keeps off larger pests, but slows air circulation, so is not needed or suitable everywhere.

However, when a roof goes wrong the thatcher says it was not him, the dealer says likewise, so where else can the blame rest, except on the reedbed, which cannot speak?

## Reed strength

*If it crumbles in your hand,*
*it will crumble on the roof.* P.W. Brockett

From the outside to the inside, various layers of the reed protect it. First there is the wax and sheen on the cuticle, described above. This protects from microbial, not ultra-violet or rat damage. Next inside is the silica skeleton, sub-epidermal. After ashing, some reeds have a latticed silica cylinder, while others just have powder. Table 6.9 shows a good skeleton may be one factor in strength, as this is usually (not always) found in strong reed.

The sclerenchyma layer is also peripheral and its lignin is the principal factor in strength. Lignin is also found in the xylem, but there is less of this, and it is less uniform than that in the sclerenchyma.

Reed walls vary in thickness. Those comprising over 50% of butt diameter (other factors being equal) give a strong reed, under 15%, a weak one. Within one area the width of the sclerenchyma band has been shown to reflect strength (e.g., Sukopp & Markstein, 1981), but this does not apply between areas. Wall width, relative wall width, sclerenchyma band width and the length, width and wall thickness of the fibres and other lignified elements are not correlated with strength, nor with a wide range of engineering measurements. Ostendorp (1995 and 1999), Binz (no date), and Binz-Reist (1989) all carried out complex mechanical tests, but only to see whether the reed will break on the bed. Endangerment to the stand could be predicted from the habit factors plus the mechanical ones. However, strength standing up in the reedbed is not the same as resistance to decay on the roof!

The cause(s) of strength and durability are therefore uncertain, though logic suggests much lignin, and a sheath of wax and silica are all part of it.

Whatever the strength of the reed while on the bed, it will change during the next few months of storage. Unfortunately the change may be in either direction! In the past few years reed from some Norfolk reedbeds have been consistently tougher and sharper on the bed; in fact seeming really good thatching reed, only to become weak afterwards. Conversely, weak reed can become fair (though not really good) during drying.

The relevant strength factor is, in fact, the susceptibility to breaks, fraying and microbial decay when on the roof, where the reed usually bears no load except that of the thatch itself. That is, the reed is on the roof, and the relevant strength is whether it crumbles while doing so, which is different to engineering strengths.

CoSIRA (the former Council of Small Industries in Rural Areas) used to assess three elements of strength, all manually measured. Hardness is how difficult it is to squash the reed, brittleness, to break it, and fibrousness, to twist it until it breaks. The last is the most important (Figure 10.8).

No bed is ever entirely strong, as individual reeds always vary. But some beds may be really strong, others, really weak (Tables 6.6–6.9 and 10.2–10.4). The better the bed, the more strong the reed, while there are always some weak shoots present, if for no other reason than because these were late-emerging 'summer' shoots which had not matured when temperatures dropped in autumn.

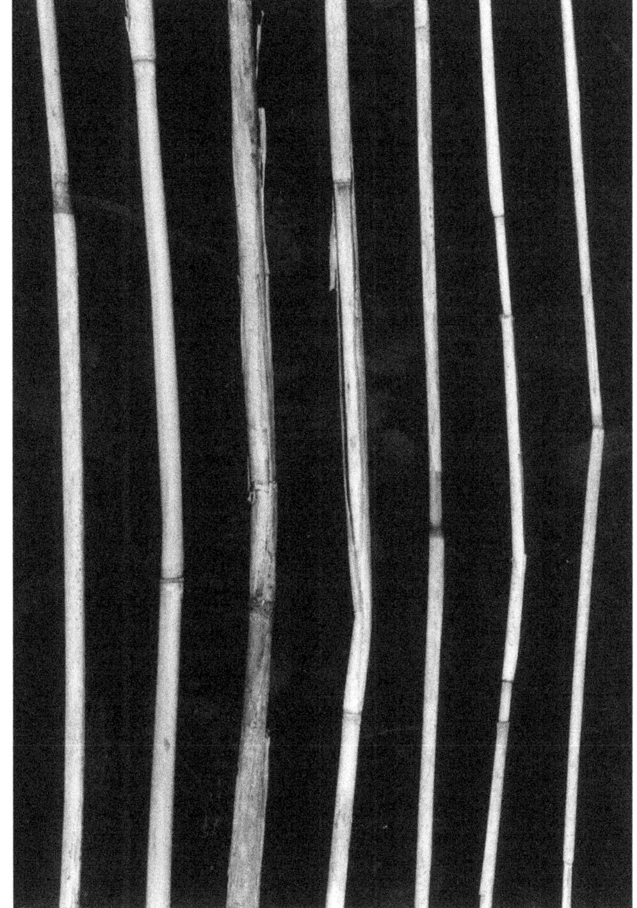

**Figure 10.8** Different types of thatching reed.

These are all fairly straight, as thatching reed should be. They are all of suitable width, but within that, vary from fine to coarse. In one (Scotland) the sheaths are loose, making the reed look larger than it is, and untidy. The two at the left are probably the best quality. The sheaths are light, and come nearly up to the next node. It is probable they are strong: but sometimes such reed is an unpleasant surprise, being weak. Those on the right either are narrow, slightly dog-legged or have longer distances between sheath tops and nodes and are probably weaker (subject to the same caveat). (In fact, appearances are correct.)

As seen from the tables, weak reed seems the 'default position' which all reeds can attain. While some reedbeds can, in some or most years, be strong, others never have been, in two decades. Strength goes up and down in the stronger beds, but trying to correlate this with environmental variables has, in twenty-five years, proved fruitless. This is primarily because there has never been funds for DNA testing, the obvious first step. That would show the genetic variation within and between beds, and see whether strong reeds are genetically different: and if they are allied. Does strength differ between beds with similar genotypes? Due to history? As noted above, managed reed is more uniform, dense and straighter than the bed's wild reed before. This managed type can persist for anything from a few years to at least 80 years after the last harvest. It can indeed return in a bed dominated by Tall Herb between the harvested and the later reedbed (same clones). Transplants check these. Therefore a bed can be profoundly influenced by events further back than memory goes. Finally, there is environmental variation, sun, temperature, rainfall pattern, flood, frost, etc.. For instance, sun and hot weather do make reed taller, and unusual dry conditions make it shorter, and so on (see Chapters 3–8). Strength, however, is unaffected by all these.

Reed grows for its own benefit, not for the benefit of the thatcher. This is important, and often forgotten. *Phragmites* is a wild plant. There has never been plant breeding, never any human alteration or adaptation for thatching or other use (apart from the simple management described above). *Phragmites* has no need of extra strength in the reed. Competition is through the growing shoots (and the roots), and as long as the reeds stay upright over the winter so that the rhizomes are aerated, they have fulfilled their purpose and can fall. In parts of the Camargue, they do exactly that. Most dead reed has fallen by May. At the other extreme is, e.g., Hickling Broad, Norfolk, where, under a previous regime, brackish-water reed could be cut every third year (triple wale), that is, after three winters it was still hard as well as upright.

Strength is useful to the thatcher, not to the plant.

## Nutrients

It is often thought, and indeed printed, that high nitrogen or other nutrients makes reed soft (low carbon : nitrogen ratio). In fact this is not the case. As already noted, reeds can dominate in wetlands unless nutrients are really short (see Chapters 4 and 6), so in a wide range of nutrient regime.

Reed strength, likewise, is independent of its nutrient content (Tables 6.6–6.9) (strength is not correlated with nutrient ratios, either). There is, though, a good correlation (always better than .001 significance) between strength and silica skeleton, with the size and complexity of the skeleton, not with the amount of ash. The skeleton may help the reed to resist decay, but it is not the only, or the principal cause.

The most nutrient-rich reeds analysed were from the Mediterranean, where large shoots live through, at least, their first winter (Chapter 3). Therefore food remains in the reed, rather than all possible food being moved to the rhizomes. Even so, there is no association between strength and nutrient content (though on only two sites).

The thatch samples include strong reed with a low nutrient content, reeds from thatch breaking down after only five years, strong, century-old reed with high nutrients, and up to 400 year-old reeds which, like the last, had especially high nutrient contents (Table 6.9). So much for high nitrogen leading to early breakdown!

There is no evidence that nutrient concentrations in thatching reed have increased since *c*. 1950 when fertilisers spread. There is partial evidence that they have decreased since the 1960s. The older thatch samples have rather higher levels, but there are too few samples to prove this. The only mature 1980s reeds with nutrient levels as high as those are from a much-fertilised Czech fish pond, much-fertilised fen dykes, and Mediterranean still-living samples. Comparable with these are shoots frost-killed while immature and still containing food.

Phosphorus levels are low in reed from phosphorus-deficient habitats, both fen and bog, where the stand is visibly nutrient-deficient (Chapter 6). In contrast, *Phragmites* can grow well and strong in chemical factory and sewage works effluent (though Professor Kickuth did check all over (the former West) Germany to find the clones best for cleaning effluent, and chose the most effective for this purpose).

Rhizome nutrient levels are even less affected by the habitat around.

Therefore, nutrient concentrations in reed relate to habitat only in extremes. In the main range of the dominant reedbeds, nutrient levels in habitat and reed are irrelevant to strength (see Chapter 4 for relevance to performance). Extremes include traumatised reed and otherwise unsatisfactory habitat. (See Bornkamm & Raghi-Atri, 1986; Klötzli, 1973; Raghi-Atri & Bornkamm, 1980; and Sukopp & Markstein, 1981, who, in deteriorating stands, find weak reed contains high nutrients.)

Clonal variation in strength undoubtedly exists, in those beds with discrete clones (Chapter 5), but this again does not help to increase strength in the weaker clones within a bed.

Strength varies annually.

Poor reed is weak. Unlike waste, where shortness is the crucial factor, here weakness is the primary character, though short, weak reed will be discarded as waste.

Poor reed may be due to:

- genotype;
- annual variation;
- many summer shoots (late shoots), both wide and narrow. They often die in summer or are frost-killed before they mature. They are usually waste. More

emerge in warm weather and when general emergence is delayed or prolonged by frost, cutting, flood, drought, insect damage, etc.. The numbers thus depend on clone, habitat, weather and management;

- shoots from the back of the rhizome in less good stands, where the back buds, so shoots, are small, leading to weak shoots;
- other shoots from small buds rising high on rhizomes;
- trauma from drastic water regime change, disturbance, ploughing, over-grazing, over-mowing all lead to weak, sparse, short shoots. Recovery takes 1–4+ years after the trauma ends. Horizontal rhizomes, which control, live 3–6(–7+) years, so, after four years most rhizomes have not suffered from the trauma, and the reeds recover;
- habitat factors operative for years or decades in the past.

The conditions for strong reed are much easier to define:

- good strong genotype;
- good management (see above);
- plenty of large buds, few small ones;
- (no traumas).

## Discussion

People often say reedbeds are altered or deteriorating because of pollution, whether this be agrochemical run-off, or worse. Given that *Phragmites* is one of the best species for cleaning pollution, this is rather a contradiction in terms. It is highly improbable that any 'normal' pollution remains uncleaned 10 m into a good reedbed (unless in a flood spreading over), and this width is irrelevant to harvesting, especially when some is left uncut anyway, for animal habitat. Extreme pollution, raw sewage, raw chemical effluent, etc., may well lower performance (the literature is divided), but such are not found in ordinary commercial beds – reedbeds purify.

*Phragmites* is a perennial – and living – wild plant. It can be modified, but it cannot be controlled. Reedbeds can be given the best conditions, but weather (e.g., extreme flood) is unpredictable, and so is the appearance of the bulldozer to make an arable field (drainage), supermarket, or housing estate.

The core East Anglian reedbeds have been managed for centuries. Because of changes in policy as much as in economics, the expertise of good management has been allowed, nay encouraged, to die out (see specialised burning, above). This is typical of the loss of heritage now beloved by the English, while they talk big about conserving it. Other reedbeds have gone in and out of production, or have been recently created, e.g., the Tay Estuary. When new beds are created, the plans may (or may not) include making an income from reed harvesting. But they never include planting the bed with the strongest, most durable reed. Reed is just reed, it is all the same. This book shows otherwise!

Reedbed creation can be for pollution purification, the present most numerous constructions (Chapter 9), it can be under an international or national agreement which includes the provision of money, e.g., the EU Bittern Project (some of whose 'new' reedbeds are on long-drained ex-reedbeds); or it may be 'just' for conservation. The two planned new huge wetlands in the English Fenland are this, massive re-creations, and no doubt, at least for a century or two while the ex-agricultural soil is full of nutrients, there will be reedbeds. Very little mention is made of the erosion of the fen peat, that enormously rich and fertile soil which has been excellent for crops since drainage; but unfortunately is now running out. Drying and ploughing oxidise peat, which then flies: and adds to the carbon dioxide in the air. The Buttery Clay below much of this is far less fertile. What a happy coincidence that wetland restoration is planned!

While the English in particular like and enjoy thatched houses, that craft and heritage will survive (apart from the ancient thatch happily being destroyed, see above). Other countries may be pleased to find a 'new' use for the reedbeds whose former thatching use has been superseded, and export much to England. All this imported reed is needed only partly because English reedbeds were destroyed in, especially, the late nineteenth and early twentieth centuries. It is more because of the development of short-strawed wheat and other grain straw which is too short for thatching. Straw has for centuries been the main thatching material in agricultural lands, but now it has to be reed, for all-too-many houses (straw thatch is the historic thatch except near large reedbeds).

## COMMENTS

Science is not all. 'Planning' and politics are also important. The historic thatching material in non-wetland areas is straw, the long (especially wheat) straw from pre-1960s farming, where two crops were gathered, the grain and the thatching straw.

Reed thatch spread when there was no straw available. On historic houses this is non-traditional. Hence controversy: how can you keep wheat straw thatch when you cannot obtain any up to standard? Or: how can you desecrate a historic building or historic area by using a different material (which also uses a different, e.g., roof pitch)? That shocking practice must be outlawed.

And indeed some heritage bodies and various councils do 'forbid' reed thatch. Law suits are in progress (and probably will continue in progress for years).

Adding to this, is cost. A thatch costs much the same for the same house, but can be expected to last about 50 years with reed, only between 15 and 25 years with wheat. Once reed has been tried, how many owners want a 5-figure (in sterling) repair more often than necessary?

It is of course possible to grow enough old varieties of wheat to get full supplies of good thatching straw. But who is going to pay for and authorise the lower yield and different grain crop? Those demanding wheat thatch? A council has even demanded a thatch should look shaggy, like a badly-thatched nineteenth century hovel or outbuilding. That decays fast: it is pointed out above that frequent thatching with free materials and non-expert thatchers can be cost-effective and beneficial. Nowadays neither the free material nor the cheap thatcher is available, and the cost of frequent thatching, prohibitive.

**Table 10.1** Characters of reed unsatisfactory for (particularly) traditional thatchers, East Anglia (modified from Haslam, 2009)

These characters were unwanted by thatchers when reed was in surplus and choice was available.

[1] The treatments decrease height, but not for the same reason. The appropriate remedy should be studied and chosen.

[2] Similarly, here and to (y), investigation is needed to find the cause of the trouble.

| Character | Possible ways to correct the reed |
|---|---|
| a) Too long. Over 10% large (1 cm wide) reeds in bundle, and/or many reeds over 2.0 m[1] | Cut, or spin off, young shoots<br>Cut single wale<br>Expose to spring frost and temperature fluctuations<br>Burn in spring<br>Lower the water level<br>Decrease nutrients and pollution<br>Cut with long stubble |
| b) Too many short and thin reeds in bundle[2] | Burn in early spring<br>Correct the water regime<br>Do not expose to spring frost (e.g. flood in spring)<br>Cut double wale<br>Remove weeds<br>Increase flow |
| c) Reeds in bundle very uneven in height and width | Burn in spring<br>Cut young colts then flood to prevent frosting<br>Raise the water level<br>Remove the weeds<br>Stabilise the management |
| d) Feather very uneven spread over half of bundle | See (c) and (e) |
| e) Too much feather | Burn in spring<br>Decrease pollution<br>Lower the water level<br>Hope for bad weather<br>Cut off feather before sale: if with fruits, sell for stuffing?<br>Biotype unsatisfactory. Replacing it is too expensive, changing it (burning, flooding, drying, etc.) is of uncertain value except where stated. Usually therefore, the bed must be put up with |
| f) Very little feather | (Objected to only as one possible indicator of double wale or older reed) |
| g) Bundle tapering too much | (i) Too many short and thin reeds, see (b) above<br>(ii) Each reed tapering too much<br>Remove weeds<br>Raise the water level<br>Improve the nutrient status |
| h) Middle of bundle wider than butt | Cut single wale and spring-burn for several years<br>Biotype unsatisfactory. Replacing it is too expensive, changing it (burning, flooding, drying, etc.) is of uncertain value except where stated. Usually therefore, the bed must be put up with<br>Lower water level |
| i) Too short | Improve water regime<br>Improve nutrient status<br>Do not burn in spring<br>Do not expose to spring frost<br>Cut double wale<br>Stabilise management<br>Remove pollution<br>Remove weeds<br>Remove grazing animals<br>If sea-flooding in past, drain out salt<br>If liable to flood with brackish water, cut above likely maximum level |

**Table 10.1** Continued

| Character | Possible ways to correct the reed |
|---|---|
| j) Too sparse | Improve water regime<br>Improve nutrient status<br>Burn in spring<br>Cut single wale<br>Cut in March, not December<br>Expose to frost and temperature fluctuations<br>Stabilise the management<br>Remove weeds and litter<br>Remove pollution by poison<br>If sea-flooding in past, drain out salt<br>Remove reedbug<br>Remove grazing animals |
| k) Shrinks during storage or on roof | Do not cut while still green (time of death varies with frost, local protection from frost, and clone)<br>Biotype unsatisfactory. Replacing it is too expensive, changing it (burning, flooding, drying, etc.) is of uncertain value except where stated. Usually therefore, the bed must be put up with |
| l) Too soft when harvested | If reed immature in early autumn, correct this in subsequent years<br>If clone dies late, cut in March, not December<br>Biotype unsatisfactory. Replacing it is too expensive, changing it (burning, flooding, drying, etc.) is of uncertain value except where stated. Usually therefore, the bed must be put up with<br>Raise the water level<br>Cut single wale (if clone liable to rapid decay)<br>If sea-flooding in past, drain out salt<br>Remove pollution<br>Correct the nutrient balance of the soil |
| m) Becomes soft during storage or soon after thatching | Any of the above factors may apply<br>Dry bundles rapidly, do not allow to be wet and warm<br>Do not use old reed<br>If local bad decay, store in another place next year and protect from infection. If on roof, replace affected part<br>Remember clonal variation in hardness and in susceptibility to decay |
| n) Broken ends jagged | Reed decayed, see (m) above |
| o) Bottom of reed too soft, rest satisfactory | Do not flood in late summer or autumn<br>Treat for 'too soft' as above<br>Cut with short stubble to get base of reed (ordinary clone)<br>Cut with long stubble to avoid base of reed (butt unhealthy) |
| p) Roots on bottom of reed | Do not flood in late summer<br>Cut with longer stubble |
| q) Reedbug infection serious. Reeds 'branch' above ground | Burn in spring to kill eggs and larvae<br>Cut with short stubble and do not allow litter to accumulate<br>Biotype unsatisfactory. Replacing it is too expensive, changing it (burning, flooding, drying, etc.) is of uncertain value except where stated. Usually therefore, the bed must be put up with |
| r) Bottom blackened | Do not flood in late summer<br>If on single wale, sell to thatcher who prefers this<br>If on old reed, treat as 'too soft'<br>Cut with long stubble |
| s) Dog-legged, bent at knots | Cut single wale for several years, remove litter<br>Burn in spring<br>Biotype unsatisfactory. Replacing it is too expensive, changing it (burning, flooding, drying, etc.) is of uncertain value except where stated. Usually therefore, the bed must be put up with<br>Lower water level |
| t) Curling or bent | Erect, or alter position of, wind breaks<br>If due to severe frost in early winter, will probably straighten later<br>Do not knock over, or bend under wheels, when young<br>If flattened by storm: if rare, ignore; if frequent, treat as 'too soft' or cut double wale<br>If due to colts emerging in June or July, cut with longer stubble |

**Table 10.1** Continued

| Character | Possible ways to correct the reed |
|---|---|
| u) Reed very rough to the touch | (i) reed too coarse, treat for this<br>(ii) Biotype unsatisfactory. Replacing it is too expensive, changing it (burning, flooding, drying, etc.) is of uncertain value except where stated. Usually therefore, the bed must be put up with |
| v) Sheaths fraying while dead reed still young | Do not cut reed until dead<br>Do not cut reed which was immature in autumn (even if completely dead)<br>Biotype unsatisfactory. Replacing it is too expensive, changing it (burning, flooding, drying, etc.) is of uncertain value except where stated. Usually therefore, the bed must be put up with |
| w) Weeds serious | Raise water level<br>Change level of pollution<br>In late winter or spring, remove weeds and litter by burning (gives selective advantage to reed), or cutting (by hand or machine). Timing to depend on method used<br>Decrease flow of water if reedbed wet<br>If these do not work, take other measures suitable for the weed species concerned |
| x) Bush invasion serious | Kill the bushes by burning, cutting, spraying, etc., and remove the dead wood<br>For prevention: raise the water level, treat as for 'Weeds serious' (w) above |
| y) Wheeltracks with little or no reed | Use heavy vehicles only when water level is low and soil is hard |

**Table 10.2** Year-to-year variations in reed strength, East Anglian marshes

Sixty samples of 15 to 30 reeds were collected from each marsh. These samples were graded on a 5-point scale (Poor, Poor–Fair, Fair, Fair–Good and Good). For the Tables, when so many samples are averaged, intermediate categories between, e.g. Poor and Poor–Fair are used. For analyses, two extremes were recognised as useful, Very Poor and Very Good, which are dropped before publication).

The grading is on three characters, devised by the (former) Council of Small Industries in Rural Areas: (i) Hardness, easiness to squash; (ii) Lack of Brittleness, easiness to break; and (iii) Fibrousness, easiness to twist without breaking. While each sample contains reeds of a range of strengths, the overall strength is usually clear (and if not, is noted as 'mixed').

All beds have weak reed in some years. Some have better reed most years, or occasionally. The strongest reed varies with bed and year, with 1995 the year most often the strongest.

| Type of bed | Strength | | | | | | | |
|---|---|---|---|---|---|---|---|---|
| | Poor | | Poor–Fair | | Fair | | Fair–Good | | Good |
| 1. Brackish (so liable to be hard). Many different clones. 1986/7–2007/8 | 1 | 3 | 5 | 3 | 7 | – | – | 3 1995/6, 2000/1, 2007/8 | 2 1999/00, 2005/6 |
| 2. Slightly brackish, good reputation, Centuries, not millennia old, 1987/8–2007/8 | 1 | 3 | 1 | 3 | 4 | 4 | 2 1995/6, 1996/7 | 1 2005/6 | |
| 3. Ancient bed, much-used, very slightly brackish, 1986/7–2007/8 | 8 | 9 | 2 | 1 | 2 1990/1, 1992/3 | | | | |
| 4. As last, but inland, 1986/7–2007/8 | 9 | 12 | | | | | | | |
| 5. As last. 1986/7–2007/8 | 8 | 9 | 1 | – | 1 1996/7 | | | | |
| 6. Bird Reserve Clone. Strong before polluted-river flooding stopped, 1986/7–2007/8 | 5 | 2 | 3 | 4 | 3 | 2 | 1 1991/2 | – | 1 1995/6 |
| 7. Same Bird Reserve. Different clone. Up to dredging, 1986/7–1995/6 | 2 | 2 | 4 | – | – | – | 1 1995/6 | | |
| 8. As last, another Poor clone | 7 | – | 2 | 1 1995/6 | | | | | |
| 9. As last, but a clone after river flooding stopped, 1996/7–2007/8 | 4 | 3 | 3 | | | 1 1996/7 | | | |
| 10. As last, another clone | 7 | 3 | – | – | 1 2001/2 | | | | |
| 11. Ancient bed, very slightly brackish, good reputation, 1987/8–2007/8 | 2 | 3 | 5 | 2 | 4 | 2 1995/6, 1996/7, 1997/8 | 3 | | |
| 12. Well inland, bed about 40 years old, 1986/7–2007/8 | 4 | 7 | 3 | 3 | 4 | 1 1988/9 | 1 1990/1 | | |

**Table 10.3** Distribution of sample strengths within reedbeds

Strength was measured for each sample as described in Table 10.2. This Table shows the number of samples of each strength in each year tested. The samples were cut low on the bed (15–35 cm above ground), and were 30–40 cm long. Both ends of the sample were tested. The record is of the average.

| Strength | 1985/6 | 1986/7 | 1987/8 | 1988/9 | 1989/90 | 1990/1 | 1991/2 | 1992/3 | 1993/4 | 1994/5 | 1995/6 | 1996/7 | 1997/8 | 1998/9 | 1999/00 | 2000/1 | 2001/2 | 2002/3 | 2003/4 | 2004/5 | 2005/6 | 2006/7 | 2007/8 |
|---|---|---|---|---|---|---|---|---|---|---|---|---|---|---|---|---|---|---|---|---|---|---|---|
| *Bed A: centuries old, and had strong reed* | | | | | | | | | | | | | | | | | | | | | | | |
| Poor | – | – | – | **44** | – | – | **32** | – | **30** | 6 | 10 | 8 | **20** | 4 | **27** | – | – | 2 | **31** | **15** | 1 | 7 | 3 |
| Poor–Fair | – | – | **8** | 11 | – | – | 14 | – | 10 | **21** | 6 | 6 | 13 | 12 | 7 | – | – | **16** | 13 | 18 | 1 | 9 | 11 |
| Fair | – | – | **8** | 5 | – | – | 12 | – | 14 | 13 | 8 | 13 | 14 | **15** | 14 | – | – | **10** | 13 | 11 | 14 | **16** | **14** |
| Fair–Good | – | – | **6** | 8 | – | – | 12 | – | 5 | 13 | 18 | 10 | 4 | **10** | 8 | – | – | **14** | 2 | 1 | 15 | **10** | **15** |
| Good | – | – | 1 | – | – | – | – | – | 3 | 6 | **24** | **23** | 7 | **17** | 2 | – | – | **14** | 3 | 7 | **27** | **17** | **17** |
| *Bed B: certainly centuries, possibly over a millennium old, and had weak reed* | | | | | | | | | | | | | | | | | | | | | | | |
| Poor | – | **48** | – | **20** | **19** | **50** | **38** | **48** | **37** | **29** | **45** | 6 | **56** | **38** | **51** | – | **49** | – | **45** | **56** | **54** | **44** | **44** |
| Poor–Fair | – | 20 | – | **15** | **26** | 21 | 19 | 8 | 12 | 12 | 9 | 13 | 16 | 9 | 8 | – | 9 | – | 7 | 3 | 4 | 12 | 9 |
| Fair | – | 2 | 3 | **21** | 14 | 7 | 7 | 3 | 9 | 15 | 5 | **22** | 5 | 5 | 1 | – | 2 | – | 5 | 1 | 2 | 3 | 2 |
| Fair–Good | – | – | – | 4 | – | – | 2 | 1 | 2 | 3 | 1 | 12 | 1 | 5 | – | – | – | – | 1 | – | – | – | 3 |
| Good | – | – | 1 | – | – | – | – | – | – | – | – | 6 | – | 1 | – | – | – | – | – | – | – | – | – |
| *Bed C: relatively new bed, in between in strength* | | | | | | | | | | | | | | | | | | | | | | | |
| Poor | 7 | 14 | **44** | 1 | **31** | 2 | **35** | 5 | **18** | **43** | 3 | **54** | **32** | 12 | **14** | **29** | 7 | **36** | **32** | **53** | **35** | **35** | **16** |
| Poor–Fair | 17 | **27** | 27 | 6 | 13 | 9 | 20 | **14** | **16** | 12 | **20** | 6 | **24** | 9 | **17** | **16** | **15** | 11 | 17 | 5 | 14 | 13 | **19** |
| Fair | **30** | 10 | 18 | **19** | 14 | **26** | 5 | **19** | **19** | 3 | **28** | – | 4 | **19** | **16** | 13 | **17** | 9 | 10 | 1 | 6 | 7 | **17** |
| Fair–Good | 8 | 2 | 6 | **18** | 4 | **34** | – | **16** | 5 | 1 | 10 | – | 2 | 11 | 8 | 1 | **13** | 4 | 1 | 1 | 3 | 3 | 4 |
| Good | 1 | 1 | – | 5 | 1 | 1 | – | 6 | 1 | – | – | – | 1 | 9 | 4 | 2 | 7 | – | 1 | – | 2 | – | 1 |

**Table 10.4** Range of strengths in some East Anglian reedbeds

The samples were graded as in Table 10.2 and 10.3. This table shows the range in different beds. All but one bed has one or more years in which strength is Poor (and that one is only marginally stronger than Poor). However, while all beds also have at least one year stronger than Poor, some only reach a little stronger, others reach Good strength. It would seem Poor is the default strength, and strengths better than that vary with both reed genotype, and some environmental conditions.

The difference between Poor and Poor–Fair is counted as one grade, that between Poor, and half-way between Poor and Poor–Fair as half a grade, in the last column. Three of the beds have strongest and weakest separated by not over 1.5 grades, four, by 3.5–4 grades: a great difference.

The beds are numbered as in Table 10.2.

| Site | Strongest year | Weakest year | Average | Difference in grades |
|---|---|---|---|---|
| 1. | Good | Poor/Poor Fair | Fair | 3.5 |
| 2. | Fair–Good/Good | Poor | Poor–Fair | 3.5 |
| 3. | Fair | Poor | Poor–Fair | 2 |
| 4. | Poor/Poor–Fair | Poor | Poor | 0.5 |
| 5. | Fair | Poor | Poor//Poor–Fair | 2 |
| 6. | Good | Poor | Fair | 4 |
| 7. | Fair–Good | Poor | Poor//Poor–Fair | 3,5 |
| 8. | Poor–Fair | Poor | Poor | 1.5 |
| 9. | Fair | Poor | Poor–Fair | 2 |
| 10. | Poor/Poor–Fair | Poor | Poor | 0.5 |
| 11. | Fair–Good | Poor | Poor–Fair | 3 |
| 12. | Fair–Good | Poor | Poor–Fair//Fair | 3 |

# Chapter 11
# Impact and distribution

## INTRODUCTION

This may seem an odd combination, impact, in fact human impact, the effects of what people do, and distribution, where *Phragmites* is found. Within the geographic and ecological limits set by climate, topography, isolation, etc., where *Phragmites* is actually found depends on the choice of people. The destruction of one of the main enormous reedbeds (Chapter 1), that in Iraq, shows that no reed is safe from such devastation.

Neglect or abandonment is often the best conservation measure. Wet places generally stay wet unless human impact makes them dry. The difficulty comes when the land all around is drained and it requires conservation management to maintain high water level.

Long-term, patterns differ. This is both because low-lying beds, coastal and a little inland are vulnerable to sea level changes, and because although reedbeds may live thousands of years, sooner or later, as *Phragmites* is a peat-builder, they come to an end (Chapter 1). Impact, though, as conservation management, can prolong their life.

There are no records of pests or diseases causing so much damage to *Phragmites* that another species took its place (as has happened with, e.g., *Epilobium hirsutum*, in Tall Herb).

## BRINGING BACK REED, OR BRINGING IT NEWLY IN

To start this depressing chapter with the good news!

In 2010 part of the Iraqi marshes was being restored, but how far the new marshes will resemble the old ecologically (within a reasonable number of centuries), and how far it will again be inhabited in the traditional way, are as yet unknown. Since 2016, though, a large part of the marshes has been listed as a UNESCO Heritage Site, and 40–60% was re-flooded. But work on the conservation and restoration of these wetlands is still hampered by agricultural schemes and the building of large dams further upstream, thus reducing the availability of water. Creation of new reedbeds is discussed by Brookhouse (1998) and Mills *et al.* (1998).

In Europe in the 1990s, the EU Bittern Project increased reedbeds, on a small scale. This interesting bird is in decline, and lives in reedbeds (a summer migrant). The EU provided money for new reedbeds suitable for this bird (RSPB, 1994 and 2000; Gilbert, 2003 and 2005). Not for the sake of *Phragmites*, please note, but for the bitterns. This is a common phenomenon. In Britain particularly, and Europe generally, those interested in preserving birdlife far exceed those interested in plants. Plants can get help only because – and how regrettably to the bird lovers – birds live on and among plants, and to preserve the birds, the habitat also must be preserved. What an unnecessary nuisance!

In Britain, the bitterns' heartlands are the coastal East Anglian Broads (man-made, e.g., Jennings & Lambert, 1951), with a centre also in the Severn estuary, and occasional elsewhere. Until nineteenth century exploitation, and twentieth century habitat loss, bitterns were frequent. Now, to have a (male) bittern booming on one's nature reserve is a cause of pride. Eighty-five per cent of breeding is on nature reserves, showing again the deplorable habitat loss. Nature reserves are only a tiny part of low, potentially wet land.

Bitterns, most luckily, prefer reedbeds harvested to give 'bare' areas of flooded stubble. This means that winter cutting (alternating areas) is not just acceptable but necessary. This contrasts with the reedbeds where 'bird lovers' prevent cutting, largely because it is not 'natural' – when in most English commercial beds the immensely rich wildlife exists only because of the management.

April to June is when bitterns have their nests of dead reed just above water *c.* 10 cm deep, in dense reedbed. The recommended reedbed size is 20–25 ha, though they can be smaller if part of a group. There should be dykes and pools on at least 20% of the area, to provide good fishing. Most birds feed within 15 m from the water's edge. Water 10–25 cm deep allows fish over a larger area, important during nesting. Bitterns eat amphibians and fish, especially eels – so sea access is needed. Some wintering pools are too small for breeding, but very important for the population as a whole. Tidal reedbeds also are used for feeding, not usually for nesting.

High water level of up to 25 cm in June can give difficulty for harvesting. If there is water control, this is easy, and may be done anyway (winter-dry). Without this depth in June means that in winter, with even deeper water, machine access may be impossible.

However, some reedbeds have been improved, and some areas of wet grassland, or even wet arable, have been turned over to reed, increasing the total area.

The 1960s reedbed in Wicken fen was described in Chapter 6, where fertiliser allowed reed to grow in a previously arable (ex-fen) area. Old Buckenham fen, fertilised, also grew a reedbed.

Projects for new wetlands are increasing, and because the soil was ex-fertilised or already has medium-high nutrient levels, reed flourishes. And, of course, some are ex-reedbeds anyway. The new Lakenheath fen is along a river, where *Phragmites* dominated prior to drainage (and in pockets since). It has now spread further, on ex-fertilised fields. In Upton marsh, a major flood protection scheme was wanted, but this was planned to also widen and increase reed and fen. Here current policies, particularly of the EU, mean conservation also is included where, a few decades ago, there would have been only destruction. Increasing marsh by widening river fringes and keeping water in (not draining it out) behind the flood walls is excellent, benefiting both reed stands and other fen and marsh communities: and their wildlife. Two large Fenland projects are described in Chapter 9, as was the great increase of small reedbeds for water purification. (And see Bardsley, 2001; Bartlett, 1999; Eurosite, 1997; Pfadenhauer, 1998; Ecosystems Applied Ecologists, 1999; English Nature, 2005; Hatcher, 1998.)

But what is all this compared with habitat destruction, both large-scale and just draining or ploughing that little patch of reeds in the corner that 'is no use to anyone' (Figure 11.1)! Large-scale drainage projects are usually termed 'reclamation', making the unwary suppose the land was formerly arable or settled, and became wet by mischance, which is unfortunate for conservation (also see Cížková *et al*., 2001)!

However, on the other side of the world, in China, native *Phragmites* is encouraged or planted in ponds which might be susceptible to alien invaders, to keep the native flora. (China, so far, is not so drained, so surplus ponds and wetlands exist. C. Zhou, pers. comm.)

**Figure 11.1**
Reed–willow (*Phragmites–Salix*) bands round streams, Hungary.
**a)** Narrow reed band.

Figure 11.1
Continued
**b)** Wide reed band.
**c)** Reeds in small valley, willow entering downstream.

## Dieback

In about 1970, from all over Europe, came reports that lake reeds (e.g., Havel lakes, Germany; Lake Constance, Switzerland and Germany) were dying back, and this caused alarm to those interested in wetlands (e.g., Ostendorp, 1989 and 1999; Hartog *et al.*, 1989; Krumscheid *et al.*, 1989; Leyrink & Hubatsch, 1993; Pier *et al.*, 1993; Schröder, 1979 and 1987; Stark & Dienst, 1989; Armstrong & Armstrong, 2001). This author, on hearing all these continental reports, promptly wrote to all organisations concerned with the fringing reed on Broadland rivers, so much appreciated by those in the (over-numerous) leisure boats, by artists, and so on. The replies, tactfully, made it clear it was just a woman scaremongering. The letters from the same organisations in about 1974 took a very different tone: ...could the international expert please tell them how to replace those lost fringes. Unfortunately, too late. To protect by fencing, etc., is expensive but possible. To completely rehabilitate plants and protect banks was and is prohibitively expensive, in waterways too lucrative to be closed.

In fact in the Broads, and no doubt on the continent, reed dieback was already an old phenomenon. George (1992) recorded reed disappearing from within, not at the fringes of, Broads in the 1940s. *Cladium mariscus* and others, though, started to dieback from the Broads in the nineteenth century (it is nutrient-medium, and more fertilisers meant more nutrient-rich water), followed by *Scirpus lacustris* and *Typha* spp. The decline was hastened by the temporary explosion of coypu (see Chapter 6) in the 1950s. Dieback started later on the firm clay and gravel, which is less easily disturbed than soft, reed peat by these animals.

Loss of reedswamp means less bird habitat, so fewer birds, and more flocking together and local damage.

The most obvious cause is disturbance, due to the explosion of leisure boats and the coming of cheap powercraft.

While, as above, England was annoyed about reed loss, the continentals sprang into action. Research and trials showed the best outcomes from (review by Ostendorp *et al.*, 1995):

- underwater fences to break waves;
- other constructions to dissipate waves;
- planting;
- stopping: recreation, gravel extraction affecting the reedswamp, and damaging construction;
- protection of banks (e.g., brushwood, reed bundles).

The consensus of causes of lake dieback is (see Table 11.1):

- wave scour, stressing shoots, and enhanced by driving algae and other plant bits away, especially by boats;
- disturbance by boats, swimming and paddling, bringing cattle to drink, picnicking and suchlike;
- constructions on the inland fringe of the reedswamp, e.g., embankments, roads and car parks;
- pollution from settlements, roads, factories, farms, etc.. This is only acceptable for *Phragmites* within its limits of cleaning power;
- disease. Not yet on a large scale in Britain, or commercially damaging anywhere for repeated years but (see Chapters 4 and 5), liable to lower yield or occasionally render one reedbed unharvestable (Ostendorp, 1999; Haslam, 1995).

The more stresses a reedbed has, the more likely it is to decline. Stresses act cumulatively. Huge amounts of reed have been lost since *c.* 1965, and this is roughly the period, in Europe, where wealth recovered after the Second World War and money could be spent on leisure pursuits which, among other things, cause disturbance to and subsequent loss of, the reedbed. Four of the five causes above are due to this cause (assuming the constructions are for leisure and simple farming), and the fifth becomes greater by stress, the primary cause. Table 11.1 sums up the processes.

As well as lake stands, and stands fringing rivers, however, there are the inland reed marshes, whose only access to open water is via their dykes. Dieback has never been seen in these. They are protected from dieback principally because they are protected against excessive recreation. There is no wave scour. Boats and swimmers are absent and even picnickers are few within the beds, etc.. Pollution in ordinary amounts is cleaned (see Chapter 7).

However, the dangers facing inland reedbeds are no less, and often more. They may be converted to grass, arable or settlement, all for 'good' economic reasons.

Reed marshes are the more likely to be intentionally removed to make money, and open water reedswamps to be lost by ignorance and negligence. (See also Brändle *et al.*, 1996; Cížková-Kancalova *et al.*, 1999; Cížková-Kancalova & Květ, 1991–1993; Dittrich & Westrich, 1990; Fuchs, 1993; Huben, 1993; Iseli, 1990; Krauss, 1993; Ksenofontova, 1989; Kühl & Neuhaus, 1993; Luft, 1993; Ostendorp *et al.*, 1995; Ostendorp, 1999; Leyrink & Hubatsch, 1993; Sukopp, 1991; van der Putten, 1993; Wöbbecke & Ripl, 1990.)

## Endurance

Malta (Chapter 8) is a primary example of endurance! In these dried islands, with overmuch disturbance, there is *Phragmites*. Less than there was in the 1920s, still less than in the pre-drainage 1830s, but there, in the remaining possible habitats, especially in bays and estuaries, there is *Phragmites*. Of course it is favoured by salt (Chapters 3 and 8) so can grow in drier and more disturbed places when brackish. Here stands have stayed, where not forcibly removed. Indeed the stands are smaller, in several or most places, and some are looking less healthy and decreasing more, but they are there, a tribute to the toughness of reed, physiologically and ecologically.

## Grazing

Grazing, for millennia, has been a major impact on reeds, wherever there were herbivores large enough to bite it, or small and with a taste for reed leaves and stems (see Chapter 6). *Phragmites* is harmed by grazing, because its growing points are at a browsable level (Chapter 3).

In deep water too – Europe not having hippopotamuses or manatee – livestock cannot graze. The relevant depth is greater for the larger animals, and depends also on substrate (firm enough to support the animal) and other conditions (e.g., White, 1788, describes cows, in hot weather, getting into ponds to be nearly covered by water).

Deep water, therefore, limits grazing. And in lightly-grazed land there is often a boundary between tall reedswamp in the water, and short grazed communities above it. Only removing grazing shows where *Phragmites* could dominate.

There is also shoot age and height. Young shoots are soft and palatable, tips of shoots remain so up to late summer, possibly until the leaves fall. In Malta goats grazed stands at about 1 m, leaving the lower leaves to continue making food, though the stems cannot flower. Sheep and ponies mostly graze younger – or if a short population, any – shoots near ground level. Cows do both, the young shoots being preferred.

There are, however, gradations even in light grazing, and grazing alone does not determine the outcome. *Phragmites* has more 'vigour', ability to grow well, under one harmful factor, if other habitat factors are good. Here, this means firstly whether the grazed area is wet enough. If flooded in winter (even by 5 cm of water, or perhaps water at soil surface) and damp in summer, it has the vigour. Secondly, if the nutrient status is high enough (see Chapter 6). To tolerate grazing, the rhizome must be able to send up many new buds (shoots), and without enough food, it cannot. Food comes from the aerial shoots, plus the minerals absorbed by the roots. If these are too few, as if the plant is too dry, it is easier to eliminate *Phragmites*. Reed has a narrower habitat range in acid than in alkaline soils, in mineral-poor than in mineral-rich ones.

These days grazing is more on nutrient-low than nutrient-rich areas in Europe, since most nutrient-rich land is cultivated intensively for pasture or arable, while moor, heath and mountain still exist, and are usually lightly grazed. This is mostly by sheep. Wet fields often have cows eating reed over the fence.

The more the replacement shoots, the better the stand, so the toleration to grazing.

In north-west Scotland reed patches, swampy and inaccessible to sheep, were drained, and so became accessible. Six years later shoots were very short and sparse, among the stumps of the previously-dominant reed. Competition with grazing-tolerant species was not an issue. Elsewhere, such competition may help to reduce *Phragmites*, it being lowered by grazing, so more susceptible to competition (see Chapter 6).

## Multiple impacts

To sum up this complex picture of interacting factors, Figure 1.5 shows where, on the scale from low to high nutrients, wet to dry, and low to high impact, *Phragmites* dominates (simplified: naturally reed occurs elsewhere as dominant, as well as in all the non-bog wetter habitats, and some of the drier ones). Taking away much of the impact, Figure 1.5b shows a simplified chart of vegetation in relation to nutrient and water regimes. This has *Phragmites* often dominant in rich fen and marsh, in wet to medium water regimes. It is also likely to be present, whether sparse or frequent, in all the other types except those of bog. And, even in bog, it is likely to be there, sparse or in patches, whenever nutrient regime improves (by run-off, flush, calcium-rich rock below, silt accumulation, etc.). (Also see Bailey *et al.*, 1998.)

Figure 11.2 looks at the kinds of fluctuating water regime that favour *Phragmites*. It occurs where the water is fairly stable. It cannot tolerate the really fluctuating levels of the riverside woodlands of the Danube, Rhine, etc.: although in any water regime, it may be present sparse, or in field-side dykes.

Figure 11.3 shows the effect of different types of management on what is a very small range of habitat. *Phragmites* could dominate in all the types of vegetation, if given suitable treatment. Rush pasture in low-nutrient hill valleys may need fertiliser as well. That on the Norfolk drained marshes just needs reed management. Complex!

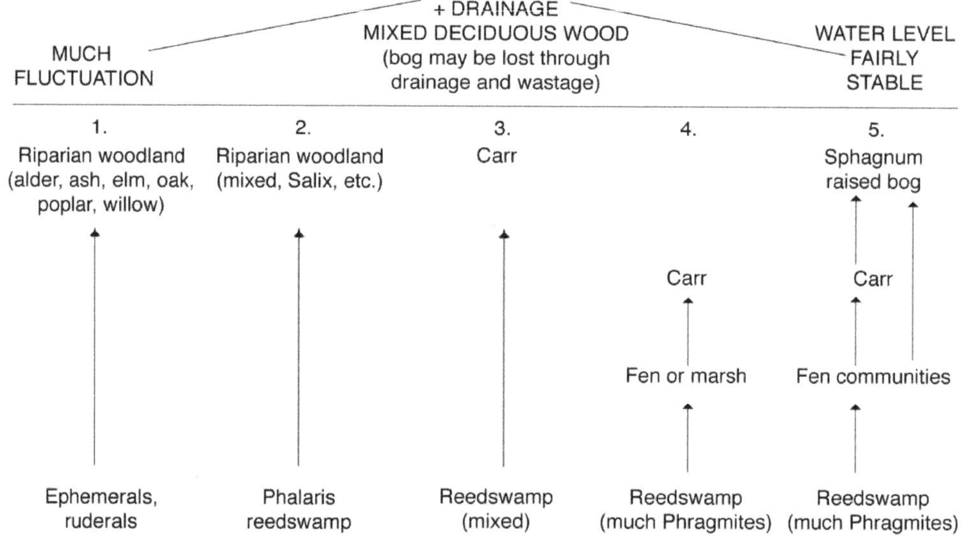


**Figure 11.2**

The effect of water level fluctuations on vegetation type (Haslam, 2003)

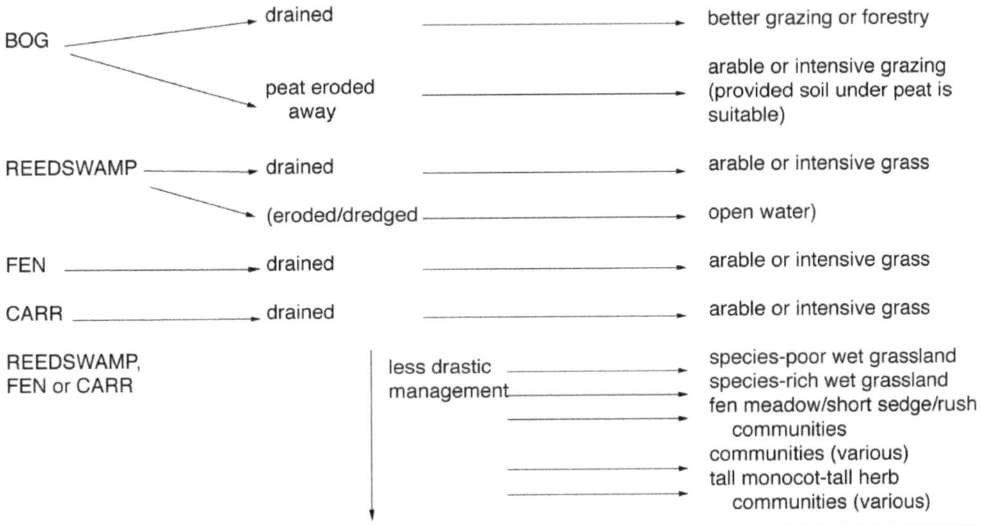


**Figure 11.3**
When peat is dried for farming (Haslam, 2003)

## Distribution

This section covers distribution within the available geographic areas (Figure 1.2). Stands vary from thousands of hectares to barely 10 shoots. Most land cannot bear *Phragmites*. It is under the sea or dry, bearing land vegetation. There is, of course, a grey area, of land able to bear *Phragmites* if other conditions are favourable, and not, if not.

There is also the question about how much *Phragmites*.

Firstly there are the dominant reedbeds, whether shallow-water reedswamps or managed, drier reedbeds. The former, of course, are much larger in area, though within England, the latter are the more important. Drainage and disturbance have removed most of the small stands, and some of the large. (It is easier to drain off a corner of a field than to do so to Iraq's reed marshes.) So the habitat is massively reduced, and, it must be presumed, also the *Phragmites* populations.

Large populations may, or may not, be recorded, e.g., Whittlesey Mere in the Fenland was, but small ones rarely are. At a guess, there are well under half – a quarter? – the reedbeds than there were two centuries ago.

Secondly, there is the often overlooked habitat in lowland streams, and in flatter areas of hill ones, where *Salix* spp. (willows or sallow: occasionally alder or poplar) fringe the stream, with a dominant *Phragmites* band outside the *Salix* (Figure 11.1). In less-impacted areas, parts of the English Midlands and East Anglia, a good deal more of (non-mountain) Hungary, and similarly across temperate Europe, these

*Salix–Phragmites*, willow-reed bands are common, and over wider areas, sparse. They used to be normal on wet streamland and have been lost by both the drainage and the ever-increasing cultivation of edges.

In Hungary these bands go up into the hills, even to the mountains, in narrow valleys or small floodplains (not turned to pasture). Wherever it is suitable for *Salix*, there also is (usually) *Phragmites* even growing in clearly dry ground by a dry stream in summer. This is possible because (Chapter 6) reed can tolerate many water regimes provided they are repetitive and give plenty of water for part of the year. Such streams, and streamside land, are flooded with snowmelt water in spring. So reed can grow! It is not as tall as in the main reedswamps (up to, e.g., 1.5 m, not 2.5–3 m), but perfectly healthy and stable. One year when there was little snow, reeds were *c.* 0.5 m shorter than in the previous year, but in the real marsh and river-lake reedbeds, height was the same in both years.

This, in length and distribution throughout much of Europe must have been, in one sense, the main *Phragmites* habitat. These were narrow bands, a few metres to 30+ m each side of the stream, and the sparser, shaded-type reeds within the *Salix* (Figure 6.6), but they wound and wandered through the countryside.

Thirdly, there is the vegetation with sparse reed (as opposed to reedbed). This of course grades both to reedbed, and to an occasional reed at the edge: reed in quantities somewhere between monodominance and absence. This gives a very wide range of habitat, of communities, and of the impact (management, disturbance, drying) on them (e.g., Wheeler, 1999).

To start with, an oddity. *Phragmites* is a field weed, as described in Chapter 8. *Phragmites* here occurs at the edges, unmanaged, growing well. In the field, regular ploughing cuts the rhizomes, or at least the upper ones. Such a cut stimulates bud growth on the rhizome pieces, and as (necessarily) these pieces have no large food store (too short), the buds and so shoots are narrow, giving healthy but short shoots, which flower sparsely in Malta, but hardly at all in England. These field reeds are limited, effectively, to fields which were drained marsh. In both England and in Malta further drying (principally 1900 to the present in Malta, post-1950 in England) leads to less, and eventually to no reed. This is the same principle as described above, the damage from ploughing was, just, tolerated, but when compounded with a worse water regime, reed declined and was lost.

Another oddity is the new field of paludiculture, described in Chapter 9. This is where, on suitable excavated peats, *Phragmites*, or other species, are grown for a commercial purpose. Because of *Phragmites*' habitat range, it can be grown on any nutrient-rich to medium peats, and its high productivity makes it valuable where a use can be found for it.

There are mixed-dominated reedswamps. In fact, from earlier records of thatching reed (e.g., Moir & Letts, 1999), a quarter of the reed bundle was, acceptably, non-reed, particularly *Typha* spp. (*T. angustifolia* more in deeper water, *T. latifolia* also in barely-flooded places). This means a mixed stand! Now, mixed is not the fashion, and 10% non-reed is almost too much.

Management has responded to fashion, and the beds are closer to monodominant. *Typha* spp. may be in discrete patches (clones, or several clones), or may be scattered throughout (commoner in drier beds). In sparser lake stands the species may be separated by habitat: *Phragmites* patches in siltier places, *Scirpus lacustris* ones in less silted, and deeper water. Here there is little overlap.

Relative sea level changes are described in Chapter 1. Without massive human intervention these override all other habitat factors in determining the occurrence of *Phragmites* in estuarine and near-coastal areas. Such a massive intervention was the drainage of much of the Zuider Zee, in The Netherlands, after the Second World War. Undersea land was drained (partly by *Phragmites*, Chapter 9), protected from flood and farmed. *Phragmites* removed from the land above sea level was subsequently fringing dykes in the same way as in old-established areas.

River fringes seldom contain *Phragmites*, not even sparse *Phragmites*. It occurs fringing nutrient-rich, sometimes nutrient-medium dykes, drains and ditches. Here *Glyceria maxima* is typically associated with silt, whether on Welsh flood plains, or in English fens, *Phragmites* being restricted to the more peat or soil places. Given time and enough disturbance and fertiliser, however, *Phragmites* replaces *Glyceria*, as on and near the Ouse Washes. Here there is an artificial habitat (dykes in drained land), becoming ever more suitable for reed as impact increases: until further drainage means the dyke, the habitat, can be done away with, and *Phragmites* is lost.

Saltmarshes are another potentially mixed habitat. In Europe, in a stable saltmarsh, *Phragmites* also is reasonably stable at the seawards side, and sometimes grading via reedbed to wet woodland behind (depending on impact) (but see Humphreys, 2005).

In the USA east and south-east coastal saltmarshes, however, there has been a near-explosion, over the past few decades, of *Phragmites* (Figure 5.1) (e.g., Weis & Weis, 2003). With inadequate DNA records (Chapter 5), the hypothesis was put that these are new and alien (in the circumstances, European) strains that for the first time are able to swallow up saltmarsh vegetation. Newer DNA records suggest this is unlikely (Chapter 5). Given the degree of impact now, it is as likely that the saltmarsh has been weakened, and for the first time is not able to repel *Phragmites* from large tracts of it. There is more evidence for this latter (also see Ailstock *et al.*, 2001). In China, in contrast, saltmarshes are, in parts, becoming saltier, and *Spartina* is spreading landwards, not the reverse. This is considered purely due to changing salt levels (C.F. Zhou, pers. comm.).

Wet woodland is another, often forgotten (Figure 6.4), habitat of *Phragmites*, but there it is, sparse in shade and patchily frequent or dominant in open glades. Short, little-flowering and flaccid, but there it stays, as long as the habitat does. Not showy, so hardly noticed: but in large areas of wet woodland, willow, alder, poplar, extending the ecological range of the species.

Tables 11.2–11.4 and Figure 11.4 show, just within Britain, the immense range of vegetation types in which *Phragmites* may be present. Even if these types have been created from reedbed by human impact, nevertheless, *Phragmites* is there.

**Calcium dominance**: *Phragmites* is there but sparse, waiting for the chance that disturbance, peat growth or drying may break the dominance and allow a reedbed (as happened at Redgrave and Lopham Fens, England, and widely elsewhere).

**Bog**: in any area of extra nutrients, there is a *Phragmites* patch. In Poor Fen, Rich Fen, edges of wet grassland, there is *Phragmites*. Not always, but often, it is there.

The habitat and vegetation range of continental varieties of *Phragmites* is just as great (see Tables 12.2 and 12.3).

Looking at the distribution of *Phragmites* means not just the large reedbeds, but the stream bands and the vast number (if not in area) of other wetland types where *Phragmites* (if counted by the stem and not by the area covered) is there, and well established and stable. Of course, vegetation relying on management may not long outlast the management, and peat-building marsh vegetation eventually builds itself out of existence, once it outgrows the water which made its habitat. But *Phragmites*, if there at all, is always ready to spring to dominance if conditions become better (as shown in the is-Simar marsh, Malta, Chapter 8).

| | |
|---|---|
| ***Phragmites***<br>Commercial reedbed | Winter-cut usually annually. (Nutrient-medium to nutrient-rich) |
| **Cladium** | Summer-cut every three to four years (calcium-influenced or -dominated) |
| **Tall grass**<br>Former marsh hay | Summer-cut, once or twice a year (nutrient-rich, wet) |
| **Medium grass-dominated**<br>Community | Summer-cut, about twice in nutrient-rich habitats, once in poorer ones. Or with light grazing |
| **Wet grassland**<br>Short grass | Grazed and/or mown more intensively than the two preceding |
| **Short sedge or rush**<br>Communities | Summer grazing or mowing. Variation in the patterns of these, together with variation in soil and water, leads to a wide variety of communities. The higher the nutrient status, the more the treatment required to keep vegetation short (but less so than for wet grassland) |
| **Tall herb**<br>Community | Develops from any of the others, if abandoned, and if nutrient-poor also dried to release nutrients through mineralisation, etc. |

These communities are created by management. This is either prolonging into drier places communities typical of wetter ones, or creating large areas of herbaceous vegetation of types otherwise restricted to local areas of, e.g. shallow soil and disturbance. (Management here includes abandonment.) In the natural succession, these habitats would be covered by carr.

**Figure 11.4** Management options in damp habitats (Haslam, 2003)
Different vegetation types created by different treatments.

**Table 11.1** Causes for reed decline (Ostendorp, 1989)

| | **Causal factor** |
|---|---|
| Direct destruction | Land reclamation<br>Recreational activities<br>Summer mowing<br>Electric fishing<br>Recovery of ammunition |
| Mechanical damage | Wave action (cargo ships, winds)<br>Floating rubbish<br>Drifting wood<br>Wash of filamentous algae<br>Drifting ice<br>Underwater gravel pits |
| Grazing | Geese, swans, coots, coypus, muskrats, grass carp<br>Horses, cattle |
| Water and sediment quality | Eutrophication (in general)<br>Sewage disposal<br>Duck manure, carp farming<br>Silting, sapropel formation, nutrient enrichment of sediments<br>Toxic effects of algal wash |
| Increased salinity (natural or through impact) | Increased salinity (natural or through impact) |
| Lake regulation and related effects | Lake regulation (in general)<br>Compensation of water level oscillations<br>Artificial rise in the water level<br>Artificial drawdown of water level<br>Bank erosion<br>Flood disaster |
| Others | Suppression by bushes<br>Replacement by *Typha* and/or *Glyceria*<br>(Disease) |

**Table 11.2** Dominants and understory groups, Britain (Rodwell, 1995)

**a) Dominant species of the swamp or fen**

*Phragmites australis*
*Glyceria maxima*
*G. fluitans*
*Phalaris arundinacea*
*Carex elata*
*C. paniculata*
*C. appropinquata*
*C. riparia*
*C. acuta*
*C. acutiformis*
*C. rostrata*
*C. aquatilis*
*C. lasiocarpa*
*C. vesicaria*
*C. pseudocyperus*
*Carex otrubae*
*Cladium mariscus*
*Scirpus lacustris* ssp. *lacustris*
*S. lacustris* ssp. *tabernaemontana*
*S. maritimus*
*Eleocharis palustris*
*Typha latifolia*
*T. angustifolia*
*Sparganium erectum*
*Acorus calamus*
*Sagittaria sagittifolia*
*Apium nodiflorum*
*Nasturtium officinale*
*Equisetum fluviatile*

**b) Understory species across swamp**

A Consistently rich and often structurally complex mixture of *Juncus subnodulosus*, *Calamagrostis canescens*, *Filipendula ulmaria*, *Iris pseudacorus*, *Valeriana officinalis*, *Lythrum salicaria*, *Lysimachia vulgaris*, *Peucedanum palustre*, *Eupatorium cannabinum*, *Galium palustre*, *Mentha aquatica*, *Calliergon cuspidatum* and *Campylium stellatum*

B Generally somewhat open and more species-poor mixtures of *Angelica sylvestris*, *Eupatorium cannabinum*, *Lythrum salicaria*, *Iris pseudacorus*, *Filipendula ulmaria*, *Galium palustre* and *Mentha aquatica* with occasional records for other species in A

C Tall and luxuriant but often species-poor mixtures, frequently with patchy local dominance, of *Filipendula ulmaria*, *Angelica sylvestris*, *Urtica dioica*, *Galium aparine*, *Oenanthe crocata*, *Epilobium hirsutum*, *Arrhenatherum elatius*, *Solanum dulcamara* and *Calystegia sepium* with rare records for other species in A

D Medium-tall and often lush but frequently species-poor mixtures of *Carex rostrata*, *Equisetum fluviatile*, *Menyanthes trifoliata*, *Potentilla palustris*, *Galium palustre*, *Cardamine pratensis* and *Agrostis stolonifera* with one or more of *Calliergon cuspidatum*, *C. cordifolium* and *C. giganteum*

E Usually open mixtures of *Carex rostrata*, *Equisetum fluviatile*, *Menyanthes trifoliata* and *Potentilla palustris*, sometimes with *Caltha palustris* or *Juncus bulbosus*

F Open and sometimes fragmentary mixtures of *Galium palustre*, *Mentha aquatica*, *Myosotis scorpioides* and *M. laxa* ssp. *caespitosa*, sometimes with locally abundant *Juncus effusus* and/or *Epilobium hirsutum*

G Open and often fragmentary mixtures of *Sparganium erectum*, *Alisma plantago-aquatica*, *Nasturtium officinale*, *Apium nodiflorum* and *Myosotis scorpioides*

H Mixtures of *Atriplex prostrata*, *Agrostis stolonifera* and *Potentilia anserina*, sometimes with species of *Armerion* or *Puccinellion* saltmarsh communities

**Table 11.3** British communities with *Phragmites australis*, according to the National Vegetation Classification (Rodwell, 1991a,b and 1995)

To show the wide range of communities, from woods to saltmarshes in which *Phragmites* can occur. Dominant and co-dominant reed communities in bold type.

|  |  |
|---|---|
| **1. Woods** | |
| W1 | *Salix cinerea* – *Galium palustre* |
| W2 | *Salix cinerea* – *Betula pubescens* |
| W3 | *Salix pentandra* – *Carex rostrata* |
| W4 | *Alnus glutinosa* – *Carex paniculata* |
| W6 | *Alnus glutinosa* – *Urtica dioica* |
| W24 | *Rubus fruticosus* – *Holcus lanatus* underscrub |
| **2. Mires** | |
| M5 | *Carex rostrata* – *Sphagnum squarrosum* |
| M9 | *Carex rostrata* – *Calliergon cuspidatum/giganteum* |
| M13 | *Schoenus nigricans* – *Juncus subnodulosus* |
| M21 | *Narthecium officinale* – *Sphagnum papillosum* |
| M22 | *Juncus subnodulosus* – *Cirsium palustre* |
| M24 | *Molinia caerulea* – *Cirsium dissectum* |
| M25 | *Molinia caerulea* – *Potentilla erecta* |
| M26 | *Molinia caerulea* – *Crepis palustris* |
| M27 | *Filipendula ulmaria* – *Angelica sylvestris* |
| M28 | *Iris pseudacorus* – *Filipendula ulmaria* |
| **3. Heaths** | |
| H5 | *Erica vagans* – *Ulex europoeus* |
| **4. Swamps** | |
| S1 | *Carex elata* (not noted in NVC, but frequent in East Anglian stands) |
| S2 | *Cladium mariscus* |
| S3 | *Carex paniculata* |
| S4 | ***Phragmites australis*** |
| S5 | *Glyceria maxima* |
| S6 | *Carex riparia* |
| S7 | *Carex acutiformis* (not noted in NVC, but occasional in East Anglia) |
| S8 | *Scirpus lacustris* |
| S12 | *Typha latifolia* (not noted in NVC, but sometimes occurs in East Anglia) |
| S13 | *Typha angustifolia* (not noted in NVC, but sometimes occurs in East Anglia) |
| S14 | *Sparganium erectum* |
| S17 | *Carex pseudocyperus* |
| S18 | *Carex obtrubae* |
| S20 | *Scirpus lacustris* ssp. *tabernaemontana* |
| S21 | *Scirpus maritimus* |
| S23 | Other water-margin vegetation (some) |
| S24 | *Peucedenum palustre*–***Phragmites australis*** |
| S25 | ***Phragmites australis***–*Eupatorium cannabinium* |
| S26 | ***Phragmites australis***–*Urtica dioica* |
| S27 | *Potentilla palustris*–*Carex rostrata* |

**Table 11.4** Extract from the description of the most complex *Phragmites*-mix Tall Herb fen, *Phragmites australis–Peucedanum palustre* Tall Herb fen of the (British) National Vegetation Classification

Selected to show the influence of management and soil (from Rodwell, 1995, also omitting references).

---

**PHYSIOGNOMY**

The *Peucedano-Phragmitetum* is a herbaceous fen vegetation community, somewhat varied in composition but often species-rich and of complex physiognomy.

Tall monocotyledons make up the major structural component and, of these, *Phragmites australis, Cladium mariscus* and *Calamagrostis canescens* are usual dominants. Less frequent but occasionally dominant are *Carex paniculata, Glyceria maxima, Typha angustifolia, Phalaris arundinacea* and *Calamagrostis epigejos*. The gross appearance of the vegetation at any particular time of year depends very much on the phenology of these species and the fen treatment, if any. In summer, unmown stands have a characteristically tall canopy, 1–2 m high, of varying density but often with a superficially uniform appearance.

Intermixed with these helophytes are a variety of tall herbaceous dicotyledons, which give the vegetation a typically colourful appearance in the flowering season. *Lysimachia vulgaris, Lythrum salicaria, Eupatorium cannabinum, Filipendula ulmaria* and *Peucedanum palustre* are constant. (Milk parsley having its major locus in this community.) Also frequent are *Valeriana officinalis, Iris pseudacorus* and *Lycopus europaeus* and, more unevenly between sub-communities, *Angelica sylvestris, Cirsium palustre, Rumex hydrous pallium, Epilobium hirsutum* and *Symphytum officinale.*

Beneath these, a layer some 60–80 cm tall, is generally dominated by sedges or rushes. *Juncus subnodulosus* and *Carex elata* are the most abundant species but *G. riparia, C. acutiformis, C. appropinquate, C. lasiocarpa, C. diandra, Molinia caerulea, Schoenus nigricans* or *Thelepteris palustris* can all attain local prominence.

Small herbaceous species are patchy throughout the community, being greater where the layers of tall and bulkier species are more open. *Mentha aquatica* is the only frequent species throughout but some of the following often occur: *Scutellaria galericulata, Hydrocotyle vulgaris, Potentilla palustris, Oenanthe fistulosa, Epilobium palustre, Cardamine pratensis, Lychnis flos-cuculi* and *Myosotis laxa* ssp. *caespitosa.* In some wetter expressions of the community, floating-leaved aquatics may be present.

A variety of sprawlers and climbers may be found sometimes binding the vegetation into an almost impenetrable tangle. Most frequent of these is *pulustre* but *Solanum dulcamara, Calystegia sepium, Vicia cracca, Galium uliginosum* and the rarer *Stellaria palustris* and *Lathyrus palustris* also occur.

Although *Calliergon cuspidatum, Campylium stellatum* and *Brachythecium rutabulum* occur in the community, over litter or patches of bare substrate, bryophytes are typically not well represented.

Seedlings and saplings of *Salix cinerea* and *Alnus glutinosa* are occasional to frequent throughout the community and, in some sub-communities, are characteristically abundant giving a scrubby appearance to the vegetation. Other generally low-frequency woody species include *Frangula almis* and *Betula pubescens* and one type of *Peucedano–Phragmitetum. Myrica gale* forms a distinctive open canopy.

Seven sub-communities are characterised.

***Carex paniculata* sub-community**

Here, the physiognomy is overwhelmingly dominated by the sometimes, enormous tussocks of *C. paniculata*; indeed, the vegetation looks like a sedge-swamp with a *Peucedano-Phragmitetum* element perched on the tussock tops.

***Glyceria maxima* sub-community**

The varied vegetation in this sub-community is characterised in general by the prominence of *G. maxima* among the taller fen species. In some stands, it is dominant and its lush growth and the lodging of the shoots tend to depress the richness of the associated flora. *Phragmites* persists, together with most of the tall dicotyledons of the community, and there may be patches of *Epilobium hirsutum* and *Urtica dioica*, some *Caltha palustris* and *Scutellaria galericulata* below and sprawling *Calystegia septum*. Other stands have a reduced cover of *G. maxima*.

***Symphytum officinale* sub-community**

Generally floristics, this is more of a central type of *Peucedano–Phragmitetum* than either the *Carex paniculata* or the *Glyceria* sub-community, although it shows some particular similarities to the latter. Often, local variation, especially among the dominants, masks the general features of the vegetation.

The sedge/rush layer is sometimes extensive, though generally species-poor and variable in its composition.

Smaller herbaceous species are not well represented, though there are usually some sprawlers with *Galium palustre, Lathyrus palustris* and *Calystegia sepium* frequent.

**Table 11.4** Continued

In general, drier underfoot than many types of the community and invasion by woody species is frequent.

**Typical sub-community**

*Phragmites* is the most frequent helophyte here and is usually dominant.

Tall dicotyledons are well represented with, in addition to the community constants, *Valeriana officinalis, Iris pseudacorus, Lycopus europaeus, Rumex hydrolapathum, Angelica sylvestris* and *Cirsium palustre. Epilobium hirsutum, Phalaris arundinacea* and *Urtica dioica* are uncommon and rarely abundant.

There is usually a prominent understorey *of J. subnodulosus* with varying amounts of *Thelypteris palustris'*, in some stands, especially where there is a shrubby element, the latter species forms a thick cover. Sedges are, however, rather infrequent.

***Cicuta virosa* sub-community**

*Phragmites* usually the most prominent helophyte although *C. mariscus* is also frequent and it may dominate. *C. canescens* is less common here than in the typical sub-community and rarely abundant; in some stand; it tends to be replaced by *Typha angustifolia.*

The sedge/rush level is well developed and species-rich.

***Schoenus nigricans* sub-community**

*Phragmites* normally dominate in this sub-community, the helophyte and tall-herb components are somewhat poorer than in other types of *Peucedano–Phragmitetum.*

Beneath, however, the sedge/rush layer is distinctive in the high frequency, and sometimes the abundance, of *Schoenus nigricans*, intermixed with *C. data* and *J. subnodulosus.* The small herb element, too, has a characteristic quality.

***Myrica gale* sub-community**

In the rather species-poor vegetation of this sub-community, *Phragmites, Cladium* or *Calamagrostis canescens* normally dominate. The constant dicotyledons of the community are all well represented but, among other tall herbs, only *Valeriana officinalis* is at all frequent. The sedge/rush layer, too, is not *rich.*

The most distinctive feature is the presence of scattered leggy bushes of *Myrica gale,* often with saplings, although small numbers of *Salix repens* are characteristic.

**HABITAT**

The *Peucedano–Phragmitetum* is generally restricted to fen peats with a moderate to high summer water table and some winter flooding with base-rich, calcareous and often oligotrophic waters.

It is a community of topogenous mires, occurring as a primary fen in open-water transitions but being especially characteristic of floodplain mires where long land complex histories of exploitation for peat and marsh crops have often led to the maintenance of the community as anthropogenic secondary vegetation.

The community is now almost entirely confined to Broadland with a few outlying stations on fragments on floodplain mires that have escaped drainage and embankment or where a high level of unenriched waters is artificially maintained.

Shallow peat extraction and other forms of soil disturbance have much altered the state of the surface, often producing very distinct local conditions for recolonisation. On the more solid peats, too, the community has been and, on a much reduced scale, still is affected by a variety of mowing treatments.

The most obvious impact of these seems to be on the quantitative balance between species within the sub-communities affected.

Local traditions of treatment may have tended to confirm local natural effects and it may be such combinations of influences that are responsible for the rather striking pattern of distribution of some of the sub-communities in particular stretches of fen. Mowing, at least winter mowing for reed has also traditionally had to be confined to drier areas of fen.

Where mowing occurred, it is likely that the treatment itself has produced some modification of soil conditions with harvesting of the crop, burning of litter, controlled unseasonal flooding and the application of fertilisers or marl. Making sense of all this is not helped by the sometimes fragmentary survival of the community, even in Broadland, and the difficulty of documenting treatment histories, especially now that most of the mowing traditions are defunct.

Sub-community which comprises the bulk of primary fen with *Phragmites* as the usual dominant over a firm, though sometimes floating, peat raft with a summer water table around the substrate surface and with often-prolonged winter flooding. In more eutrophic conditions, *Glycerin maxima* can retain its pre-eminence over *P. australis* in primary fen as well as in swamp. With the increased dereliction of dykes and consequent restriction of free water movement since the 1940s, this *G. maxima*-dominated primary fen has become even more restricted in its distribution.

Table 11.4 continued overleaf...

**Table 11.4** Continued

The *Cicuta* sub-community also occurs as primary fen around some broad margins, suggesting a combination of extreme wetness with lack of throughput may play some part in determining the distinctive nature of the associates. If the community has been accessible for the harvesting of various fen crops and there is little doubt that much *Peucedano–Phragmitetum has* been maintained as 'secondary fen' by a range of cutting regimes which have repeatedly set back the invasion of woody species possible with a lower water table.

Typical variant of the *Cicuta* sub-community and the *Schoenus* sub-community, are generally found only on drier peats and then only where there has been a tradition of mowing.

It is easy to make suppositions about the effects of mowing on these sub-communities of the *Peucedano–Phragmitetum* but much more difficult to be precise. Even when mowing was a normal part of the Broadland rural economy (and at one time or another most of the area occupied by the communities was regularly cut) stretches of fen seem to have been subjected to a complex pattern of treatments which might vary from year to year and from one compartment to the next and even within compartments according to need, available labour, weather conditions and changes of ownership, etc.

The *Peucedano-Phragmitetum* has traditionally provided four types of crop: *Phragmites* for reed thatch, *Cladium* for sedge thatch, *G. maxima* for green fodder, hay or litter and more mixed vegetation, often with abundant *Juncus subnodulosus* and/or *Molinia caerulea* also for litter. The most obvious general effect of regularly mowing for any of these crops was to set back repeated and natural invasion of woody species and so maintain the community, albeit in modified forms as secondary fen.

This would probably have been aided by keeping dykes in good repair for water circulation and to control flooding, both of which could maintain the substrate in a moister condition than might otherwise have been the case. Now much of this variation is masked by the more uniform advance of scrub and carr over former mowing mash, but it seems that in general impacts have not disrupted the edaphically influenced variation within the community. One effect that does seem to be discernible is an increase in species-richness associated with the regular but infrequent summer mowing for sedge.

The complex of open water transition and floodplain mires in Broadland, where the *Peucedano–Phragmitetum* forms the major fen element, has a rich invertebrate fauna which has long attracted attention. Although there is no absolute coincidence between the community and organised animal assemblages, it has provided food plants or distinctive physiognomic niches for a wide variety of species, some of which are now virtually restricted to its own range, others of which have become extinct with the advance of fen reclamation.

**Zonation**

In Broadland, the *Peucedano–Phragmitetum* generally occupies the middle zone of the floodplain mires between the open water of the rivers and broads and the valley sides. In some places, it forms part of a complete and fairly clear sequence from swamp through fen to scrub and woodland, including vegetation developed over both the colonised margins of the deep peat cuts of the broads and the intact alluvial deposits behind. Such zonations have, however, been much confused especially on these drier solid substrates towards their landward limit, by the various traditional mowing treatments and shallow peat digging, and are now further complicated by neglect and the spread of woody vegetation over the abandoned fen compartments and dykes.

Such sustained summer mowing, especially on drier peats, seems to be one of the most effective ways of breaking the cycle of maintenance of secondary fen by converting the *Peucedano–Phragmitetum* to another vegetation type. Systematic grazing by stock, especially when combined with embankment and drainage, is probably another. On the grazed levels of the Yare, the community appears to develop into the *Holco–Juncetum* under such treatment. More drastic disturbance of drier peats may result in oxidation of the organic deposits and a release of nutrients with an increase in nitrophilous species. Such a development is perhaps already seen within the *Glyceria* and *Symphytum* sub-communities but it may also transform the *Peucedano–Phragmitetum* into the tall-herb fens of the *Phragmites–Urtica* community or the *Epilobium hirsutae.*

The vegetation of shallow peat diggings is often further complicated by their isolation from the main paths of water movement through the broads and dykes and by the possible lateral seepage of water at the junctions of the flood-lain mire with the valley sides. It is in such positions that the *Cicuta* sub-community can occur in striking mosaics with a carpet of the *Carex rostrata–Calliergon* fen. These mosaics are the locus of a number of Broadland rarities, including *Liparis loeselii*, *Anagallis tenella*, *Drosera anglica*, *Parnassia palustris* and *Hypericum elodes.*

**Distribution**

The most extensive tracts of the community occur in the middle reaches of the Yare, Bure, Ant and Thurne valleys in Broadland, east Norfolk. The *Symphytum* sub-community has been recorded from Wicken and Woodwalton Fens in Cambridgeshire and rather similar vegetation has been described from Catcott Heath in Somerset and the Ham fens in Kent.

Even in those areas which have been designated as nature reserves, the ongoing processes of terrestrialisation and recolonisation by shrubs and trees are reducing the extent of the *Peucedano–Phragmitetum* or modifying its floristics. Although a patchwork of fen and scrub may be valuable in some respects (for breeding or migrant passerines), for example, the generally open character of former areas of mowing marsh has, in many places, been lost and some species characteristic of regularly managed fen have declined or become extinct.

# Chapter 12
# Can reeds diagnose habitat?

## Introduction

Some plant species have a narrow habitat range, so if they are present, their habitat is known: *Parnassia palustris*, the beautiful Grass of Parnassus is one. It occurs with low, but not very low nutrients, in short or patchily short vegetation, it being short is killed by shade, and frequents wet but not flooded conditions. It is in part of the phytosociological entity *Parvocaricetum*, in Poor Fen. That is, in its core western European habitat! Seeing *P. palustris* implies all this, in England (see Haslam, 1994).

*Phragmites*, however, has a very wide habitat range, and can extend to unlikely extremes. This means less can be deduced by reed alone. Pelechaty (2004) in Poland looked at many lakes with reed. There were many different species associations, naturally enough, since the lakes ranged from the low-nutrient, almost bog, *Lobelia dortmanna* community to the nutrient-rich. It was the associated species which indicated the nutrient status. Within the wide range of dominant *Phragmites*, it is necessary to look not just at the *Phragmites* but at the habitat around, the other species, and indeed the habit of the *Phragmites*, for any diagnosis. This diagnosis can be only what is obvious anyway: brown, boggy lake water within a bog, curiously enough, bears a boggy community! But there are also interesting deductions to be made. Apart, that is, from the obvious one that if *Phragmites* is there, then also is access to water.

"THE PLANT IS ALWAYS RIGHT." This statement, by A.S. Watt (ed. in Fiala K & Květ J. 1971), is vital for all ecology. The plant *is* always right: and its corollary is; the botanist may be wrong. If the plant is there (other than having been transplanted last week), then all its needs are met, the habitat is suitable for the plant. There are, of course, as with all in ecology, peculiarities, often due to human impact, which must be looked for, before diagnosis is assured. One of the oddest this author has met was to be shown round a fen, and to find *Berula erecta*, which in England is calcium-demanding, in a nutrient-rich fen regularly flooded from a river (Ouse Washes). I commented on the oddity. A few weeks later the information came: the wall by which the *Berula erecta* was growing was made of limestone imported from outside the Fenland. The plant was indeed always right. The botanist was nearly wrong, having considered whether the range of *B. erecta* in this region could occasionally be extended. The answer is 'No'.

As *Phragmites* has such a wide range anyway, a range confirmed by numerous occurrences, not just a single case, anything as extreme as this site has not been noted. However, there are occasions when thought is needed. The Hungarian willow–reed bands up in the mountain valleys are one such instance. These are (Chapter 6) dry. *Phragmites* needs access to water. Yes, but Hungary has winter snow for several months (usually three to four where this feature was studied), and melting snow means heavy flows that wet the streams and the low ground around them. Access to water, there is. Again, the dry small stands in Malta, surely very peculiar. Yes, but they are remnants of larger wetter populations, able to persevere – ever-decreasing – in dry conditions because firstly, plants can hold on in unfavourable conditions they cannot invade, and secondly, salt and a lack of dryland competition are both favourable factors, enabling the plants to do better than if there had been freshwater or competition.

Thirdly, an incompletely answered oddity: the influence of steam trains. From west to east (England to Hungary), *Phragmites* can be found high on railway banks, well up above water (Figure 6.6), where also other tall plants can grow (though with continuous or discontinuous stands reaching downslope to wet habitats). Why? Obviously, decades after steam trains ran, ash and clinker are still in the soil. Exactly what chemical is even more favourable than salt, allowing *Phragmites* to grow several metres above ground level, is unfortunately not known. It, or they, obviously exist, and if also occurring in nature, would allow *Phragmites* to grow in much drier places than expected.

## Deductions from plant habit

The habit of *Phragmites* is highly variable, as already seen (and summarised in Table 1.1, also in Table 12.1). Variation is from genetic as well as environmental variables, so both must be considered. If, for instance, in a nutrient-poor lake with many small bays with reed stands, one stand is very much taller, why is that? The bays are all similar. The area between the lakewards limit of deep water or scouring waves, and the landwards limit of stones or grazing may vary, but the bay habitat itself is superficially similar. So what makes the reed of one bay so much larger? First, obvious genetic differences can be checked. Dates of flowering and fruiting? Shape of panicles? Other? Second, the habitat: is there much more silting in the large-reed bay? (In low-nutrient areas, silt, with its available minerals, makes a great difference to nutrient regime.) Is the substrate in the tall stand silty, in the others, just stony with little 'earth' even below where the rhizomes are? Could the stand in the other bays be reached by grazing livestock in early summer, and the large one be the only one ungrazed? (What are the livestock? Cattle can graze much higher and in deeper water than sheep.) (Mowing is discussed by Güsewell, 1998.)

When all such habitat factors have been exhausted, a more detailed study of the habit, and the possible genetic factor, can be made. This comes last because it takes the longest. Factors suggesting a (so far) missed habitat variable include the tall-reed bay having a better population type (optimal versus sub-optimal or if worse, sub-optimal versus restricted or depauperate, Chapter 4), the correlation graph between basal stem width and height, and with leaf width, being the same, the inflorescence shape and colour (not necessarily size) being the same, the pattern (not the date) of leaf drop, being the same, and so on (Chapters 4 to 8, also using Table 1.1). If all fail, a tall genotype is likely. It can be checked only by DNA (preferably) or transplant of the tall and short to the same habitat, and waiting for four years (when different genotypes will still be different, but quite possibly more like each other than in the lake).

The conclusion is that reed stands differ both genetically and with habitat, and quite a lot can be observed about habit differences, and the differences in habitat they indicate.

Table 12.1 summarises the more simple and obvious habit differences that can be used to diagnose habitat differences. This is far less complex than the example above. It is important to know diagnosis of habitat can be simple or complex!

Tables 11.2–11.4 and 12.1–12.3 illustrate habitat and *Phragmites* behaviour, distribution, and associate species. They show both the wide range and some of the responses to habitat variation.

What reason is there for the occurrence of a reed patch here, and not there? In high-nutrient places the answer is too often random competition, the willows became established here, and *Phragmites* was reduced to sparse, flaccid shoots (the *Salix* became established because there were locally suitable patches causal for the *Salix*, but random for the *Phragmites*), or past patchy heavy grazing allowed other species to grow and then compete, also patchily, or... (see Chapter 11 for reed dieback).

## Deductions from associated species

Each species has its own habitat range, which differs, in a major or minor way, from every other species. Consequently looking at a list of species from any one place can give a very accurate picture of the water, nutrient, management, etc., status, since the range in which all these species can occur is very narrow.

However, as always, there is complexity. An exact and excellent list compiled for one locality may be only approximate and inexact for another.

Species show geographic variation. One area may be in the centre of distribution, where the species is tough and widespread (able to occur in a wide habitat range), another may be peripheral, where the habitat range is narrower (so more useful as a monitor), and yet a third may be outside its geographic range, e.g., *Phalaris arundinacea* in the south Mediterranean, so its absence means nothing, the 'I wonder why x is not there?' has no ecological significance. In addition, there are regional differences in geology, topography and large-scale management, which lead to variations. (See Clevering *et al.*, 2001; Davies, 1945; Frolik, 1941; Griffiths, 1932; Kulczinski, 1949; Moss, 1913; Newbould, 1960; Ratcliffe & Walker, 1958; Regel, 1947; Rose, 1950 and 1953; Ssymank & Hauke, 1998; Clapham, 1940; Harding, 1992; Kassas, 1951.)

Therefore good lists of associate species give a good habitat diagnosis and assessment, provided only that someone has been to do the research in that particular area, eco-habitat, already. Tables 6.1, 6.2, 11.2, 12.2 and 12.3 give some examples where this has been done (also see Hejny & Segal, 1998). Even here there is a snag to be watched out for. As mentioned above, it is possible for a community to continue, after conditions have changed. The community could no longer become established. It can no longer renew itself, but long-lived plants can still live out their natural life span (e.g., *Carex paniculata*, probably a few centuries), and those species which depend on the long-lived ones (e.g., *Rosa canina* for the last) can also persist. However, closer study will reveal this. The annual or short-lived and frailer plants will disappear. So when the species list, though impressively long and correct, nevertheless leaves out this group of species (e.g., *Galium uliginosum*): warning bells should ring. They may be species fastidious in water regime, nutrient regime, grazing management, other impact, or whatever. If the short-in-life-span or frail-in-narrowness-of-range are absent, checking for the status of the community is essential.

Similarly, if a whole group of species appears, whether newly in time, or erratically in space, this needs study. Newly in time means changing conditions (regrettably, mostly from drying, raising nutrients – perhaps in consequence – or increasing impact). Newly in space means altered conditions across the habitat, water regime changes cutting off calcium-rich waters from part, more grazing on the drier places, pollution reaching in from an increasingly polluted river, and suchlike. These all need investigating. It is the plants which have alerted the researcher that something is happening, the plants that have 'diagnosed' change, and indicated what sort of change it is.

## Conclusion

Because of its very wide habitat range, *Phragmites* is easily picked out as present, particularly when dominant. This is in itself an assessment, of a wet (a damp) place, with suitable nutrient regime and without great human impact.

This wide habitat range, though, means reed presence does not diagnose much more. To get further, either the reed must be in an extreme form, or there is a good selection of associated species, and the habitat range of the group is already known.

**Table 12.1** Reed features characteristic of different habitat

| **Feature** | **Possible habitat** | **Likelihood of conclusion being correct** |
|---|---|---|
| Monodominant, sparse | Flooded | High |
| Monodominant, very dense | Not flooded, and burnt and/or frosted | Fair |
| Optimal population | Good or excellent habitat | Very high |
| Sub-optimal population | Something wrong | High |
| Restricted population | Bud production hindered by poor habitat (e.g. nutrients, dryness, competition) | High |
| Depauperate population | Shoot growth and bud size hindered by grazing or similar 'draining' habitat | High |
| Reeds 3–10 m high | Damp or flooded | Very high (the reverse is untrue) |
| Reeds miniature, leaves soft | Dry, trampled | Fair |
| Reeds miniature, leaves sharp | Dry, trampled, salty | High |
| Leaves semi-horizontal, flaccid, pale | Shade | Very high |
| Leaf blades ribbed | Calcium dominance | Fair |
| Leaf blades yellow early, stiff | Calcium dominance | Fair–high |
| Leaf blades yellow, normal | Root competition, serious early flood or drought. Genetic | Each, low to fair |
| Leaf blades brown early | Nutrient-poor (other). Genetic | Each, low to fair |
| Violet on stem internodes | Tall reed (individually) | High |
| *Legehalme* present | Genetic. And space on damp soil or quiet water. And brackish water | High |
| Rhizomes oval | Good and damp/wet habitat | Very high |
| Water roots present | Wet | High |
| Water roots abundant | Flooded, nutrient-rich | High |
| Litter mat 20+ cm thick | Not harvested, little or no flooding | High |

**Table 12.2** Vegetation pattern in reedswamp, The Netherlands (Mook & van der Toorn, 1982)

**a) The Biesbosch**

The high herbaceous layer (height: >150 cm) was composed of *Phragmites australis* and *Typha angustifolia*, the amount of coverage by both decreasing in the indicated direction, the low-growing plants (height: <100 cm) increasing in the same sense with a corresponding increase in the total number of species. On the basis of the dominant species in the herbaceous and moss layers, the following zones can be distinguished:

Zone 1 Vegetation of *Stratoites aloides* with *Utricularia vulgaris* interspersed with *T. angustifolia* and *Phragmites*. The habitat consists of open water with a depth of about 80 cm

Zone 2 Dense reed vegetation composed of only a few water plants with limited cover and vitality, occurring along open water with a depth of about 15 cm

Zone 3 Reed vegetation with *T. angustifolia,* a few water plants (e.g. *Hydrocharis morsus-ranae)* and the moss *Calliergonella cuspidata*, occurring adjacent to Zone 2 or along open water with a depth roughly the same as that of Zone 2

Zone 4 Reed vegetation with *T. angustifolia*, the well-developed moss layer showing *C. cuspidata* in combination with *Pallavicinia lyellii*. Species-rich reed stand halfway between open water and *legakker*. Water level close to soil surface

Zone 5 Reed vegetation with *T. angustifolia,* the almost complete moss cover showing *P. lyellii*, *Sphagnum fimbriatum*, and *S. squarrosum*. Species-rich vegetation in which the reed is less strongly developed than in the preceding zones. Water level a few centimetres below soil surface

Zone 6 Vegetation with highly developed moss layer in which *S. fimbriatum* and *S. squarrosum* dominate. Open and species-rich herb layer in which reed occurs with limited cover. Water level to 10 cm below soil surface

Zone 7 Reed vegetation with *R. ficaria*, *C. amara*, *P. trivialis*, and *Anthriscus sylvestris*, occurring on levées; MHWL at -13 cm

**b) Zuidland**

As in The Biesbosch, the number of plant species increases with the soil elevation, and in the same direction there is a shift in the species from high (>150 cm) to less high (100–150 cm). On the basis of dominance and composition of the vegetation, the following zones can be distinguished:

Zone 1 *Phragmites* mixed with *Scirpus maritimus* f. *compactus,* which is strongly developed. Occurs in back-swamps lying in the transitional area between the *Phragmites* belt and the *Scirpus maritimus* belt; MHWL around soil surface

Zone 2 Reed vegetation with weakly developed and vegetative *S. maritimus*. Adjacent to Zone 1, with corresponding water level and drainage

Zone 3 Reed vegetation with other vegetation of little vitality, including *Atriplex hastata*, *Agrostis stolonifera*, and *Cochlearia officinalis*. Occurs in back-swamps; MHWL at -14 cm

Zone 4 Reed vegetation with abundant *A. stolonifera* and rather well developed *Rumex crispus*. Occurs on levées; MHWL at -10 to -18 cm

Zone 5 Reed vegetation with abundant *A. stolonifera* and C. *officinalis*. Occurs on levées; MHWL at -17 to -26 cm

Zone 6 Vegetation of *Elytrigia pungens* and *Althaea officinalis* mixed with *Phragmites*. Occurs on levées; MHWL at -31 cm

Zones 2 to 5 can be assigned to the *Phragmites* consociation.

Zone 1 forms a transition to the *Scirpus maritimus compactus* consociation with which the *Halo–Scirpetum maritimi* is synonymous.

Zone 6 corresponds with the *Althaeo–Calystegietum sepium* association, (subass. with *Cochlearia officinalis*). This association is found on drift zones in reed vegetations occurring in the mesohaline area.

It is striking that in this area the reed is the only representative of the class Phragmitetea. Phytocenologically, the plant is characterized here on the one hand by *Asteretea tripolii* species and on the other by Artemisietea species.

A negative influence of reed on other species is demonstrated by the negative correlation between the amount of light absorbed by the reed and the coverage of species accompanying reed.

MHWL = Mean High Water Level

**Table 12.3** Reedbed variation with habitat hands of different reedbed areas

These are individual sites, and data may not be applicable in other sites.

**1. Nutrient status – Example: Part of East Anglia, England**

The nutrient range of each species is different. Many overlap, and of these, the more the species present, the narrower the possible nutrient regime, and so the better the diagnosis.

There is not only one balance of nutrients. Not only does pH vary, but so does calcium, magnesium, phosphate, nitrate, potassium, ammonium, and so on. And balance is vital. This table shows a high calcium sequence, with calcium dominance forming part of the low nutrient ranges. Starred species are those most characteristic of calcium dominance or influence. b) shows some calcium influence, but not nearly as much.

The species listed are those found in or beside reed beds. There are in their most frequent habitat, but may occur in others. Hence the importance of a long species list.

a) Very low *Anagallis tenella**, *Betula* spp.*, *Drosera* spp.*, *Eriophorum angustifolium*, *Molinia caerulea*, *Myrica gale*, *Pinguicula vulgaris**, *Potentilla erecta*, *Schoenus nigricans**, *Sphagnum* spp.*

b) Low *Betula* spp., *Briza media**, *Cirsium palustre*, *Cladium mariscus*, *Eriophorum angustifolium*, *Galium uliginosum**, *Genista anglica*, *Gymnadenia conopsea**, *Juncus articulatus*, *J. subnodulosus*, *Lotus pedunculatus*, *Mentha aquatica*, *Menyanthes trifoliata*, *Molinia caerulea*, *Myrica gale*, *Parnassia palustris**, *Potentilla erecta*, *Sanguisorba officinalis*, *Schoenus nigricans**, *Succisa pratensis*, *Valeriana dioica*

c) Medium *Calamagrostis canescens*, *Carex elata*, *C. paniculata*, *Cirsium palustre*, *Cladium mariscus*, *Filipendula ulmaria*, *Galium palustre*, *Glyceria maxima*, *Iris pseudacorus*, *Juncus effusus*, *Lychnis flos-cuculi*, *Lysimachia vulgaris*, *Lythrum salicaria*, *Rumex hydrolapathum*

d) High *Calamagrostis canescens*, *Caltha palustris*, *Carex acutiformis*, *C. paniculata*, *Cirsium arvense*, *C. palustre*, *Filipendula ulmaria*, *Galium palustre*, *Glyceria maxima*, *Iris pseudacorus*, *Juncus effusus*, *Lychnis flos-cuculi*, *Lysimachia vulgaris*, *Lythrum salicaria*, *Rumex hydrolapathum*

e) Very high *Arrhenatherum elatius*, *Carex acutiformis*, *C. riparia*, *Cirsium arvense*, *Epilobium hirsutum*, *Galium aparine*, *Iris pseudacorus*, *Juncus effusus*, *Lythrum salicaria*, *Phalaris arundinacea*, *Urtica dioica*

(NB: *Juncus effusus*, in moorland etc. situations grows also in fairly low-nutrient regimes.)

**2. Water regime – Example: The Netherlands (Kalenberg, The Biesbosch, Zuidland) (van der Toorn, 1972)**

a) Open water (*c.* 80 cm deep) with sparse plants. *Stratotes aloides*, *Typha angustifolia*, *Utricularia vulgaris*

b) Dense reed, with few and poor associates. Water 15 cm deep, alongside open water

c) Reedbeds with *Typha angustifolia*, *Hydrocharis morsus-ranae*, *Calliergon cuspidata* and a few others. By (b) or by open water *c.* 15 cm deep

d) Reedbed with *T. angustifolia*, a well-developed moss layer of *C. cuspidata* and *Pallavicini lyellii*. Species-rich. Water close to but below, ground level

e) *Sphagnum* reedland with *T. angustifolia*, an almost complete moss cover of *P. lyellii*, *Sphagnum fimbriatum*, *S. squarrosum*. Species-rich. Reed less well developed than in the above. Water level a few centimetres below ground level. A complete *Sphagnum* cover, in the absence of flooding, drying, or disturbance, becomes self-sufficient and leads to peat bog. Reed can grow well while roots and rhizomes are in the high-nutrient, calcium moderately-rich high pH fen soil below

f) *Sphagnum* reedland, with a highly developed layer of (mostly) *S. fimbriatum* and *S. squarrosum*. The beginning of raised bog. No nutrients except those in rain reach the bog. Open and species-rich herb layer, in which reed occurs with limited cover. Water level to 10 cm below ground level

**3. Water regime – Example: Comparative water depths in four nutrient-poor Scottish lochs (principal species occurring with *Phragmites*) (Spence, 1964)**

| Water depth cm | Lindores | Kilconuhar | Carlingwark | Clunie |
|---|---|---|---|---|
| 130–100 | *Chara* spp., *Nitella* spp., *Potamogeton praelongus* | No reed | No reed | No reed |
| 60–100 | *Littorella uniflora* | No reed | No reed | No reed |
| 20–60 | *Potamogeton filiformis*, *Typha angustifolia*, *Polygonum amphibium*, *Littorella uniflora*, *Nuphar lutea* | *Phragmites* society, *Scirpus tabernaemontana* | *Typha angustifolia* No reed | No reed |

Table 12.3 continued overleaf...

**Table 12.3** Continued

| Water depth cm | Lindores | Kilconuhar | Carlingwark | Clunie |
|---|---|---|---|---|
| 0–20 | *Typha angustifolia*, *Eleocharis palustris*, *Littorella uniflora* | *Phragmites* society, *Phragmites australis–Galium aparine* | *Glyceria maxima*, *Typha latifolia*, *Carex rostrata*, *Sparganium erecta* | *Nuphar lutea*, *Polygonum amphibium*, *Menyanthes trifoliata*, *Littorella uniflora–Lobelia dortmanna* |
| 0– -20 | *Glyceria maxima*, *Iris pseudacorus*, *Littorella uniflora* | *Glyceria maxima*, *Ranunculus lingus*, *Phragmites–Galium* | *Glyceria maxima*, *Typha latifolia*, *Carex rostrata*, *Sparganium erectum* | *Phalaris arundinacea–Filipendula ulmaria–Iris pseudacorus* |
| -20+ | No reed | No reed | No reed | No reed |

It is important to note:
*Phragmites* goes deeper into some waters; depth penetration depends on other habitat factors.
When *Phragmites* is the only tall monocot present in water, it is not suffering from competition.
At the dry end of the range, competition is effective (see Chapter 7) and, with less nutrients, *Phragmites* succumbs more easily. (Other tall species are present, so grazing has not removed all.)

**4. Grazing regime – Example: Effect of grazing at edge of Scottish lochs (Spence, 1964)**

| Water depth cm | Ungrazed | Grazed |
|---|---|---|
| +10 | *Phragmites* (80 cm high) | *Phragmites* (20 cm high) |
| | *Phragmites* | |
| | *Phragmites–Myrica gale* | *Carex nigra–C. panica* sward |
| -20 | *Myrica gale* (*Calluna vulgaris*) | *Molinia caerulea* (*Calluna vulgaris*) |

More intensive grazing would remove the *Phragmites*. Once the rhizomes are drained of food, by constantly sending up new shoots which are eaten before they can make food, they die.

# Chapter 13
# Remarkable reeds in marvellous marshes

*What you know, you can value;*
*What cannot be justified, will not be protected.*

IMCG Restoration Manual Draft, 2006

*They went quickly down until the willows were*
*left behind and the turf had given place to the*
*cold black soil where the thin grass straggled...*
*Then the soil grew softer and began to stick;*
*rivulets of water were in it, and became pools;*
*the pools grew larger and tall reeds were*
*growing from the water; and suddenly, before*
*their feet, there was nothing else. Reeds and*
*water filled the world, and the Mere had imposed*
*itself. Nothing moved; nothing broke the stillness*
*but the lap of water on the reeds.*

Neill, 1952

...Even in winter, when the summer-grazed grass is long and straggly and of a dull browny green, and the damp soil looks black and cold. How well this author, too, remembers those years studying winter reedbeds! That straggly, tussocky grass showing winter was come indeed, its presence showing botanically that there is grazing. Livestock graze and keep down reed until the ground is too wet, both for livestock and for the grass. Even then the reedbed imposes its own fascination, its eerie other-worldness, the sense of moving to a different time and place. A sense greater than the sense of cold and wet! Or, rather, the cold and wet are gathered up into the sense of the reedbed.

The reedbed is human size (unless 3–10 m high), and this is part of its uniqueness. A wood dominates a person, a person dominates a heath. But a reedbed, tall enough to overtop a person, though only by 0.5 m or so, that is different to either. Above is the sky alone, and in Norfolk particularly the skies are marvellous. Working all day, surrounded by the reeds, no sound but that of reeds and birds, no view beyond a few metres except looking up to the blue sky and remarkable white clouds: an experience not to be forgotten.

Reedbeds hide the researcher. Once, in the Admiralty Marsh at Rosyth, which had open access and this author had permission to visit, she saw, in mid-morning, a policeman in the distance. Obviously nothing to do with her, so she returned into the demonstration pattern of genotypically different recent clones. When moving between them, she saw the policeman a couple of times and wondered vaguely why any such should stay in a marsh. In late afternoon, heading back to the car, a dozen burly Admiralty Police converged upon her. They had spent 'all day' searching for the spy they had glimpsed! The author had spent her usual enjoyable solitary day almost always within the beds. The Police had spent a frustrating day searching in the open around the beds (and not calling out).

So, even for people, the reedbed is not always as peaceful as it appears. The researcher expects total peace and quiet, not interference from other people. One Bank Holiday this author, in Woodwalton Fen, Huntingdonshire, more in than out of the reedbeds, yet saw, far off, occasional people. Disgusted after having seen half a dozen, she left in early afternoon, unable to tolerate not being alone. That is the bad side, that the sense of otherness is addictive! Of course there is a worse side. Fugitives can hide in the reedbeds, and being smoked out by fire is the most effective means of capture.

We, who go down to reedbeds, for research, harvesting or maintenance, working alone, feel almost a magic, refreshing, renewing and for the researcher, intensifying the wish to learn. In deplorable 'science-speak', this is known as the 'spiritual societal effect', a phrase calculated to end it.

Solitary? Peaceful? Other-worldly? Only for people! For the reed it is an intense but silent battle for survival. There is no peace in the reedbed for *Phragmites*. There is a struggle for existence. There may or may not be a struggle by the plant for the existence of the reedbed, or that may be done by the, e.g., water regime alone. There is certainly a struggle between the shoots on a single plant with small shoots frequently succumbing to food competition from large ones (Chapters 3 and 4).

In Europe, the struggle for community existence, in the hazy band 'twixt land and water', is usually won by *Phragmites*. In the Americas, Australasia, Africa and much of Asia, it is usually won by other reedswamp species. Even in Europe there is more than just the wide reedswamp. There is the isolated clump, the hectares of Poor Fen, the fringes around dykes and drains, the patches in bogs and green fields and, in particular, the willow-reed bands enclosing streams which have not been drained, both flatter lowland, and on gentle slopes up in (fertile, wooded) mountains. These bands have dominant reed, but often 1–1.5 m high, not the taller reedbed of the lakes, or the giant reedbeds of lakes in some warm-summer, more eastern areas, like Turkey, Iraq, North Africa and the Danube Delta.

The willow-reed bands have the reeds outside the willows. The willows get (or before drainage and regulation, used to get) the scour from the stream, which they, but not dominant reed, can tolerate. The reed grows out further from the stream, so having less water, so being shorter: like the reed on railway banks. In wet woodland, whether this is a narrow band of sallow or willow, or a genuine large wood, *Phragmites* is also shorter, flaccid and plainly not 'healthy' (Figure 6.4). But if 'unhealthy' means doing poorly and heading for extinction, that is certainly not so. *Phragmites* stays, whatever its appearance, for however long the wood stays. There is never a 100% canopy in such woods. The wide-ranging rhizomes can produce many shoots in sunlit glades, where the reeds are also taller, shaded for only part of the day; while in more shade shoots are sparser, and are rare in the heaviest shade.

The influence of light interacts with the structure of the *Salix* wood or carr and the behaviour of *Phragmites*. The *Phragmites* stays.

The influence of the other widespread (well, for light, in the formerly widespread woods and bands) damage factor – grazing – is different. In the wood, *Phragmites* stays, because light is always adequate, taking the wood as a whole. This is not the case with grazing. This also can be mild, but it can be more severe, and at its heaviest, eliminates reed. Grazed shoots must be replaced, if the plant is to photosynthesise and live. Replacements are few if minerals, so food, are in short supply, so grazing kills reed quicker in nutrient-poor habitats. But death comes whether quick or slow, as it does not, in a shaded wet woodland.

In the disappearing populations in Malta (Chapter 8), the study is more like a Treasure Hunt. Is it still here? Can it be seen? Is it hidden by looking like a grass tuft? But, when found, there is much of interest. How is a nation's *Phragmites* lost? Well, long-continued drying and disturbance, does it, very very slowly, over decades and even a small number of centuries. How fascinating, though, is the great range of *Phragmites* behaviour in the small area left! Dwarf and prickly, *legehalme*, 3 m high reed, field reed, reed growing through tarmac, varied genotypes, crystal-ball Egyptian introduction, and the grass-like remnants at the edge of stone wall at the 'rivers' edge', likely to be washed off in the next decade.

Then, what are the uses of reeds and reedbeds? Whole human populations can live in, and in extreme cases, nearly entirely on, the reedbed. The extreme variety of uses (Chapter 9) from pens to spears, medicine and baskets is most surprising, for a plant seldom very obvious to courts, governments and cities. These prefer to be sited on dry land (Chapters 1 and 11).

Uses continue, for thatch in England and those countries now supplying England's shortage of reed. The recent use of constructed wetlands for water purification is still expanding, and in Europe this normally uses reed. And how many people, gazing at a reedswamp, think, "There is my water storage?" And probably even fewer think, "There is my carbon storage, taking carbon from the air, we must not let it be destroyed because of all the carbon that would be let into the air." The Fenland, and similar agricultural areas of The Netherlands and elsewhere, attest to that. Every year peat oxidises and vanishes, by maybe a metre in 25–100 years, a huge amount, when spread over hectares. What about peat burned as fuel for power stations? For crofters, peat fuel is a small, sustainable loss, but power stations use huge amounts. So do gardeners for horticulture and new peat is bought each year to counter oxidation. (If every flower bed needed peat just the once, it would still be bad, not evil.)

*Phragmites* is so important in so many ways, and since its recorded uses started early in Ancient Egypt, it must be supposed it has been so since *Homo sapiens* – if not Neanderthal man – first invaded Europe, and in Africa before that. Though early man in Africa would probably not have had so many uses. There are better plants there for spears, pipes, and thatch, for example (and it took some time for early man to develop sophisticated technology). From spears to carbon storage is a long jump, but given millennia, over-population and increasing environmental damage, so it has happened. Reeds have remained important to people, for their benefit to people (see Lefeuvre, 1998).

The earliest English description of a reedbed is that in Felix's *Life of St Guthlac* (Colgrave, 1956), AD 730. Guthlac lived in the Isle of Crowland (701–716). This is the wet Anglo-Saxon period, when the Roman drainage had fallen into disrepair also. His fens had dense reedswamp, isolated reeds, thickets, presumably on small islets, where houses also could be built, and open water, pools large enough to have waves recorded. Guthlac travelled by boat, but feared foes coming on foot. 'Reedswamp', at the time, could be several species (Chapter 1), but that it was *Phragmites* is shown by documents being placed, for collection, in the top of broken reeds. *Phragmites* it must be, others could not support the weight. The water was muddy, presumably from peat disturbed by huge numbers of fish and fowl.

These days much the same can be found in, say, the Hungarian Lake Tisza or Lake Velenzia, except there is little land fringing the lake reedswamps: agriculture, housing, roads, recreation all fill that space. However, there is a great psychological difference. Hermits, and to some extent monks and nuns, went into the wetlands so their prayers would purify and sanctify the haunt of demons. (When they got hallucinatory fevers of the malarial type, the presence of demons was assured.) *Beowulf*, the Anglo-Saxon epic, deals with the monsters of wetland. Now, such reedswamps are full of light, both actually and figuratively. The actual light has obviously always been there. The eyes perceiving doom and gloom are fewer.

After the Anglo-Saxons, came another drier climatic optimum – that of mediaeval times. In 1250 Matthew Paris could write that the Fenland was recently transformed from a haunt of devils to fertile meadows and fields (more from climate than drainage, though there was drainage).

Wetlands, including reedbeds, are odd in that they attract a great deal of talk from national and international organisations, but hardly any money for research or conservation. Most vegetation types have both (e.g., ancient oak woods) or neither (e.g., river plants). There is plenty on the value of wetlands, but where is the money even to fence out livestock, let alone stop the abstraction of the underground water which keeps wetland wet? Dry a reedbed and in due course, a woodland is there instead.

As discussed in Patten (1990) non-consumptive values of wetlands now tend to involve higher aspirations, philosophy, beauty, learning, spiritual and humanitarian concerns. Not the fearful abode of all things evil, which could be combated only with strong spirituality.

Losing a wetland means losing its associated species, its flora (Chapter 12), and its fauna (Chapter 4), and a varied, prolific and interesting fauna it is, too (e.g., George, 1992). It would take centuries, if that, to replicate the fauna in a newly wetted field turned reedbed, with different soil and treatment, and isolated from older marshes. Too many people think it is possible to 'restore' a reedbed in new surroundings. To create (or restore) a reedbed is. The vigour and versatility of reed means it can grow in most wet places (see earlier chapters). To restore the birds is more difficult. Some of them, yes, they will come whenever they see a habitat. Not all though, and certainly not all the invertebrates. George (1992) for Broadland describes the richness and biodiversity, but also the differences, and often substantial differences, between the various marshes and Broads. Those are very difficult to restore, particularly when many of them are still to be found in particular marshes only, and no one knows the reasons for the different biodiversity: was it the mowing of 500 years ago? That, by itself, is a silly question, but the integrated management over the past 700–800 years could well be the cause, and one not possible to emulate (without *extremely* accurate details of today's habitat; and even then there is the question of dispersal).

Is it possible that reedbeds – or vegetation with sparse reeds, or *Phragmites–Salix* bands – could be restored to anything like their former quantity? Would it be desirable if it could be done? The desirability is the easier question, and the answer varies with the person answering. More reedbeds would be good, for the values already described, for harbouring fauna, and for spiritual values for people. However, they are bound to be limited, since wetland diseases are, to put it mildly, unwanted. Then, space on this earth is limited, and will become more limited, as an already far too large human population becomes even more so. This space will have to be used primarily for the immediate benefit of people. Will food become less important than other future uses for harvested reed? That is unknown, but unlikely. Only thirdly, unfortunately, is the ethical concern for the conservation of natural heritage. Some people may talk differently, but how many would actually let children die for the sake of a new reedbed (not even for the sake of an ancient one)? Very few! If, though, some plague halved the world's population, many areas in Europe would revert to wetland, partly reedbed, without any direct action.

On whether increased reedbeds are possible now, in current conditions, the answer seems that all things are possible, but this is unlikely without a profitable, commercial use. Such restoration means using land specialised for another purpose.

Agricultural land on drained wetland can be reverted, as seen at Wicken Fen (Chapter 11) and many other recent 'restoration' projects. Wicken Fen is a good example, however, because the reedbed is on a former sedge-mix area, lower-nutrient, and with *Phragmites* only sparse or locally dominant. How far are reedbeds more valuable than other types of wetland? Is it right to allow reedbed where, traditionally, there was Poor Fen? Even worse, should another existing wetland type be destroyed to make a reedbed? With the overall nutrient status of waters becoming higher, it is the lower-nutrient wetland types that are in greater need of protection, and the least likely to develop on abandoned land. That some people derive greater pleasure from reedbeds is no reason to ignore other types!

Even where land is set aside for reedbed, there is another quickly-diminishing essential: water. Reedland is wetland! Where low-lying ground is drained, drains can be blocked, and water will return. When it is abstracted, it is gone. There are now concerns whether former fen and marsh can become wet again, because its water is taken away upstream. Wet woodland, yes; but flooded vegetation? Many small 'wetlands' still there and fenced, have been severely dried in the last half-century because abstraction licenses, granted on the assumption water extraction would remain constant, have been used for vastly increased amounts. Retrospective law without compensation is always difficult: and who wants to pay the compensation?

So water loss means not only new wetland may be impossible, but that existing wetlands are lost. It should be remembered that reedswamp uses more water than any other type of vegetation, so needs more water available! And if wetlands are dry enough for woodland, and are not intensively managed, woodland they will become.

To be effective, EU and similar-project plans to increase reedbeds must also provide the water.

Can such projects increase total reedbeds? They increase those where drying is from drainage, certainly. Prevent loss in abstracted areas? Doubtful. The formula 'Conserve the best, forget the rest' means loss of back-up and surrounding habitat, with eventual degradation also of the 'best'.

These dubious thoughts could be overturned by the as yet unknown effects of unknown climate change. As already argued (Haslam, 2008), the main effect on European wet habitats has been that from increasing temperature over the past three centuries.

Far too many floods ascribed to climate change are in fact due to land management. It is easy to prevent rain water upstream from soaking into the soil:

- intensive agriculture;
- firming soil;
- settlement;

- roads;
- straightening, channelling and lining rivers.

It is then even easier to say downstream floods are due to climate change. So far not enough is known about the rainfall pattern (amount of rain, heaviness of rain, each day of the year) since 1700 to say there is a change in western Europe (rather than a recent minor alteration which could be cyclic). Rises in sea level of course greatly alter the configuration of estuaries and other muddy coasts, and therefore reedbeds there (Chapter 1).

Nevertheless, *Phragmites* is a fascinating and unique species. It is one of the few extant plants to build peat and, unlike *Sphagnum*, to build it so that it destroys its own habitat. Reedswamp, by building peat, changes the habitat from shallow water to wet wood (sparse reeds continue into wet woodland, though not into dry). Just a grass, a wild grass, yet how complex is its behaviour, how adaptive, how sensitive to habitat and environment, how able to persist under difficulties! Its underground rhizome, inadequately known even though studied more than most species, is so very large and complex, and so crucial for the behaviour of the plant. While the aerial shoots photosynthesise, and the root tips absorb minerals and water, it is the rhizome which controls how many, where, what size, what purpose and at what level in the substrate the buds arise.

This part of the plant, well underground, the best protected from the vagaries of the habitat above, determines the future of the plant. It determines whether there shall be

- few, small buds which form short aerial shoots;
- many 20–50 m *legehalme* shooting the plant across a bare marsh, covering the ground with reeds quicker than other species can invade and shade;
- many branch horizontal and curving vertical rhizomes with even more wide buds to form dominant reed in good conditions;
- few rhizomes and high-rising narrow buds giving sparse, short shoots in over-dry or grazing-damaged places;
- an advancing margin little can resist, or sparse rhizomes slipping around.

All these are determined by the horizontal rhizome, acting on signals received from above and below. How exactly does this work? Surely most interesting, but a specialisation of study quite different to that of the ecologist!

Exactly how does, to use an extreme, a *Juncus subnodulosus* mat decrease *Phragmites*' bud production, narrow these buds greatly around soil surface during emergence, make them emerge a fortnight later than those just outside, yellow and die a fortnight earlier in autumn? What complex chemistry, signals and response!

A plant whose viable shoots can vary from 5 cm to over 10 m high and whose genotypes include ones specialised, e.g., for salt, and no doubt other extremes, which can remain in place (with apparently many of the same clones) for several thousand years.

## What a plant!

## *Voilà Phragmites! Vive Phragmites!*

# Bibliography

(Information taken from the author's research is not separately referenced. Most of the basics are in the publications listed here, which include further evidence.)

**Adams JB & Bate GC. 1999.** Growth and photosynthetic performance of *Phragmites australis* in estuarine waters: a field and experimental evaluation. *Aquatic Botany,* **64**(3–4): 359–368.

**Ailstock MS, Norman CM & Bushmann PJ. 2001.** Common reed *Phragmites australis*: control and effects upon biodiversity in freshwater non-tidal wetlands. *Restoration Ecology*, **9**(1): 49–59.

**Anglo-Saxon Poetry (transl. RK Gordon). 1954 edn**. London: Dent.

**Anthony PA. 2000.** The macro fungi and decay of roofs thatched with water reed, *Phragmites australis*. Mycological Research, **103**(10): 1346–1352.

**Arber A. 1934.** *The Gramineae*. Cambridge: University Press.

**Armstrong R. 2004.** Survey of SPA interest at Castle Loch and Hightae Lochs. Edinburgh: Scottish Natural History. SNH Commissioned Report 24.

**Armstrong J & Armstrong W. 2001.** An overview of the effects of phytotoxins on *Phragmites australis* in relation to die-back. *Aquatic Botany*, **69:** 251.

**Armstrong J, Armstrong W & Beckett PM. 1992.** *Phragmites australis*: Venturi and humidity-induced pressure flows enhance rhizome aeration and rhizosphere oxidation. *New Phytologist*, **120:** 197–207.

**Asaeda T, Rajapakse L, Monatunge J & Sahora N. 2006.** The effect of summer harvesting of *Phragmites australis* on growth characteristics and rhizome resource storage. *Hydrobiologia*, **553:** 327–335.

**Athie D & Cerri CC, eds. 1987.** *The use of macrophytes in water pollution control.* Oxford: Pergamon Press.

**Ausden M. 2005.** The effects of cattle grazing on tall-herb fen vegetation and molluscs. *Biological Conservation*, **122**(2): 317–326.

**Bahrman N & Gorenflot R. 1983.** Apport des proteins foliares solubles dans l'interprétation du complexus polyploids du *Phragmites australis* [Cav.] Trin. ex Steudel. *Revue Generale Botanique*, **90:** 177–184.

**Bailey RG, José PV & Sherwood BR, eds. 1998.** *United Kingdom Floodplains*. Otley: Westbury Academic & Scientific Publishing.

**Bakker D, Jonker JJ & Smits H. 1960.** Land reclamation in Holland; bringing the polders into production. *Span*, **3**: 143–151.

**Bardsley L. 2001.** *Wetland restoration manual. Tranche 1 and 2*. Newark: The Wildlife Trusts.

**Bartlett J. 1999.** *An investigation into the potential for the recreation of a brackish lagoon at Hook Lake*. Southampton: Southampton Institute.

**Bastlová DI, Bastl M, Cíñková H & Květ J. 2006.** Plasticity of *Lythrum salicaria* and *Phragmites australis* growth characteristics across a European geographical gradient. *Hydrobiologia*, **570:** 1.

**Bayly IL & O'Neill TA. 1972.** Seasonal ionic fluctuations in a *Phragmites* communis community. *Canadian Journal of Botany*, **50**(11): 2103–2109. The National Research Council of Canada.

**Bibby CJ & Lunn J. 1982.** Conservation of reedbeds and their avifauna in England and Wales. *Biological Conservation,* **23**: 169–180.

**Binz HR. [no date].** Measurements of reed rigidity: some theoretical and methodological aspects. Research Report.

**Binz-Reist H. 1989.** Mechanische Belastbarkeit natürlicher Schilfbestände durch Wasser, Wind und Treibzeug. *Veröffentl. Geobot. Inst. ETH,* Zürich, **101:** 1–536.

**Bittmann E. 1953.** Das Schilf. *Angewandte Pflanzensoziologie*, **7**.

**Björk S. 1963.** Chromosome geography and ecology of *Phragmites communis. Södra Sveriges Fiskeriförening*, 1961–1962.

**Björk S. 1967.** Ecological investigations of *Phragmites communis*: studies in theoretic and applied limnology. *Folia Limnologica Scandinavica,* **14**.

**Blackmore RD. 1869.** *Lorna Doone*. London: Blackie.

**Borg J. 1927.** *Descriptive Flora of the Maltese Islands*. Malta: Government Printing.

**Bornkamm R & Raghi-Atri F. 1986.** Uber die Wirkung unterschiedlicher Gaben von Stickstoff und Phosphor auf die Entwicklung von *Phragmites australis* (cav.) Trin. ex Steudel. *Archivum Hydrobiologia,* **105:** 423–441.

**Bosman MTM. 1985.** Some effects of decay and weathering on the anatomical structure of the stem of *Phragmites australis* Trin. ex Steudel. *AWA Bulletin N.S.*, **6:** 165–170.

**Brändle R, Cíñková H & Pokorny J, eds. 1996.** *Adaptation Strategies in Wetland Plants: links between ecology and physiology*. Grangärde, Sweden: Oppolus Press.

**Brix H & Cíñková H. 2001.** Introduction: *Phragmites*-dominated wetlands, their functions and sustainable use. *Aquatic Botany*, **69:** 87–88.

**Brix H, Sorrell BK & Lorenzen B. 2001.** Are *Phragmites*-dominated wetlands a net source or net sink of greenhouse gases? *Aquatic Botany*, **69:** 313–324.

**Bromfield WA. 1841–2.** Curious form of the Common Reed. *Phytologist*, **1:** 146.

**Brookhouse J. 1998.** Creating reedbeds for wildlife. *English Nature, Enact*, **6**(2): Supplement, 5–6.

**Burdick DM & Konisky RA. 2003.** Determinants of expansion for *Phragmites australis*, common reed, in natural and impacted coastal marshes. *Estuaries,* **26**(2B): 407–416.

**Buttery BR. 1959.** An investigation into the competition between *Glyceria maxima* (Hartm.) Holmb., and *Phragmites communis* Trin. in the region of Surlingham Broad, Norfolk. PhD Thesis, University of Southampton.

**Buttery BR & Lambert JM. 1965.** Competition between *Glyceria maxima* and *Phragmites communis* in the region of Surlingham Broad. I. The competition mechanism. *Journal of Ecology*, **53:** 163–181.

**Buttery BR, Williams WT & Lambert JM. 1965.** Competition between *Glyceria maxima* and *Phragmites communis* in the region of Surlingham Broad. II. The fen gradient. *Journal of Ecology*, **53:** 183–196.

**Caffrey JM & Beglin T. 1996.** Bankside stabilisation through reed transplantation in a newly constructed Irish canal habitat. *Hydrobiologia*, **340:** 349–354.

**Chambers RM, Meyerson LA & Saltonstall K. 1999.** Expansion of *Phragmites australis* into tidal wetlands of North America. *Aquatic Botany*, **64**: 3.

**Carter SP. 2003.** Reedbeds as refuges for water voles (*Arvicola terrestris*) from predation by introduced mink (*Mustela vison*). *Biological Conservation*, **111**(3): 371–376.

**Chapman VJ. 1960.** *Saltmarshes and salt deserts of the world*. London: Leonard Hill.

**Cížková H, Brix H & Herben T. 2001.** Ecology of *Phragmites* populations in the changing landscape. *Folia Geobotanica*, **35**. Uppsala, Sweden: Opulus Press.

**Cížková H, Brix H, Kopecky J & Lukavská J. 1999.** Organic acids in the sediments of wetlands dominated by *Phragmites australis*: evidence of phytotoxic concentrations. *Aquatic Botany*, **64**(3–4): 303–316.

**Cížková H, Istvánovics V, Bauer V & Balázs L. 2001.** Low levels of reserve carbohydrates in reed (*Phragmites australis*) stands of Kis-Balaton, Hungary. *Aquatic Botany*, **69**: 209–216.

**Cížková H, Strand JA & Lukavaská J. 1996.** Factors associated with reed decline in a eutrophic fishpond, Rozmberk (South Bohemia, Czech Republic). *Folia Geobotanica Phytotaxonomia*, **31**: 73–84.

**Cížková-Kancelova H & Květ J. 1991–1993.** Effects of interactions between eutrophication and major environmental factors on the ecosystem stability of reed vegetation in European land/water ecotomes (EUREED).

**Cížková-Kancelova H, Květ J, Lakarska J & Priban K. 1991–1993.** Physiological mechanisms of reed decline in eutrophic habitats: role of carbon balance. *Hydrobotany Report, Academy of Sciences of the Czech Republic*. Institute of Botany, Section of plant ecology at Třebon. 26.

**Clapham AR. 1940.** The role of bryophytes in the calcareous fens of the Oxford District. *Journal of Ecology*, **28**: 71–80.

**Clayton WD. 1967.** Studies in the Gramineae: XIV. *Kew Bulletin*, **21**: 111–117.

**Clayton WD. 1968.** The correct name of the common reed. *Taxononomica*, **17**: 168–169.

**Clevering OA & Lissner J. 1999.** Taxonomy, chromosome numbers, clonal diversity and population dynamics of *Phragmites australis*. *Aquatic Botany*, **64**(3–4): 185.

**Clevering OA, Brix H & Lukavská. 2001.** Topographic variation in growth responses in *Phragmites australis*. *Aquatic Botany*, **69**: 89–108.

**Colenutt S. 2003.** Managing priority habitats for invertebrates. *Habitat Section 25: Reedbeds*. Peterborough Bug Life.

**Colgrave B, ed. 1956.** Felix (author). *Life of St Guthlac*. Cambridge: University Press.

**Cooper FW. 1972.** The thatcher and the thatched roof. In *The Reed*. Norfolk Reedgrowers Association, 58–69.

**Cooper PF & Findlater BC, eds. 1990.** *Constructed wetlands in water pollution control*. Oxford: Pergamon Press.

**Cope TA. 1982.** *Flora of Pakistan, 143*. Karachi, Pakistan: University of Karachi.

**Couwenberg J. 2007.** Biomass energy crops on peatlands: on emissions and perversions. *IMCG Newsletter*, **2007**/3: 12–14.

**Curran PL. 1969.** Fertility of *Phragmites communis* Trin. *Irish Naturalists Journal*, **16**: 242.

**Darby HC. 1968.** *The draining of the Fens*. Cambridge: University Press.

**Darby HC. 1983.** *The changing Fenland*. Cambridge: University Press.

**Davies EG. 1945.** Figyn Blaen Brefi: A Welsh upland bog. *Journal of Ecology*, **32**: 147–166.

**Defoe D. 1724.** *A tour through the whole island of Great Britain*. 1959 edn. London: Everyman, Dent.

**Dittrich A & Westrich B. 1990.** Erosions Erscheinungen und Schilfrückgang in der Flachwasserzone des Bodensees. *Landschaftsentwicklung und Umweltforschung*, TU Berlin, **71**: 86–93.

**Djerbrouni M. 1992.** Variabilité morphologique, caryologique et enzymatique chez quelques population de *Phragmites australis* [Cav.] Trin. ex Steudel. *Folia Geobotanica Phytotaxonomia*, Praha, **27**: 49–59.

**Durska B. 1970.** Changes in the reed (*Phragmites communis* Trin.) conditions caused by diseases of fungal and animal origin. *Polskie Archivum Hydrobiologica*, **17**(30): 373–396.

**Dvorak J & Imhof G. 1998.** The role of animals and animal communities in wetlands. In D.F. Westlake, J. Květ, & A. Szczepanski, eds. *The Production Ecology of Wetlands*. Cambridge: University Press, 211–317.

**Dykyjová D. 1970.** Comparative biometry of *Phragmites communis* ecotypes and its significance to investigation of reed stands productivity. In *Productivity of terrestrial ecosystem production processes. PT-PP/IBP. Rep.*, **1**: 105– 109. Praha: Czechoslovak Acad. Sci.

**Dykyjová D. 1971.** Ekomorfózy a ekotypy rákosu obecného *Phragmites communis* Trin. [Ecomorphoses and ecotypes of *Phragmites communis* Trin.]. *Preslia*, **43**: 120–138 [Engl. sum.].

**Dykyjová D. 1979.** Selective uptake of mineral ions and their concentration factors in aquatic higher plants. Reports. *Folia Geobotanica Phytotaxonomia* Praha, **14**: 267–325.

**Dykyjová D & Hradecká D. 1976.** Production ecology of *Phragmites communis*, 1. Relations of two ecotypes to the microclimate and nutrient conditions of habitat. *Folia Geobotanica Phytotaxonomica*, Praha, **11**: 23–61.

**Dykyjová D & Květ J. 1970.** Comparison of biomass production in reedswamps communities growing in South Bohemia and South Moravia. In *Productivity of terrestrial ecosystem production processes. PT-PP/IBP Rep.* **1**: 71–78, Praha, Czechoslovak Acad. Sci.

**Dykyjová D & Květ J. 1978a.** Mineral nutrient economy in wetlands of the Trĕbon Basin Biosphere Reserve, Czechoslovakia.

**Dykyjová D & Květ J, eds. 1978b.** Pond littoral ecosystems, structure and functioning. *Ecological Studies*, **28**. Berlin: Springer-Verlag.

**Dykyjová D & Olehlova B. 1998.** Mineral economy and cycling of minerals in wetlands. In D.F. Westlake, J. Květ & A. Szczepanski, eds. *The Production Ecology of Wetlands*. Cambridge: University Press, 318–366.

**Dykyjová D, Véber K & Pribán K. 1971**. Productivity and root/shoot ratio of reedswamp species growing in outdoor summer hydroponic culture. *Folia Geobotanica Phytotaxonomia*, **6**: 233–254.

**Ecosystems Applied Ecologists. 1999.** Re-creation options for River Severn/Avon floodplain wetlands: final report. Peterborough: *English Nature Research Report*, **692**.

**English Nature. 2005.** Wetland HAP Steering Group. Getting wetter for wildlife: guidance on habitat restoration and creation. Peterborough: English Nature.

**Esselink. 2000.** The effects of decreased management on plant-species distribution patterns in a salt marsh nature reserve in the Wadden Sea. *Biological Conservation*, **93**(1): 61–76.

**Eurosite. 1997.** The creation, restoration and use of reedbeds. *Proceedings of the 28th Eurosite Workshop*, Carentan.

**Fiala K. 1970.** Rhizome biomass and its relation to shoot biomass and stand pattern in eight clones of *Phragmites communis* Trin. In *Productivity of terrestrial ecosystem production processes. PT-PP/IBP Report*, **1**: 95–98. Praha: Czechoslovak Academy of Sciences.

**Fiala K. 1973**. Seasonal changes in the growth and total carbohydrate content in the underground organs of *Phragmites communis* Trin. *IBP/PT-PP Report*, Trĕbon, **3**: 107–110.

**Fiala K & Květ J. 1971.** Dynamic balance between plant species in south Moravian reedswamps. In E. Duffey & A.S. Watt, eds. *The scientific management of animal and plant communities for conservation.* Oxford: Blackwell Scientific Publications, 241–269.

**Fiala K, Dykyjová D, Květ J & Svoboa J. 1968.** Methods of assessing rhizome and root production in reed-bed stands. In M.S. Ghilarov *et al.*, eds. *Methods of Productivity Studies in Root Systems and Rhizosphere Organisms*. Nauka: Leningrad, 36–47.

**Friday L, ed. 1997.** *Wicken Fen. The making of a wetland nature reserve*. Colchester: Harley Books.

**Frolik AL. 1941.** Vegetation on the peat lands of Dane Country, Wisconsin. *Ecology Monographs*, **11**: 117–140.

**Fuchs C. 1993.** The beetle *Donacia clavipes* as possible cause for the local reed decline at Lake Constance (Untersee). *Limnologia Aktuell*, **5**: 49–60.

**Fürtig K, Pavelic D, Brunold C & Brändle R. 1999.** Copper- and iron-induced injuries in roots and rhizomes of reed (*Phragmites australis*). *Limnologica*, **29**(1): 60–63.

**Fürtig K, Rüegsegger A, Brunold C & Brändle R. 1996.** Sulphide utilization and injuries in hypoxic roots and rhizomes of common reed (*Phragmites australis*). *Folia Geobotanica Phytotaxonomia*, **31**: 143–151.

**Gaevskaya NS. 1966.** *The Role of Higher Aquatic Plants as Food for Animals in Freshwater Reservoirs.* Nauka: Moscow.

**Gainey PA. 1997.** Trembling sea-mat: baseline distribution in England and species action plan. Peterborough: English Nature. Research Report 225.

**Gee GW, Fayer MJ, Rockhold ML & Campbell MD. 1992.** Variation in recharge at the Hanford Site. *Northwest Sci.*, **66**: 237–250.

**George M. 1992.** *The land use, ecology and conservation of Broadland.* Chichester: Packard Press.

**Gilbert G. 2003.** Nestling diet and fish preference of bitterns *Botaurus stellaris* in Britain. *Ardea*, **91**(1): 35–44.

**Gilbert G. 2005.** Nesting habitat selection by bitterns *Botaurus stellaris* in Britain and the implications for wetland management. *Biological Conservation*, **124**(4): 547–553.

**Godwin H. 1941.** Studies in the ecology of Wicken Fen. IV. Crop-taking experiments. *Journal of Ecology*, **29**: 83–106.

**Godwin H. 1981, 1978 and 1986.** *History of the British Flora.* Cambridge: University Press.

**Godwin H & Bharucha FR. 1932.** Studies in the ecology of Wicken Fen. II. The fen water table and its control of plant communities. *Journal of Ecology*, **20**: 157–159.

**Godwin H & Mitchell GF. 1938.** Stratigraphy and development of two raised bogs near Tregaron, Cardiganshire. *New Phytologist*, **37**: 425–454.

**Godwin H & Newton L. 1938.** The submerged forest at Borth and Ynyslas, Cardiganshire. *New Phytologist*, **37**: 333–344.

**Godwin H & Turner JS. 1933.** Soil acidity in relation to vegetational succession in Calthorpe Broad, Norfolk. *Journal of Ecology*, **21**: 235–263.

**Gordon-Gray KD & Ward CJ. 1971.** A contribution to knowledge of *Phragmites* (Gramineae) in South Africa, with particular reference to Natal populations. *Journal of South African Botany*, **37**: 1–30.

**Gorham E & Pearsall WH. 1956.** Production ecology. III. Shoot production in *Phragmites* in relation to habitat. *Oikos*, **7**: 206–214.

**Granéli W. 1989.** Influence of standing litter on shoot production in reed, *Phragmites australis* [Cav.] Trin. ex Steudel. *Aquatic Botany*, **35**(1): 99–110.

**Gratton C. 2005.** Restoration of arthropod assemblages in a *Spartina* salt marsh following removal of the invasive plant *Phragmites australis*. *Restoration Ecology*, **13**(2): 358–372.

**Griffiths BM. 1932.** The ecology of Butterby Marsh, Durham. *Journal of Ecology*, **20**: 105–128.

**Grosch UA. 1978.** Die Bedeutung der Ufervegetation für Fisch und Fischerei dargestellt am Beispiel Berlins. *Arb. Dtsch. Fisch.-Verb.*, **25**: 1–15.

**Grundling P-L, Linström A, Grobler R & Engelbrecht J. 2008.** The Tevreden Pan wetland complex of the Mpumalanga Lake District (South Africa). *IMCG Newsletter* 2008, **1**: 18–21.

**Gunnison D & Barko JW. 1989.** The rhizosphere ecology of submersed macrophytes. *Water Resources Bulletin*, **25**: 193–202.

**Güsewell S. 1998.** Does mowing in summer reduce the abundance of common reed (*Phragmites australis*)? *Bulletin of the Geobotanical Institute ETH*, **64**: 23–35.

**Gustafsson A & Simak M. 1963.** X-ray photography and seed sterility in *Phragmites communis* Trin. *Hereditas*, **49**: 442–450.

**Hafliger P, Schwarzlander M & Blossey B. 2006.** Comparison of biology and host plant use of *Archanara geminipunctata, Archanara dissoluta* and *Arenostola phragmitidis* (*Lepidoptera: Noctuidae*), potential biological control agents of *Phragmites australis* (*Arundinacea:Poaceae*). *Annals of the Entomological Society of America,* **99**(4): 683–696.

**Hammer DA, ed. 1989.** *Constructed wetlands for wastewater treatment.* Chelsea, Michigan: Lewis Publishers.

**Hanganu J, Mihail G & Coops H. 1999.** Responses of ecotypes of *Phragmites australis* to increased seawater influence: a field study in the Danube Delta, Romania. *Aquatic Botany*, **64**(3–4): 351–358.

**Harding M. 1992.** Redgrave and Lopham Fens: A case study in change due to groundwater abstraction. Report to English Nature. Suffolk Wildlife Trust.

**Hardy EM. 1939.** Studies of the post-glacial history of British vegetation. V. The Shropshire and Flin Maelor mosses. *New Phytologist*, **38**: 364–396.

**Hartog C den, Květ J & Sukopp H. 1989.** Reed. A common species in decline. *Aquatic Botany*, **35**: 1–4.

**Hartsendorf T & Rolletschek H. 2001.** Effects of NaCl-salinity on amino acid and carbohydrate contents of *Phragmites australis. Aquatic Botany*, **69**: 195–208.

**Haslam SM. 1968.** The biology of reed (*Phragmites communis*) in relation to its control. *Proceedings of the 9th British Weed Control Conference*, 1968, 392–397.

**Haslam SM. 1969a.** Stem types of *Phragmites communis* Trin. *Annals of Botany*, **33**: 127–131.

**Haslam SM. 1969b.** The development and emergence of buds in *Phragmites communis* Trin. *Annals of Botany*, **33**: 289–301.

**Haslam SM. 1969c.** The development of shoots in *Phragmites communis* Trin. *Annals of Botany*, **33**: 695–709.

**Haslam SM. 1969d.** *The Reed.* Norfolk Reedgrowers Association, 50. (1972 – Reprint with Further Chapters by Other Authors.) Revised edition 2009. Norwich: British Reed Growers Association.

**Haslam SM. 1970a.** The development of the annual population in *Phragmites communis* Trin. *Annals of Botany*, **34**: 571–591.

**Haslam SM. 1970b.** The performance of *Phragmites communis* Trin. in relation to water supply. *Annals of Botany*, **34**: 867–877.

**Haslam SM. 1970c.** Variation of population type in *Phragmites communis* Trin. Phragmites communis. *Annals of Botany*, **34**: 147–158.

**Haslam SM. 1971a.** Community regulation in *Phragmites communis* Trin. I. Monodominant stands. *Journal of Ecology*, **59**: 65–73.

**Haslam SM. 1971b.** Community regulation in *Phragmites communis* Trin. II. Mixed stands. *Journal of Ecology*, **59**: 75–88.

**Haslam SM. 1971c.** The development and establishment of young plants of *Phragmites communis* Trin. *Annals of Botany*, **35**: 1059–1072.

**Haslam SM. 1971d.** Shoot height and density in *Phragmites* stands. *Hydrobiolica*, **12**: 113–119.

**Haslam SM. 1972.** *Phragmites communis* Trin. Biological Flora of the British Isles 128. *Journal of Ecology,* **60**: 585–610.

**Haslam SM. 1973a.** Some aspects of the life history and autecology of *Phragmites communis* Trin. *Polskie Archiwum Hidrobiologia*, **20**: 79–100.

**Haslam SM. 1973b.** The management of British wetlands. I. Economic and amenity use. *Journal of Environmental Management*, **1**: 303–320.

**Haslam SM. 1973c.** The management of British wetlands. II. Conservation. *Journal of Environmental Management*, **1**: 345–361.

**Haslam SM. 1975.** The performance of *Phragmites communis* Trin. in relation to temperature. *Annals of Botany*, **39**: 881–888.

**Haslam SM. 1978.** *River Plants*. Cambridge: University Press.

**Haslam SM. 1979.** Infra–red colour photography and *Phragmites communis* Trin. *Polskie Archiwum Hidrobiologia*, **26**: 65–72.

**Haslam SM. 1982.** *Vegetation in British Rivers*. London: Nature Conservancy Council. 2 Vols.

**Haslam SM. 1987.** (With Harding JPC, Spence DHN. Authors not listed on title page). Methods for the use of Aquatic Macrophytes for assessing water quality 1985. *Methods for the examination of waters and associated materials*. Her Majesty's Stationery Office, London.

**Haslam SM. 1989.** Early decay of *Phragmites* thatch: an outline of the problem. *Aquatic Botany*, **35**: 129–132.

**Haslam SM. 1990a.** *Phragmites* culm strength and thatch breakdown: Some difficulties. *Landschaftsentwicklung und Umweltforschung*, Technische Universität Berlin, **71**: 58–77.

**Haslam SM. 1990b.** *River pollution: an ecological perspective*. London: Belhaven Press.

**Haslam SM. 1991.** *The Historic River*. Cambridge: Cobden of Cambridge Press.

**Haslam SM. 1994.** *Wetland habitat differentiation and sensitivity to chemical pollutants (non open water wetlands)*. Her Majesty's Inspectorate of Pollution, London.

**Haslam SM. 1995.** A discussion of the strength (durability) of thatching reed (*Phragmites australis*) in relation to habitat. Reed research report, Department of Plant Sciences, Cambridge. (Annual updates follow.) Norwich: British Reed Growers Association.

**Haslam SM. 2003.** *Understanding wetlands: fen, bog and marsh*. London: Taylor & Francis.

**Haslam SM. 2006.** *River Plants*. 2nd edn. Cardigan: Forrest Text.

**Haslam SM. 2008.** *The Riverscape and the River*. Cambridge: University Press.

**Haslam SM & Bone T. 2020.** *Reed—on the edge*. River Friend Series. Cambridge: Tina Bone UK.

**Haslam SM & Borg J. 1998.** *The river valleys of the Maltese Islands*. Malta: Foundation of International Studies.

**Haslam SM, Klötzli F, Sukopp H & Szczepanski A. 1998.** The management of wetlands. In D.F. Westlake, J. Květ & A. Szczepanski, eds. *The Production Ecology of Wetlands*. Cambridge: University Press, 405–464.

**Hatcher J. 1998.** *Wetland creation and strategic planning in East Anglia*. Sandy, Bedfordshire: RSPB.

**Hauber DP, White DA, Powers SP & De Francesch FR. 1991.** Isozyme variation and correspondence with unusual infrared reflectance patterns in *Phragmites australis* (Poaceae). *Plant Systomatics and Evolution*, **178**: 1–8.

**Hegi G. 1906.** *Illustrierte Flora von Mittel-Europa*. München: J.F. Lehmans Verlag.

**Hejny S & Segal S. 1998.** General ecology of wetlands. In D.F. Westlake, J. Květ & A. Szczepanski, eds. *The Production Ecology of Wetlands*. Cambridge: University Press, 1–77.

**Henson S. 2001.** National key sites for water voles in Norfolk. *Norfolk Bird and Mammal Report*, **35**(2): 304–308.

**HMSO London. 1994.** *Biodiversity. The UK Action Plan.*

**Hofmann K. 1991.** The role of plants in subsurface flow constructed wetlands. In: C. Ernier & B. Guterstam, eds. *Ecological engineering for waste-water treatment*. Gothenburg: Botskogen, 248–259.

**Hoi H. 2001.** *The ecology of reed birds*. Vienna: Austrian Academy of Sciences Press.

**Holdgate MW. 1954.** The vegetation of some springs and wet flushes on Tarn moor, near Orton, Westmorland. *Journal of Ecology*, **43**: 80–89.

**Holdgate MW. 1955.** The vegetation of some British upland fens. *Journal of Ecology*, **43**: 389–408.

**Hradecka D. 1973.** Common reed (*Phragmites communis* Trin.) in South Bohemia, South Moravia and South Slovakia. Morphology of the Inflorescences and Flower Wraps. Ecosystem Study on Wetland Biome in Czechoslovakia. *Czechoslovak IBP/PT-PP Report*, Trĕbon, **3**: 47–53.

**Hubbard CE. 1968.** *Grasses*. Harmondsworth: Penguin.

**Humphreys J. 2005.** *The ecology of Poole Harbour*. Amsterdam: Elsevier.

**Hürlimann H. 1951.** Zur Lebensgeschichte des Schilfs an den Ufern der Schweizer Seen. *Beitr. Geobot. Landesaufn. Schweiz*, **30**.

**Ilbagi H. 2006.** Common reed (*Phragmites communis*) is a natural host of important cereal viruses in the Trakya region of Turkey. *Phytoparasitica*, **34**(5): 441–448.

**Ingram HAP, Barclay AM, Coupar AM, Glover JG, Lynch BM & Sprout JL. 1980.** *Phragmites* performance in reed beds in the Tay estuary. *Proceedings of the Royal Society of Edinburgh*, **788B**: 89–107.

**Isambaev AI. 1964.** The underground stolons of common reed in different ecological environments. *Trudy Institute Botanica Alma-Ata*, **19**: 185–201.

**Iseli C. 1990.** Die Geschichte der Schilfröhrichte am Bielersee und Folgerungen für den praktischen Schilfschutz. *Land. Umwelt*, TU Berlin, **71**: 212–228.

**Jankovska V. 1998.** Flood plain biodiversity. In Prach *et al. Floodplain ecology and management: the Lun̆nice River in the Trĕbon Biosphere Reserve, Central Europe*. Amsterdam: SPB Academic Publishing.

**Jaweir JK & Al-Kenzawi MA. 2009.** Freshwater research in the marshes of Southern Iraq. *FBA News*, **45**: 2–3.

**Jennings JN & Lambert JM. 1951.** Alluvial stratigraphy and vegetational succession in the region of the Bure Valley Broads. *Journal of Ecology*, **39**: 106–119.

**Johnson S. 1755.** *A Dictionary of the English Language*. London: Strahern.

**Joosten H. 2007.** Peatlands, biofuels, energy: an introduction. *IMCG Newsletter*, **2007/3**: 7–8.

**Kaart A. 2007.** Carbon stocks in peatlands: a vital gap in the carbon market. *IMCG Newsletter*, **2007/3**: 20.

**Kassas M. 1951.** Studies in the ecology of Chippenham fen. II. Recent history of the fen from evidence of historical records, vegetational analysis, and tree-ring analysis. *Journal of Ecology*, **39**: 19–32.

**Kiendl J. 1953.** Zum Wasserhaushalt des *Phragmitetum communis* und des *Glycerietum aquaticae*. *Berichte der Deutschen Botanischen Gesellschaft*, **66**: 246–262.

**Kingsley C. 1866.** *Hereward the Wake*. London: Eversley.

**Kirby JJH & Rayner ADM. 1988.** Disturbance, decomposition and patchiness in thatch. *Proceedings of the Royal Society of Edinburgh*, **94B**: 245–253.

**Kirby JJH & Rayner ADM. 1989.** *Aspects of the decomposition, mechanical strength and anatomy of water reed (*Phragmites australis*) used in thatching*. Bath: University of Bath.

**Kiviat E & Hamilton E. 2001.** *Phragmites* use by Native North Americans. *Aquatic Botany*, **69**: 341–358.

**Klötzli F. 1973.** Conservation of reed beds in Switzerland. *Polskie Archivum Hydrobiologica*, **20**: 231–237.

**Koppitz H. 1999.** Analysis of genetic diversity among selected populations of *Phragmites australis* world-wide. *Aquatic Botany*, **64**(3–4): 209–222.

**Koppitz H, Dewender M, Ostendorp W & Schmieder K. 2004.** Amino acids as indicators of physiological stress in common reed *Phragmites australis* affected by an extreme flood. *Aquatic Botany*, **79**(4): 277–294.

**Koppitz H, Kühl H, Hesse K & Kohl J-G. 1997.** Some aspects of the importance of genetic diversity in *Phragmites australis* [Cav.] Trin. ex Steudel for the development of reed stands. *Botanica Acta,* **110**: 217–223.

**Krasovskii LI. 1962.** The biomass of the underground parts of the reed (*Phragmites communis*). *Bot. Zh. Kyyiv*, **47**: 673–677.

**Krauss M. 1993.** Die Rolle des Bisams (*Ondatra zibethicus*) beim Röhrichtrückgang an der Berliner Havel. *Limnologie Aktuell*, **5**: 49–60.

**Królikowska J, Priban K & Smid P. 1998.** Microclimatic conditions and water economy of wetlands vegetation. In D.F. Westlake, J. Květ & A. Szczepanski, eds. *The Production Ecology of Wetlands*. Cambridge: University Press, 376–404.

**Krumscheid P, Stark H & Peintinger M. 1989.** Decline of reed at Lake Constance (Obersee) since 1967 based on interpretations of aerial photographs. *Aquatic Botany*, **35**(1): 57–62.

**Krzakowa M. 1996.** Genetic diversity of *Phragmites australis* [Cav.] Trin. ex Steudel revealed by electrophoretically detected differences in peroxidases. In C. Obinger, U. Burner, R. Ebermann, C. Penel & H. Greppin, eds. *Plant Peroxidases: Biochemistry and Physiology*. University of Geneva, 184–189.

**Krzakowa M & Drapikowska M. 2000.** Sexual reproduction of *Phragmites australis* (Poaceae) in Baczkowski Pond (Poznan, Poland) revealed by peroxidase polymorphism. *Biologica Bulletin of Poznan*, **37**(1): 43–53.

**Krzakowa M, Kolodziejczak M, Drapikowska M & Jakubiak H. 2003.** The variability of reed (*Phragmites australis*) [Cav.] Trin. ex Steudel (Poaceae) populations expressed in morphological traits of panicles. *Acta Societatis Botanicorum Poloniae*, **72**( 2): 157–160.

**Ksenofontova T. 1989.** General changes in the Matsalu Bay reedbeds in this century and their present quality. *Aquatic Botany*, **35**(1): 111–120.

**Kühl H & Neuhaus D. 1993.** The genetic variability of *Phragmites australis* investigated by random amplified polymorphic DNA. *Limnologia Aktuell*, **5**: 9–18.

**Kühl H, Koppitz H, Rolletschek H & Kohl J-G. 1999.** Clone specific differences in a *Phragmites australis* stand. I. Morphology, genetics and site description. *Aquatic Botany*, **64**(3–4): 235–246.

**Kluczynski S. 1949.** Peat bogs of Polesie. *Memorialakitica Acaydemie Scientifica Cracovia*, B, **15**.

**Květ J. 1973a.** Mineral nutrients in shoots of reed (*Phragmites communis* Trin.). *Polskie Archivum Hydrobiologica*, **20**: 137–147.

**Květ J. 1973b.** Shoot biomass, leaf area index and mineral content in selected South Bohemian and South Moravian stands of common reed (*Phragmites communis* Trin.). Results of 1968. Ecosystem Study on Wetland Biome in Czechoslovakia. *Czechoslovak IBP/PT-PP Report*, Třebon, **3**: 93–95.

**Květ J & Westlake DF. 1998.** Primary Production in Wetlands. In D.F. Westlake, J. Květ & A. Szczepanski, eds. *The Production Ecology of Wetlands*. Cambridge: University Press, 78–168.

**Květ J, Svoboda J & Fiala K. 1969.** Canopy development in stands of *Typha latifolia* L. and *Phragmites communis* Trin. in South Moravia. *Hidrobiologia*, **10**: 63–75.

**Lakatós G. 1983.** Accumulation of elements in biotecton forming on reed (*Phragmites australis*) in two shallow lakes in Hungary. *Proceedings of the International Symposium of Aquatic Macrophytes, Nijmegen*, 117–122.

**Lakatós G. 1989.** Composition of reed biotecton (periphyton) in the Hungarian part of Lake Fertö. *BFB-Bericht*, **71**: 125–134.

**Lakatós G & Bíró P. 1991.** Study on chemical composition of reed-periphyton in Lake Balaton. *Biologisches Forschungs-institut für Burgenland*, Illmitz, **77**: 157–164.

**Lakatós G, Acs E & Buchtijarova LN. 1991.** Study on reed-periphyton in Lake Velence. *Annales Historico-Naturales Musei Nationalis Hungarici*, Budapest, 187–197.

**Lakatós G, Vörös L & Entz B. 1982.** Qualitative and quantitative studies on the periphyton/biotecton or reed in Lake Balaton. *BFB-Bericht,* **43**: 40–61.

**Lambert JM. 1946.** The distribution and status of *Glyceria maxima* (Hartm.) Holmb. in the region of Surlingham and Rockland Broads, Norfolk. *Journal of Ecology*, **33**: 230–267.

**Lambert JM. 1951.** Alluvial stratigraphy and vegetational succession in the region of the Bure Valley Broads. III. Classification, status and distribution of communities. *Journal of Ecology,* **39**: 149–170.

**Lefeuvre JC. 1998.** Reedbeds as living assets. English Nature, *Enact*, **6**(2): Supplement, 3–4.

**Lessmann JM, Brix H, Bauer V, Clevering OA & Comín FA. 2001.** Effect of climatic gradients on the photosynthetic responses of four *Phragmites australis* populations. *Aquatic Botany*, **69**: 109.

**Letts JB. 1993.** *Smoke-blackened thatch (SBT): a new source of late medieval plant remains from southern England.* Oxford Environmental Archaeology Unit.

**Letts JB. 1999.** *Smoke Blackened Thatch.* London: English Heritage.

**Leyrink L & Hubatsch H. 1993.** Die Uferzerstörung an den Netteseen. *Limnologia Aktuell*, **5**: 131–140.

**Lippert I, Rolletschek H & Kohl J-G. 2001.** Photosynthetic pigments and efficiencies of two *Phragmites australis* stands in different nitrogen availabilities. *Aquatic Botany*, **69**: 359–361.

**Lissner J, Schierup H-H, Comín FA & Astorga V. 1999a.** Effect of climate on the salt tolerance of two *Phragmites australis* populations. I. Growth, inorganic solutes, nitrogen relations and osmoregulation. *Aquatic Botany*, **64**(3–4): 317–334.

**Lissner J, Schierup H-H, Comín FA & Astorga V. 1999b.** Effect of climate on the salt tolerance of two *Phragmites australis* populations. II. Diurnal $CO_2$ exchange and transpiration. *Aquatic Botany*, **64**(3–4): 335–350.

**Löve A & Löve D. 1961.** Chromosome numbers of central and north-west European plant species. *Op. bot. Soc. bot. Lund.*, **5**: 1–581.

**Luft G. 1993.** Langfristige Veränderung der Bodensee Wasser-stände und mögliche Auswirkungen auf Erosion und Ufer Vegetation. *Limnologie aktuell*, **5**: 61–76.

**Lunn J. 1979 and 1980.** *Reedbed Surveys*. Sandy: Royal Society for the Protection of Birds.

**Matyuk IS. 1960.** Some types of reed growths in the delta of the river Volga. *Bot. Zh. SSSR*, **45**: 1681–1687.

**Mauchamp A, Blanch S & Grillas P. 2001.** Effects of submergence on the growth of *Phragmites australis* seedlings. *Aquatic Botany*, **69**: 147.

**McCartney M. 1999.** Hydrological implications of developing reedbeds in the Llyn Ystumllyn SSSI wetland. CCW Contract Science Report 267.

**McDougall DSA. 1972.** Marsh management for reed and sedge production. In *The Reed.* Norfolk Reedgrowers Association, 49–57.

**McKee J & Richards AJ. 1996.** Variation in seed production and germinability in common reed (*Phragmites australis*) in Britain and France with respect to climate. *New Phytologist*, **133**(2): 233–243.

**Mendelssohn IA, Sorrell BK, Brix H, Schierup H-H, Lorenzen B & Maltby E. 1999.** Controls on soil cellulose decomposition along a salinity gradient in a *Phragmites australis* wetland in Denmark. *Aquatic Botany*, **64**(3–4): 381–398.

**Micovski B. 1998.** Reedbeds and fish traps. *Enact*, 100–111. Peterborough: English Nature.

**Mills S, Taylor D & Wetton J-P. 1998.** Reedbeds in Somerset and Normandy – their creation and use. *Enact*, 7–9. Peterborough: English Nature.

**Misra RD. 1938.** Edaphic features in the distribution of aquatic plants in the English Lakes. *Journal of Ecology*, **26**: 411–452.

**Mitsch W, ed. 1994.** *Global wetlands: old world and new*. Amsterdam: Elsevier.

**Mitsch WJ & Gosselink JG. 1993.** *Wetlands*. 2nd edn. New York: Van Nostrand Reinhold.

**Moir J & Letts J. 1999.** *Thatch*. English Heritage Research Transactions, 5.

**Mook JH. 1967.** Habitat selection by *Lipara lucens* Mg. and its survival value. *Archives Néerlandia de Zoologie*, **17**: 469–549.

**Mook JH. 1971.** Influence of environment on some insects attacking Common Reed (*Phragmites communis* Trin.). *Hidrobiologia*, **12**: 305–312.

**Mook JH & van der Toorn J. 1982.** The influence of environmental factors and management of stands of *Phragmites australis*. II. Effects on yield and its relationships with shoot density. *Journal of Ecology*, **19**: 501–517.

**Moshiri GA, ed. 1993.** *Constructed wetlands for water quality improvement*. Boca Raton, Florida: Lewis Publishers.

**Moss CE. 1913.** *Vegetation of the Peak District*. Cambridge: University Press.

**Neil R. 1952.** *Moon in Scorpio*. London: Hutchinson.

**Neori A, Reddy K, Cíñková-Kancelova H & Agami M. 2000.** Bioactive chemicals and biological-biochemical activities and their functions in rhizospheres of wetland plants. *The Botanical Review*, **66**: 350–378.

**Neuhaus D, Kühl H, Kohl J-G, Dörfel P & Börner T. 1993.** Investigation on the genetic diversity of *Phragmites* stands using genomic fingerprinting. *Aquatic Botany*, **45**: 357–364.

**Newbould PJ. 1960.** The ecology of Cranesmoor, a New Forest valley bog. I. The present vegetation. *Journal of Ecology*, **48**: 361–383.

**Nikolajevsky VG. 1971.** Research into the biology of the Common Reed (*Phragmites communis* Trin.) in the USSR. *Folia Geobotanica Phytotaxonomia*, **6**: 221–230.

**Novotny V & Olem H. 1994.** *Water quality: prevention, identification and management of diffuse pollution*. New York: Van Nostrand Reinhold.

**Nuttall PM. 1997.** *Review of the design and management of constructed wetlands*. London: CIRIA.

**Ondok JP. 1971.** Horizontal structures of some macrophyte stands and its production aspects. *Hidrobiologia* (Bucharest), **12**: 47–55.

**Ostendorp W. 1989.** 'Die-back' of reeds in Europe – a critical review of literature. *Aquatic Botany*, **35**(1): 5–26.

**Ostendorp W. 1990.** Ist die Seeneutrophierung am Schilfsterben schuld? *Landesumwelt*, TU Berlin, **71**: 121–140.

**Ostendorp W. 1993.** Reed bed characteristics and significance of reeds in landscape ecology. *Akta Limnologica*, **5**: 149.

**Ostendorp W. 1995.** Estimation of mechanical resistance of lakeside *Phragmites*-reeds. *Aquatic Botany*, **51**: 87–101.

**Ostendorp W. 1999.** Susceptibility of lakeside *Phragmites*-reeds to environmental stresses: examples from Lake Constance-Untersee (SW-Germany). *Limnologica*, **29**(1): 21–27.

**Ostendorp W, Iseli C, Krauss M, Krumscheid-Lankert P, Moret J-L, Rollier M & Schanz F. 1995.** Lake shore deterioration, reed management and bank restoration in some Central European lakes. *Ecological Engineering*, **5**: 51–75.

**Ostendorp W, Tiedge E & Hille S. 2001.** Effect of eutrophication on culm architecture of lakeshore *Phragmites* reeds. *Aquatic Botany*, **69**: 177–194.

**Pallis M. 1916.** The structure and history of Plav; the floating fen of the delta of the Danube. *Journal of the Linnean Society* (Botany), **43**: 233–290.

**Pallis M. 1958.** An attempt at a statement concerning a vital unit as shown by the reed in the delta of the Danube. (M. Pallis).

**Parmenter J. 1995.** *The Broadland fen resource survey*. Norwich: Broad's Authority and English Nature.

**Patten BC, ed. 1990.** *Wetland and shallow continental water bodies. 1. Natural and human relationships*. The Hague: SPB Academic Publishing.

**Pauca-Comanescu M, Clevering OA, Hanganu J & Gridin M. 1999.** Phenotypic differences among ploidy levels of *Phragmites australis* growing in Romania. *Aquatic Botany*, **64**(3–4): 223–234.

**Pazourková Z. 1973.** Caryology of some forms of *Phragmites communis* Trin. Ecosystem Study on Wetland Biome in Czechoslovakia. *Czechoslovak IBP/PT-PP Report*, Třebon, **3**: 59–62.

**Peake JF. 1960.** A saltmarsh at Thornham, in north-west Norfolk. *Transactions of the Norfolk and Norwich Naturalists Society*, **19**: 56–62.

**Pelechaty M. 2004.** Can reed stands be good indicators of environmental conditions of the lake littoral? A synecological investigation of *Phragmites australis* dominated phytocoenoses. *Polish Journal of Environmental Studies*, **13**(2): 177–183.

**Pellegrin D. & Hauber DP. 1999.** Isozyme variation among populations of the clonal species *Phragmites australis* (Cav.) Trin. ex Streudel. *Aquatic Botany*, **63**(3): 241–259.

**Peverly JH, Surface JM & Wang T. 1992.** Growth and trace metal absorption by common reed (*Phragmites australis* [Cav.] Trin.) in wetlands constructed for landfill leachate treatment. Prepared for Intecol Wetlands Conference, Ohio.

**Pfadenhauer J. 1998.** Leitlinien für die Restaurierung süddeutscher Moore. (Guidelines for mire restoration in southern Germany). *Natur und Landschaft*, **74**: 18–29.

**Phillips J. 1930.** Some important vegetation communities in the central province of Tanganyika Territory (formerly German East Africa). *Journal of Ecology*, **18**: 193–234.

**Piccolo A. 1994.** Interactions between organic pollutants and humic substances in the environment. In N. Senesi & T.M. Miamo, eds. *Humic substances in the global environment and implications on human health*. The Netherlands: Elsevier Science.

**Pier A, Dienst M & Stark H. 1993.** Dynamics of reed belts at Lake Constance (Untersee and Uberlinger See) from 1984 to 1993. *Limnologia aktuell*, **5**: 141–148.

**Planter M. 1970.** Elution of mineral components out of dead reed *Phragmites communis* Trin. *Polskie Archivum Hydrobiolica*, **17**: 357–362.

**Prach K, Jeník J & Large ARG. 1998.** *Floodplain ecology and management: The Lužnice River in the Třebon Biosphere Reserve, Central Europe*. Amsterdam: SPB Academic Publishing.

**Preston CD, Pearson DA & Dinis ID. 2002.** *New Atlas of the British Flora*. Oxford: University Press.

**Putten WH van der. 1993.** Effects of litter on the growth of *Phragmites australis*. *Limnologia aktuell*, **5**: 19–22.

**Raghi-Atri von F & Bornkamm R. 1980.** On the stiffness of stems of the reed plant (*Phragmites australis* [Cav.] Trin. ex Steudel) at different levels of nutrients added. *Archivum Hydrobiologie*, **90**: 90–105.

**Ranwell DS, Bird EFC, Hubbard JCE & Stebbings RE. 1964.** Tidal submergence of chlorinity in Poole Harbour. *Journal of Ecology*, **52**: 627–642.

**Ratcliffe DA & Walker D. 1958.** The Silver Flowe, Galloway, Scotland. *Journal of Ecology*, **46**: 407–445.

**Reddy KR & Smith WH, eds. 1987.** *Aquatic plants for water treatment and resource recovery*. Orlando, Florida: Magnolia Publishers.

**Regel C. 1947.** The bogs and swamps of White Russia. *Journal of Ecology*, **35**: 96–104.

**Rejmankova E. 1973.** Chlorophyll content in leaves of *Phragmites communis* Trin. Ecosystem Study on Wetland Biome in Czechoslovakia. *Czechoslovak IBP/PT-PP Report*, Trěbon, **3**: 143–145.

**Retare W, Weisner SEB, Strand JA & Granéli W. 2001.** Phenotypic plasticity in *Phragmites australis* as a functional response to water depth. *Aquatic Botany*, **69**: 127.

**Ridley HN. 1923.** The distribution of plants. *Annals of Botany NS*, **37**: 1–29.

**Ridley HN. 1930.** *The dispersal of plants throughout the world*. Ashford: L. Reeve and Co. Ltd.

**Roberts J. 2000.** Changes in *Phragmites australis* in south-eastern Australia: a habitat assessment. *Folia Geobotanica*, **35**: 353–362.

**Rodewald-Rudescu L. 1974.** Das Schilfrohr–*Phragmites communis* Trin. *Die Binnengewässer*, Stuttgart, **27**: 1–302.

**Rodwell JS, ed. 1991a.** *Woodlands and scrub*. Cambridge: University Press.

**Rodwell JS, ed. 1991b.** *Mires and heaths*. Cambridge: University Press.

**Rodwell JS, ed. 1995.** *Aquatic communities, swamps and tall herb fens*. Cambridge: University Press.

**Rolletschek H, Hartzendorf T, Rolletschek A & Kohl J-G. 1999a.** Biometric variation in *Phragmites australis* affecting convective ventilation and amino acid metabolism. *Aquatic Botany*, **64**(3–4): 291–302.

**Rolletschek H, Rolletschek A, Kühl H & Kohl J-G. 1999b.** Clone specific differences in a *Phragmites australis* stand. II. Seasonal development of morphological and physiological characteristics at the nature site and after transplantation. *Aquatic Botany*, **64**(3–4): 247–260.

**Romero JA, Brix H & Comín FA. 1999.** Interactive effects of N and P on growth, nutrient allocation and NH4 uptake kinetics by *Phragmites australis*. *Aquatic Botany*, **64**(3–4): 369–380.

**Rose F. 1950.** The East Kent fens. *Journal of Ecology*, **38**: 292–302.

**Rose F. 1953.** A survey of the ecology of British lowland bogs. *Proceedings of the Linnean Society*, London, **164**: 186–211.

**RSPB (Royal Society for the Protection of Birds). 1994.** *Reedbed management for bitterns*. Sandy, Bedfordshire.

**RSPB (Royal Society for the Protection of Birds). 2000.** *Butterbump*. Sandy, Bedfordshire: English Nature.

**Rubec CDA & Overend RP, eds. 1987.** *Proceedings Symposium 1987*. Wetlands/peatlands. Alberta, Canada: Edmonton.

**Rudescu L, Niculescu C & Chivu IP. 1965.** *Monografia stufului den delta Dunarii*. Editura Academiei Republicii Socialiste, Romania.

**Rychnovská M & Smíd P. 1973.** Preliminary evaluation of transpiration in two *Phragmites* stands. Ecosystem Study on Wetland Biome in Czechoslovakia. *Czechoslovak IBP/PT-PP Report*, Trěbon, **3**: 111–119.

**Salmon DJ. 2004.** *Cornwall and Isles of Scilly reedbed survey*. Truro: Cornwall Wildlife Trust.

**Schieferstein BB. 1999.** Ecological and Molecular Biological Investigations on Reed (*Phragmites australis* [Cav.] Trin. ex Steudel) in Lakes of Northern Germany. An Overview. *Limnologica*, **29**(1): 28–35.

**Schröder R. 1979.** The decline of reed swamps in Lake Constance. *Symposia Biologica Hungaria*, **19**: 43–48.

**Schröder R. 1987.** Das Schilfsterben am Bodensee–Untersee, Beobachtungen, Untersuchungen und Gegenmaßnahmen. *Arch. Hydrobiol, Suppl.* **76**: 53–99.

**Seidel K. 1956.** *Unsere Flechtbinden.* Hydrobiologische Anstalt der Max-Planck-Gesellschaft. Coburg: Emil Patzschke.

**Seidel K. 1966.** Reinigung von Gewässern durch höhere Pflanzen. *Naturwissenschaften*, **12**: 289–297.

**Shimp JF, Tracy JC, Davis LC, Lee E, Huang W & Erickson LE. 1993.** Beneficial effects of plants in the remediation of soil and groundwater contaminated with organic materials. *Critical Review of Environment, Science and Technology*, **23**: 41–73.

**Siriwardena GM. 2001.** *Bird indicators of sustainability for the water industry.* Thetford: British Trust of Ornithology.

**Soetaert K, Hoffmann M, Meire D, Starink M, van Oevelen D, van Regenmortel S & Cox T. 2004.** Modelling growth and carbon allocation in two reed beds (*Phragmites australis*) in the Scheldt Estuary. *Aquatic Botany*, **79**(3): 211–234.

**Somerset County Council. 1998.** *Reedbed construction guidelines for environmental improvement applications: experience from the Somerset Levels and Moors, UK and the Parc des Maraide du Cotentin et du Bessin, France*. Life Project 92-1/UK/026, Taunton.

**Spence DHN. 1964.** The macrophytic vegetation of freshwater lochs, swamps and associated fens. In J.H. Burnett, ed. *The Vegetation of Scotland.* Edinburgh: Oliver & Boyd, 306–425.

**Ssymank A & Hauke U. 1998.** Landscape ecology of calcareous fens (*Caricion davallianae*) and the *Cladietum marisci* in the lowlands of NE-Germany and their relevance for nature conservation in the European Union Habitats Directive. *Phytocoenologia*, Stuttgart, **28**(1): 105–142.

**Stålberg N. 1939.** Lake Vättern. Outlines of its natural history, especially its vegetation. *Acta Phytogeogr. Suec.*, **XI**: 1–52.

**Stark H & Dienst M. 1989.** Dynamics of lakeside reed belts at Lake Constance (Untersee) from 1984 to 1987. *Aquatic Botany*, **35**(1): 63–70.

**Sukopp H. 1971.** Effects of recreational activities on littoral macrophytes. *Hydrobiologia*, **12**: 331–340.

**Sukopp H. 1991.** Röhrichte unter dem Einfluss von Großstädten, Rundgespräche der Kommission für Ökologie. *Ökologie der oberbayerischen Seen*, **2**: 63–72.

**Sukopp H & Markstein B. 1989.** Changes of the reed beds along the Berlin Havel 1962–1987. *Aquatic Botany,* **35**(1): 26–27.

**Szczepanski A. 1969.** Biomass of underground parts of the reed *Phragmites communis* Trin. *Bull. Acad. Pol. Sci. Sér. Sci. Biol.*, **17**: 245–246.

**Szczepanski A. 1970.** Methods of morphometrical and mechanical characteristics of *Phragmites communis* Trin. *Polskie Archivum Hydrobiologica*, **17**: 329–335.

**Täckholm V & Täckholm G. 1941.** *Flora of Egypt.* Cairo: Fouad I University.

**Tallantire PA. 1954.** Old Buckenham Mere. *New Phytologist.* **53**: 131–139.

**Thesiger W. 1964.** *The Marsh Arabs.* London: Longmans.

**Toorn J van der. 1972.** Variability of *Phragmites australis* [Cav.] Trin. ex Steudel in relation to the environment. *Van Zee Tot Land,* **48**. 's-Gravenhage.

**Toorn J van der & Mook JH. 1982.** The influence of environmental factors and management on stands of *Phragmites australis*. I. Effects of burning, frost and insect damage on shoot density and shoot size. *Journal of Ecology*, **19**: 477–499.

**Trémolières M, Roech U, Klein JP & Carbiener R. 1994.** The exchange process between river and groundwater on the central Alsace floodplain (Eastern France): II. The case of a river with functional floodplain. *Hydrobiologia*, **273**: 19–36.

**Tscharntke T. 1999.** Insects on common reed (*Phragmites australis*): community structure and the impact of herbivory on shoot growth. *Aquatic Botany*, **64**(3–4): 399–410.

**Vajda Z. 1998.** Hungary's heron haven. *Enact*, 12–14. Peterborough: English Nature.

**Verhoeven JTA, ed. 1992.** *Fens and Bogs in The Netherlands, vegetation, history, nutrient dynamics and conservation*. Kluwer Academic Publishers.

**Votrubová O, Soukup A, Pecháková A & Paul J. 1997.** Anatomy of common reed and its importance for survival in eutrophic habitats. *Acta Univ. Car. Biol.*, **41**: 233–244.

**Vymazal J, ed. 2001.** *Transformation of nutrients in natural and constructed wetlands*. Leiden: Backhuys.

**Vymazal J, Brix H, Cooper PF, Green MB & Haberl R, eds. 1998.** *Constructed wetlands for waste water treatment in Europe*. Leiden: Backhuys.

**Ward NC. 1999.** A study into the preferred habitat of Cetti's warbler *Cettia cetti*, with regard to invertebrate prey. Seale-Hayne Faculty of Agriculture, Food and Land Use, Plymouth University, Plymouth.

**Wazny J & Wytwer T. 1963.** Badania nad edpornasia Trzciny (*Phragmites communis* Trin.) na dzialanie grzybow niszczacyats drewno. *Folia for polonaise*, **B5**: 171–194.

**Weaver JE & Himmel WJ. 1930.** Relation of increased water content and decreased aeration to root development in hydrophytes. *Plant Physiology*, **5**: 69–92.

**Weber HC & Kendzior B. 2006.** *Flora of the Maltese Islands*. Weikersheim: Margraf Publishers.

**Weis JS & Weis P. 2003.** Is the invasion of the common reed, *Phragmites australis*, into tidal marshes of the eastern US an ecological disaster? *Marine Pollution Bulletin*, **46**(7): 816–820.

**Weisner SEB. 1987.** The relation between wave exposure and distribution of emergent vegetation in a eutrophic lake. *Freshwater Biology*, **18**: 537–544.

**Weisner SEB. 1991.** Within-lake patterns in depth penetration of emergent vegetation. *Freshwater Biology*, **26**: 133–142.

**Weisner SEB & Strand JA. 1996.** Rhizome architecture in *Phragmites australis* in relation to water depth: implications for within-plant oxygen transport distances. *Folia Geobotanica Phytotaxonomia*, **31**: 91–97.

**Westlake DF. 1963.** Comparisons of plant productivity. *Biol. Rev.*, **38**: 385–426.

**Westlake DF. 1968.** Methods used to determine the annual production of reedswamp plants with extensive rhizomes. In M.S. Ghilarov *et al.*, eds. *Methods of Productivity Studies in Root Systems and Rhizosphere Organisms*. Nauka, Leningrad, 226–234.

**Westlake DF, Květ J & Szczepanski A, eds. 1998.** *The Production Ecology of Wetlands*. Cambridge: University Press.

**Wheeler BD. 1999.** The biodiversity value of wet woodlands in comparison to open herbaceous wetlands. *Contract Science Report 336*. Countryside Commission for Wales.

**White G. 1788.** *The Natural History of Selborne*. London 1977: Penguin edn.

**Wichtmann W & Joosten H. 2007.** Paludiculture: peat formation and renewable resources from rewetted peatlands. I*MCG Newsletter*, **2007**(3): 24–28.

**Windham L. 1999.** Microscale spatial distribution of *Phragmites australis* (common reed) invasion into *Spartina patens* (salt hay)-dominated communities in brackish tidal marsh. *Biological Invasions*, **1**: 137–148.

**Winter M & Kickuth R. 1985.** Elimination of nutrients (sulphur, phosphorus, nitrogen) by the root zone process and simultaneous degradation of organic matter. *Utrecht Plant Ecology*, News Report, **4**: 123–140.

**Winter M & Kickuth R. 1989.** Elimination of sulphur compounds from waste water by the root zone process. I. Performance of a large-scale purification plant at a textile finishing industry. *Water Research*, **23**: 535–546.

**Wöbbecke K & Ripl W. 1990.** Untersuchungen zum Röhrichtrückgang an der Berliner Havel. *Landschaftsentwicklung und Umweltforschung*, TU Berlin, **71**: 94–102.

**Wood H. 1999.** *Leap forward for wildlife: promoting biodiversity through Local Environment Agency Plans.* Sandy, Bedfordshire: RSPB (Royal Society for the Protection of Birds).

**Worrall P. 1998.** Water treatment or conservation? *Enact*, 15–17. Peterborough: English Nature.

**Zeidler A, Schneider S, Jung C, Melchinger AE & Dittrich P. 1994.** The use of DNA fingerprinting to ecological studies of *Phragmites australis* [Cav.] Trin. ex Steudel. *Botanica Acta*, **107**: 237–242.

**Zhu XY, Wang SM & Zhang CL. 2003.** Composition and characteristic differences in photosynthetic membranes of two ecotypes of reed (*Phragmites communis* L.) from different habitats. *Photosynthetica*, **41**(1): 97–104.

**Zonneveld IS. 1960.** De Brabantse Biesbosch. Een studie van bodem en vegetatie van een zoetwatergetijdendelta. PhD Thesis, Landbouwhogeschool Wageningen, NL.

# Index


## A

aeration (*and see also oxygen*) 51, 122, 173–6

aerial shoots, stems, reeds 27, 28, 32, 37, 38, 44, 46, 48, 49, 51, 67, 75–78, 92, 115, 117, 119, 132, 139, 197, 223. Tables: 3.1, 4.6, 4.8, 4,10, 6.3

Africa, North, African 3, 7, 15, 38, 89, 90, 93, 133, 142, 218, 220. Table: 3.2

agriculture 132, 164, 173, 220, 222. Table: 6.3

agricultural 155, 168, 170, 185, 191, 219, 222

alder, *see Alnus glutinosa* 13, 122, 125, 126, 199, 201

Alfred, King (The Great) 149

*Alnus glutinosa* 122, 125, 154

America, North (*and see Canada*) 3, 6, 7, 147, 150, 153, 157, 163, 164, 218

USA 3, 84, 85, 87, 90, 201. Table 6.8

archaeology 10, 151, 161

*Archanara* and *Arenostola* spp. Reedbug (*also see gall*) 62, 63, 64, 71

*Archanara dissoluta* 71

*Archanara geminipunctata* 71. Table: 4.9

*Arenostola* spp. 71

Arkata, River 141, 142

*Arrhenatherum elatior* 112

*Arundo donax* 2, 112, 128, 135, 136–138, 140, 141, 143, 144, 146, 159

Australia 3, 5. 6, 7, 93, 112

*Azotobacter* 154

## B

*Beowulf* 149, 158, 220

*Berula erecta* 209

*Betula* spp., birch 10, 13, 125. Tables: 1.6, 6.1

Birch, see *Betula*

birds (habitat, nest, net, reserve, sanctuary, song, trap, watching) 22, 72, 134, 137, 138, 142, 145, 150, 153, 159, 162, 163, 177, 192, 195, 217, 221. Tables: 4.13, 10.2

Bittern 101, 153, 161, 165, 185, 192. Table: 4.13

Blade (of *Phragmites*) 7, 23, 28, 32, 44, 45–47, 50, 51, 62, 94, 104. Table: 12.1

of *Tribe oryzeae* [Rice] 5

bog (boggy) 3, 9, 10, 12, 13, 38, 48, 70, 71, 104, 105, 109, 121, 124, 125, 131, 149, 158, 159, 160, 161, 172, 183, 198, 202, 209, 218. Tables: 1.2, 6.1, 12.3

competition 131. Tables: 4.4, 4.7

invasion 124, 172

peat 8, 9, 125, 148. Tables: 1.2, 2.3, 6.5

brackish 1–3, 21, 44, 84, 102, 103, 109, 112, 133, 134, 144, 176, 182, 196. Tables: 1.6, 10.2

Braughing 41

Breck fens 152. Tables: 4.2, 4.6, 6.3

Bridge of Reeds *facing page to Chapter 1*, 2

Britain 1, 3, 4, 7–9, 21, 22, 24, 32, 35, 36, 44, 56, 62, 84, 99, 102, 113, 152, 153, 157, 165, 175, 177, 192, 195, 202. Tables: 1.2, 1.3, 3.2, 11.2

Broadland, Broads (*also see Norfolk Broads*) 11, 72, 130, 152, 159, 172, 173, 175, 192, 194, 195, 221. Tables: 1.4, 1.6, 4.13, 4.14, 6.5

buffer strips Table: 9.3

Bugibba 134

burn, burned, burning, burnt (*also see fire*) (Chapters 4 and 11) 8, 9, 56, 57, 58, 71, 99, 113, 128, 130, 132, 151, 160, 162, 164, 174, 176, 179, 184 (power stations, 219). Tables: 5.2, 10.1

## C

*Calamagrostis* spp. 70, 130

calcium 13, 25, 48, 60, 63, 69, 100, 104, 105, 125–127, 130, 131, 134, 141, 149, 173, 174, 202,209, 212. Table: 6.2

*Calystegia*, C. *sepia*, *C. sepium* 129

Camargue, France 128, 172, 182

Cambridgeshire (Cambs) 100

Canada, Canadian 152, 155, 164. Table: 3.2

Canadian Wildlife Service 164

carbon 164, 182, 219, 220. Table: 5.2

storage, loss 164,

*Carex paniculata* 44, 126–128, 131, 212. Table: 12.3

*C. davalliana 161*

*C. riparia* 20. Table 1.6

carr 70, 121, 125, 126, 131, 219. Tables: 1.6, 4.13

Cavenham fen 27, 56, 63, 64, 100. Tables: 4.1, 4.2, 4.5, 4.6, 4.7

China 3, 150, 167, 193, 201

*Cladium mariscus* (fen sedge) 3, 8, 13, 48, 100, 122, 130, 148, 169, 173, 177, 195, 222

*Claviceps* spp. 71

cleaning (*and see purification*) 153, 157, 161, 183, 184, 195

Cley, Mill, Norfolk 89, 173. Table: 6.1

climate 3, 9, 15, 40, 43, 49, 51, 52, 55, 83, 93, 102, 112, 113, 128, 144, 155, 176, 191, 220, 222, 223. Tables: 1.1, 1.2, 1.3

clone, clonal. 7, 44, 64–67, 83–85, 87–90, 92, 99, 101–103, 109, 114, 124, 135, 145, 174, 179, 182–184, 201, 218, 224. Tables: 5.1, 5.2, 6.6

*Clostridium* 154

competition (Chapter 7) 22, 24, 36, 37, 40, 41, 58, 60, 62, 63, 65, 66, 69, 70, 84, 89, 110, 121, 122–131, 141, 144, 153, 154, 179, 182, 197, 210, 211, 218. Tables: 4.4, 4.7

conservation, conserve, conserving Preface, 146, 151, 152, 156, 160–165, 171, 174, 184, 185, 191, 193, 220–222

coypu 153, 195. Table: 4.13

culture 13, 14, 149, 150, 167, 170. Table: 6.3

cut, cutting, cutter 2, 7, 8, 14, 25, 38, 40, 49–51, 57, 58, 67, 69, 99, 105, 112, 114, 126, 127, 130, 134, 145, 150, 151, 158, 160, 162, 171, 172, 174–176, 179, 182, 184, 192, 200, 212. Tables: 1.6, 2.4, 4.8, 5.2, 6.2

cuticle 180

Czech Republic (Czechoslovakia) 7, 47, 71, 91, 93, 94, 153, 183.Tables: 1.5, 3.2, 4.9, 5.4, 6.8

## D

damage (from animals, frost, gall, trampling, etc.) 7, 32, 40, 44, 46, 50, 51, 58, 62, 63, 70, 71, 106, 107, 111, 126, 127, 134, 153, 157, 163, 169, 171, 177, 179, 180, 184, 191, 195, 200, 219, 220, 223. Tables: 4.4, 4.9

Danube, River [and Delta] 6, 29, 37, 38, 106, 150, 170, 198, 218. Table: 4.15

*Deightoniella arundinacea* 71

Delaware estuary [and marshes] 84, 85. Table: 5.1

development 2, 6, 22–24, 34, 35, 37, 38, 40, 43, 46, 48, 58, 61, 101, 103, 122, 123, 125, 135, 141, 146, 162, 173, 185. Tables: 1.5, 2.1

die back 8, 112, 128, 144

diseases 71, 122, 133, 158, 159, 171, 191, 195, 221. Tables: 1.1, 5.2

dispersal, dispersed, dispersing 21, 22, 34, 39, 43, 46, 101, 111, 145, 221

distribution, of *Phragmites* [and shoots] (Chapters 1, 8 and 11) 3, 4, 12, 15, 61, 66–69, 112, 134, 147, 191, 200, 202, 211. Table: 10.3

DNA 5, 8, 182, 201, 211

dormancy, dormant 7, 24, 39, 49, 50, 55–57, 113, 171

drainage 3, 22, 100, 112, 149, 152, 155–159, 163, 164, 172, 184, 185, 193, 196, 199, 200, 201, 218, 220, 222

drought 24, 25, 58, 109, 111, 112, 126, 132, 144, 173, 184

dry, drying 6, 9, 10, 12, 13, 23, 25, 27, 38, 48, 52, 69, 70, 73, 100, 102, 103, 109–111, 121, 122, 124–126, 128, 129, 131–138, 141–146, 151, 152, 160, 162–164, 171, 172, 175–178, 180, 182, 185, 191, 192, 197, 199, 200, 202, 210, 212, 219, 221–223

Dunfermline 42

Dutch (*see Netherlands, The*)

Dweijra 134, 136

## E

East Anglia 36, 49, 56, 63, 84, 90, 104, 121, 129, 130, 152, 173, 175, 184, 192, 199. Tables: 6.2, 10.1, 10.2, 10.4

ecotype (*and see genotype*) 89, 91–94, 155. Table: 5.4

Egypt, Egyptian 2, 6, 51, 62, 101, 141–143, 145, 150, 157, 159, 219, 220

emergence 39, 40, 43, 45, 50, 52, 55–58, 60, 62, 69, 83, 89, 99, 101, 111–113, 123, 129, 130, 132, 136, 174,184, 224. Tables: 4.4, 4.8, 6.4

emergent 239

endurance (Chapter 8) 133, 196

England 4, 27, 29, 33, 41, 43, 47, 50, 57, 58, 60–64, 68, 69, 88, 100, 107, 108, 109, 110, 125, 128, 149, 150, 152, 157, 160, 162, 167, 169, 171, 172, 178, 185, 195, 199, 200, 202, 209, 210, 219. Tables: 1.4, 4.4, 4.6, 4.7, 4.8

*Epilobium hirsutum* 122, 191

Eriswell fen 43, 76. Tables: 4.5, 4.6, 6.1

erosion 150, 159, 173, 185. Table: 9.3

eutrophic (*and see nutrient-rich*) 9, 13, 106, 164

evapotranspiration 124, 160

**F**

fen sedge (*see Cladium mariscus*) 130, 177, 242. Table: 11.4

fence 14, 131, 148, 150, 152, 195, 197, 220, 222

fertilise, fertiliser(s) 50, 102, 104, 105, 173, 183, 192, 193, 195, 198, 201. Table: 6.2

fire 38, 57, 150, 151, 167, 170, 177, 218

fish, fishing 13, 14, 72, 93, 94, 135, 148, 150, 153, 161–163, 173, 183, 192, 220. Tables: 5.4, 6.3

flood, flooding, flooded 1, 9, 22–24, 29, 38, 47, 52, 58, 70, 72, 73, 88, 103, 105, 107, 109–112, 122, 124–127, 129, 130, 132, 133, 144, 145, 149, 152, 153, 155, 157, 159, 160, 162, 164, 172–174, 176, 182, 184, 191–193, 197, 200, 201, 209, 222, 223. Tables: 4.11, 12.1

fodder 14, 148, 150, 152

Fontana 141, 143

frost(s), frosty 7, 8, 22, 24, 40, 49, 50–52, 55, 56, 58, 60, 62–64, 68, 112, 113, 126, 147, 182–184. Tables: 2.1, 3.1, 4.2, 4.4, 5.4

**G**

*Galium aparine* 129, 131
*G. uliginosum* 212

gall(s) (*and see Lipara, stem miners*) 40, 51, 71. Tables: 4.9, 4.10

genotype (*and see ecotype*) 3, 37, 38, 40, 43, 46, 51, 67, 88, 91, 93, 95, 102, 109, 110, 112–114, 123, 138, 142, 144, 161, 182, 183, 184, 211, 219, 224

German, Germany 6, 90, 93, 157, 183,194

germinate, germination 21–25, 103, 123, 125–127, 160

Ghadira 134, 137

Ghajn il-Kbir 144

Ghajn Tuffieha 141

giant reed (giant *Phragmites* or *Arundo donax*) 6, 14, 62, 102, 150, 216

glacial (*and see Ice Age*) 9, 148, 152. Tables: 1.3, 1.5

*Glyceria maxima* 122, 148, 152. Tables: 1.6, 11.1, 11.4, 12.3

Gnien il-Kbir 141, 143, 144

gravel pit(s) 21

grazing, grazed 25, 37–40, 43, 50, 58, 61, 69, 70, 72, 88, 104, 105, 112–114, 122, 126, 128, 131, 144, 149, 150, 152, 163, 184, 196, 197, 210–212, 217, 219, 223

Guthlac, St. 149, 220

## H

harvesting (*and see cutting*) 47, 49, 51, 63, 73, 99, 100, 103, 113, 114, 121, 122, 129, 130, 144, 148, 150, 151, 160–163, 172–177, 179, 182, 184, 192, 195, 218, 221

Hereward 149, 159

Hickling, Norfolk 103, 182

history 9, 43, 72, 83, 99, 100, 126, 151, 161, 170, 182. Tables: 1.4, 1.5, 1.6

Horsey Table 4.4

hover (*see plav*)

Human impact (*see Impact*)

humidity 72

Hungary 28, 93, 106, 110, 111, 153, 159, 193, 199, 200, 210. Table: 6.8

## I

Ice Age (*and see glacial*) 9. Table: 1.2

ice, snow 9, 73, 111, 200, 210

Icklingham fen 29, 43, 56, 58, 68, 100. Tables: 2.2, 4.1, 4.2, 4.4, 4.5, 4.6, 6.1, 6.2, 6.4

Ijsselmeer 160

Impact, human 1, 3, 9, 21, 70, 112, 121, 122, 125–127, 129, 131, 132, 141, 146, 152, 191, 196–202, 209, 212

infra-red colour photography 83, 85, 87–89. Tables: 1.1, 5.1, 5.2

Iraq 1, 13, 62, 148, 157, 191, 199, 218

isozymes 83, 90

Israel, Israeli 93

## J

*Juncus subnodulosus* 70, 129, 223

## L

Lakenheath 193

leaves, leaf, leafy 3, 5, 7, 22–24, 31, 38, 39, 44–48, 51, 55, 62, 71, 83, 88, 91, 93–95, 102, 104, 105, 112, 122, 163 [leaflets], 211. Table: 1.1

*legehalme* 6, 28, 39, 44, 65, 66, 67, 102, 105, 139, 145, 219, 223. Tables: 3.1, 12.1

long runners 6, 28

light 37, 44, 50, 70, 88, 107, 125–127, 130, 154, 179, 219, 220

*Lipara* spp. Tables: 4.9, 4.10

- *L. lucens* 71. Tables: 4.9, 4.10
- *L. pullitarsis* 71. Tables: 4.9, 4.10
- *L. rufitaris* 71. Tables: 4.9, 4.10
- *L. similis* 71. Tables: 4.9, 4.10

litter, litter mat 7, 46–48, 56, 63, 64, 68–73, 113, 121, 124, 125, 127, 130, 132, 137, 142, 150, 171. Tables: 4.4, 4.8, 12.1

*Lobelia dortmanna* 209

Lochluichart 42

## M

macrolepidoptera Table: 4.14

Maghtab 135, 138, 144

Majorca Table. 3.2

Malta, Maltese (Chapter 8) Acknowledgements, 7, 8, 22, 24, 32, 36, 40, 41, 43, 44, 55–58, 60–63, 94, 101, 102, 109, 110, 112, 128, 133–146, 163, 196, 197, 200, 202, 210, 219. Tables: 2.2, 6.8

mammals 72, 153, 163. Tables: 4.12, 4.13

management 7, 36, 49, 71, 83, 99–101, 127, 129, 130, 131, 144, 147, 156, 157, 162–164, 172, 174–176, 179, 182, 184, 191, 192, 198, 200–202, 211, 212, 221, 222. Tables: 9.3, 11.4

Marfa 135, 139

Marsalforn 134, 136, 145, 146

Marsascala 134, 135

Marsaxlokk 134, 135

mediaeval 135, 150, 153, 170, 178, 220

Mediterranean 1, 128, 133, 144, 183, 211

metals 154, 155, 156. Table: 9.3

Mgarr ix-Xini 134, 137

Mgarr 141

minerals, mineralisation 21, 48, 50, 51, 104, 109, 125, 131, 132, 154, 156, 160, 164, 175, 197, 210, 219. Table: 6.5

mowing (*see cut, cutting*)

musical instrument, pipe 2, 158

*Myrica gale* 154. Table: 12.3

## N

National Vegetation Classification Tables: 11.3, 11.4

Netherlands, The 1, 6, 7, 9, 22, 56, 57, 62, 84, 90–93, 101, 102, 111, 125, 152, 171, 175, 201, 219. Tables: 5.3, 5.4, 12.2, 12.3

New Forest 43, 104. Table: 4.5

New Jersey 3, 87

nitrate 155

nitrogen 63, 88, 154, 156, 182, 183. Tables: 2.3, 5.4, 9.2

Norfolk Broads (*see also Broads, Broadland*) 9, 152, 168

nutrients (*and see eutrophic*) 9, 10, 22–25, 38, 40, 44, 50, 69, 72, 83, 84, 92, 103–105, 107, 109, 111, 112, 122, 123, 126, 128, 131, 141, 149, 154, 155, 161, 173–176, 182, 183, 185, 197, 202, 209, 212. Tables: 4.7, 6.2

nutrient-poor 4, 9, 104, 114, 161, 210, 219. Tables: 6.1, 6.5, 12.1

nutrient-rich 9, 22, 58, 63, 100, 109, 124, 125, 129, 131, 164, 183, 195, 197, 200, 201, 209. Tables: 6.1, 6.2

## O

Old Buckenham fen 173, 192. Table 1.6

oxygen, oxygenation (*and see also aeration*) 24, 39, 73, 109, 153, 154, 156, 176. Table: 9.2

## P

Pakistan 3, 62
paludiculture 200
paper 150, 155
peat 160–162, 164, 172, 173, 175, 185, 191, 195, 199–202, 219, 220, 223. Tables: 1.2, 12.3
pests (*also see Archanara spp., gall*) 63, 71, 126, 171, 180, 191. Tables: 1.1, 5.2
*Peucedanum palustre* Tables: 1.6, 11.4
*Phalaris arundinacea* 6, 148, 152, 211
phosphate, phosphorus 25, 63, 88, 106, 154, 156, 183. Tables: 2.3, 6.2
*Phragmites* (Chapter 1) Acknowledgements, Preface, 21, 22, 25, 27, 30, 36, 38, 48, 55, 60, 62, 65, 67, 70–73, 84, 85, 88, 89, 95, 99–114, 121–138, 141–144, 146–154, 158–162, 164, 165, 173, 182–184, 191–193, 195–202, 209–212, 218–224. Tables: 1.6, 3.1, 4.5, 6.1, 6.2, 11.4
  *Phragmites australis* Preface, 3–8, 21. Tables: 1.1, 1.6, 4.9, 4.10, 4.11, 5.3, 5.4, 11.3, 11.4
  *Phragmites communis* 5. Table: 4.2
  *Phragmites effusa* var. 6
  *Phragmites flavescens* var. 6, 7
  *Phragmites humilis* ssp. 6
  *Phragmites isaica* var. 6
  *Phragmites karka* 3, 7
  *Phragmites forma picta* 6
  *Phragmites pseudodonax* ssp. 6
  *Phragmites stolonifera* var. 6
  *Phragmites variegatum* var. 6
  *Phragmites pumila* sub var. 6
  *Phragmites stenophylla* var. 6
physiology (eco-) 73
phytoplankton 72
pipe (*see musical*)
*Platycephala planifrons* (reed bug) 71
plav, plaur, hover 37, 38, 172
ploidy 83, 89–91
Poland 71, 109, 209. Table 3.2
polder 22, 84, 160
pollen 5, 9, 34, 170
  analysis 9, 151
  zone 3, 11
pollutant, pollution, pollute(d) 134–136, 141, 146, 155, 157, 174, 184, 185, 195, 196, 212. Table: 9.3
*Populus* spp. 154
*Pucciniella phragmitidis* 71
purification, purifying (*and see cleaning*) 90, 101, 155, 161, 165, 184, 185, 193, 219, 220

## R

radiocarbon 10

Ramla 134–136, 145

Ranworth Tables: 4.4, 6.1

recreation 161–163, 185, 195, 196, 220

Redgrave fen 69, 202. Tables: 4.4, 6.1

reed (*see Phragmites*). Ancient name for, first, reed-like plants—reedswamp; later including wheat straw prepared for thatching; finally, narrowing to *Phragmites* (though 'wheat reed' for long straw, is still a thatching term). Old English (hreod), Old Saxon (hriad), Middle Dutch (ried, riet), Old High German (riot), German (ried, schilf). Reed(s), aerial shoot of *Phragmites*, particularly when mature.

reed grass (*see Glyceria maxima, Phalaris arundinacea)*

*Reed*, Anglo-Saxon poem 2

reedbed Marsh dominated by *Phragmites* where the cut stems can be used as *Phragmites* for thatching, plaster, etc. May be almost solely *Phragmites*, with an understory of short plants—discarded—or with up to a quarter or so of other tall monocotyledons. Meaning is overlapping with reedswamp, but a reedbed is, or has been, used by people. (Chapter 4) Preface, 1, 3, 6, 9, 10, 12–15, 21, 22, 24, 25, 27, 38, 43, 47, 55–84, 88, 95, 99, 10–103, 105, 107, 109, 112, 113, 121, 122, 124, 126, 127, 130–133, 144, 147–150, 152, 153, 157–162, 164, 165, 168, 170, 172–177, 179, 180, 182–185, 191–193, 195, 196, 199, 200–202, 217–223. Tables: 1.6, 10.3, 10.4, 12.3

reedbug (*see Archanara, etc.*)

reedswamp Marsh, wetland, dominated by tall vaguely reed-like vegetation. In a general sense, may include marshes dominated by tall *Carex* spp., *Glyceria* spp., *Phalaris arundinacea, Phragmites australis, Sparganium erectum, Typha* spp., etc. In this book the term is used for *Phragmites* marshes, which may, or may not be used intensively by people. 1–3, 7–9, 13, 21, 37, 83, 113, 121, 123, 146, 148, 153, 157, 160, 163, 164, 171, 195–197, 199, 200, 218–220, 222, 223. Table: 1.6, 12.2

*Rhizobium* 154

rhizome, rhizomes 1, 7, 8, 21, 32, 33, 35, 37–39, 43, 44, 46–51, 56–58, 62, 65–70, 91–94, 100, 102, 103, 105, 106, 109, 111, 113, 122, 123–125, 128, 129, 132, 134, 144, 145, 150, 152, 171, 174–176, 182–184, 197, 200, 210, 219, 223. Tables: 1.1, 6.4, 6.5
- horizontal (and lateral) 23–24, 28, 29, 37, 38, 44, 48, 49, 65, 67, 69, 102–105, 107, 109, 125, 132, 179, 184, 223. Table: 3.1
- oblique 28
- oval Table 12.1
- vertical 28, 29, 37–39, 43, 44, 48, 49, 55, 57, 68, 69, 109, 125, 223. Table: 3.1

root(s) 1, 8, 22, 24, 28, 31–33, 36, 37, 39, 48, 50, 65, 69, 70, 91, 92, 94, 105, 106, 110, 122, 123, 125, 128, 129, 131, 132, 144, 150, 153–156, 176, 182, 197, 223. Tables: 3.1, 9.1, 12.1

Rosyth 64–66, 217

Romania [Roumania] 6, 30, 62, 93, 102. Table 3.2

Runnymede 158, 159

## S

Salini (Salina) 135, 139, 144–146. Table: 4.1

salinity, saline, salt, salty 1–3, 6, 23, 73, 84, 88, 90, 101–104, 109, 112, 133, 134, 139, 144, 196, 201, 210, 224. Tables: 5.4, 9.3 [Saltmarsh 21, 132, 134, 137, 201. Table 1.6]

*Salix* spp., *S. alba, S. cinerea, S.* ◈ *babylonica* 5, 122, 125–126, 154, 193, 199, 200, 211, 219, 221. Table: 1.6

sallow (*see also Salix*) 13, 125–127, 131, 158, 199, 219

Santa Maria [Marija] 141

Santon Downham Table: 6.1

*Schoenus nigricans* 127, 161. Tables: 6.1, 11.4

Scotland 42, 64–66, 105, 157, 181, 197. Tables: 6.1, 6.3, 6.5, 12.3

season, seasonal 38, 43, 50, 51, 58, 91, 95, 101, 103, 110–113, 121, 123, 128, 145, 154, 160, 163, 167, 174

sedges (*and see Carex*) 3, 13, 44, 48, 100, 125, 126, 130, 156, 167, 173, 174, 177, 222

seed(s), seedlings (Chapter 2), 6, 9, 13, 21–26, 39, 47, 83, 84, 123, 125,–127, 145, 148, 150, 160, 161. Table: 2.3

shade, shading 22, 24, 25, 44, 47, 50, 57, 63, 65, 66, 70, 84, 88, 90, 92, 102, 107, 109, 114, 121–132, 135, 145, 174, 200, 201, 209, 219, 223

sheath 5, 6, 23, 24, 28, 32, 44, 46, 50, 51, 55, 99, 105, 180, 181. Table: 6.8

silica, skeleton 179, 180, 182. Tables: 6.6, 6.9

Simar 134, 138, 144–146, 202

Somerset Levels 10, 13, 149, 156, 159, 173

Southampton 102

Spain 93

*Sphagnum* spp. 8, 124, 125, 223. Tables: 6.1, 12.3

stem miners *Phragmitoecia costoneae, Schoenobius gigantellus, Chilo phragmitellus, Rhizedra lutosa, Lasioptera arundis, Lasioptera hungarica, Giraudiella inclusa* (the last three are also gall-formers) Table: 4.11

*Stenotarsonemus* [*Steneotarosenemus*] *phragmitidis* 71. Table: 4.9, (Bibliography 230)

substrate 11, 24, 25, 37, 72, 91, 95, 104, 123, 124, 128, 149, 153–156, 159–161, 172, 174, 197, 210, 223

summer shoots 42, 57, 63, 183. Table 4.3

Sweden, Swede, Swedish 5–7, 22, 25, 62, 91, 93, 109, 171

Switzerland, Swiss 22, 38, 194

## T

Tall herb(s) 27, 56, 62, 107, 108, 112, 122, 125, 126, 128–130, 160, 172, 182, 191. Table: 11.4

taxonomy (Chapter 1) 1–20

temperature 22, 24, 44, 47, 49, 51, 52, 56, 60, 61, 68, 72, 112, 113, 130, 154, 160, 174, 175, 181, 182, 222. Table: 2.2

thatching. Materials, management, methods on the bed, quality strength, nutrients, storage placing on roof, decay, ultra violet, weak and waste 49, 52, 90, 99, 103, 105, 113, 130, 146, 147, 150, 160, 161, 164, 165 (Chapter 10, 167–190). Tables: 6.8, 6.9, 10.1

Tisza, River, Lake 159, 220

transplant 6–8, 38, 83, 90, 91, 93, 94, 101, 111, 133, 157, 161, 182, 209, 211

Turkey 62, 171, 218

*Typha* spp. 3, 148, 153, 154, 173, 195, 200, 201. Table: 11.1

*T. angustifolia* 200. Table: 12.3

*T. latifolia* 156, 200. Table: 12.3

## U

Uganda 25
*Urtica dioica* 112, 163
US, USA, United States of America (*see America*)
uses (of reed) (Chapter 9, 147–166; and Chapter 10, 167–190) 15, 130, 146, 219, 220–222
*Ustilago grandis* 71

## V

variation (especially Chapters 1, 5, 6, 8, 11, 12) 3, 6–8, 13, 15, 40, 51, 60, 61, 66–68, 83–85, 87, 89–91, 93, 95, 97, 99, 101–103, 105–107, 109, 111–114, 146, 174, 182, 183, 210,–212. Tables: 1.1, 6.6, 10.2, 12.3
Velenzia, Lake 220

## W

Wangford Table: 2.2
water (fresh), water regime 60, 70, 88, 103–105, 106, 109–113, 122, 130, 157, 160, 172, 174, 179, 182, 184, 198, 200, 212, 218. Tables: 1.6, 5.4, 12.3
Waveney, River 84, 88. Table: 1.6
weather 7, 21, 50, 56, 57, 62, 63, 95, 105, 114, 123, 176, 182, 184, 197
wheat, straw 55, 168, 171, 178, 185, 186
Whittlesey Mere 199
Wicken Fen 9, 100, 131, 173, 192, 222
willow (*see Salix*)
woodland 3, 109, 126, 130, 132, 160, 161, 198, 201, 218, 219, 221–223. Table: 11.3
wet 1, 3, 8, 9, 22, 38, 48, 52, 70, 73, 103, 105, 109, 110, 125, 126, 128–130, 132, 141, 142, 144, 145, 148, 152, 156, 157, 160, 161, 163, 164, 172, 177, 178, 191–193, 197, 198, 200–202, 209, 210, 212, 217–223. Table: 12.1
wetland(s) Preface, 2–3, 6, 12, 13, 48, 70, 72, 112, 121, 122, 132, 141, 147, 148, 149, 153–157, 161–165, 167, 173, 182, 185, 191–194, 202, 219, 220–22. Table 9.2, 9.3
Woodwalton Tables: 4.4, 6.2, 6.4

## Z

Zuider Zee 22, 160, 201
Zuidland Table 12.2

www.ingramcontent.com/pod-product-compliance
Ingram Content Group UK Ltd.
Pitfield, Milton Keynes, MK11 3LW, UK
UKHW061954290726
14090UKWH00021B/1216